Lecture Notes in Mathematics 1514

Editors

Ulrich Krengel
Institut für Mathematische Stochastik
Universität Göttingen
Lotzestr. 13, W-3400 Göttingen, Germany

Karin Richter
Volker Warstat
Martin Luther Universität Halle-Wittemberg
Fachbereich für Mathematik und Informatik
Postfach, O-4010 Halle, Germany

Mathematics Subject Classification (1980): 28D05, 34C35, 58F03, 58F11, 60G10

ISBN 3-540-55444-0 Springer-Verlag Berlin Heidelberg New York
ISBN 0-387-55444-0 Springer-Verlag New York Berlin Heidelberg


Printed in Germany

Typesetting: Camera ready by author/editor
Printing and binding: Druckhaus Beltz, Hemsbach/Bergstr.
46/3140-543210 - Printed on acid-free paper

U. Krengel K. Richter V. Warstat (Eds.)

# Ergodic Theory and Related Topics III

Proceedings of the International Conference
held in Güstrow, Germany, October 22-27, 1990

Springer-Verlag
Berlin Heidelberg New York
London Paris Tokyo
Hong Kong Barcelona
Budapest

## Introduction

In the eighties, Horst Michel organized two conferences "Ergodic theory and Related Topics I and II" held in 1981 at Vitte (Hiddensee), GDR and in 1986 at Georgenthal (Thuringia), GDR. These conferences succeeded in bringing scientists from the East and the West together. Ergodic theorists from Austria, CSSR, France, FRG, GDR, Great Britain, Greece, Japan, the Netherlands, Poland, USA, USSR, and Vietnam discussed their recent results in measure-theoretic and topological dynamical systems as well as connections to other fields. A third conference was in the planning when Horst Michel, his wife Jutta, his younger daughter Kathrin and his mother died in a tragic car accident in December 1987. His colleagues all over the world lost a good friend.

Horst Michel was born in a little town in Thuringia. He studied mathematics at the University of Leipzig. As an assistant at the Technical College Ilmenau and at the University of Halle, he worked on iteration groups of real valued functions using methods of functional analysis. His thesis (1961) dealt with "Continuous and monotone iteration groups of nondifferentiable real valued functions". He then turned to the study of measure theoretical properties and of the classification of special groups of measure preserving transformations. Stimulated by articles of K. Jacobs, H. Furstenberg, and W. Parry, he explored the class of totally ergodic dynamical systems with quasidiscrete spectrum, in particular their embeddability into a flow. After 1970, he became interested in topological dynamics and studied so-called configuration spaces on special lattices. A list of his publications appears in "Kongreß und Tagungsberichte der Martin-Luther-Universität Halle-Wittenberg 1989/54".

Ipse abiit e vita. Remanebunt opera studiumque viri valde estimati in scientias mathematicas posita.

The idea of having a third conference was not given up. Horst Michel's students Karin Richter-Häsler and Volker Warstat organized it, and it was held in October 1990 in Güstrow, GDR, although the political events of 1989-90 caused various difficulties quite different from those of previous years. Fortyfive colleagues from 9 countries participated. This volume contains those results which are not published elsewhere. We thank all the participants of the conference for contributing towards its success, all the authors for their good cooperation with the editors, and the Martin-Luther-University at Halle-Wittenberg for sponsoring the conference.

Our special thanks go to all colleagues who offered their advice in preparing these Proceedings, especially Prof. M. Denker from Göttingen.

Göttingen and Halle
March 1992

Ulrich Krengel
Karin Richter-Häsler
Volker Warstat

Dedicated to the memory of

HORST MICHEL

(1934–1987)

# Table of Contents

# Symbolic dynamics for angle-doubling on the circle
# I. The topology of locally connected Julia sets

Christoph Bandt and Karsten Keller
Fachbereich Mathematik
Ernst-Moritz-Arndt-Universität
D-O-2200 Greifswald, Germany

## 1 Introduction

The study of the dynamics of complex polynomials leads to some problems which belong to topology, combinatorics and number theory rather than complex function theory. Douady and Hubbard used trees to study Julia sets [6, 7], and Thurston [15] introduced invariant laminations of the circle. Our point is to show how symbolic dynamics can be used to strengthen and clarify their results. We restrict ourselves to quadratic polynomials although some of our results extend to polynomials or to invariant factors of shift spaces [1].

The basic concepts are simple. We consider the circle $T = R/Z$ and the angle-doubling map $h: T \to T$, $h(\beta) = 2\beta \bmod 1$. Fix $\alpha \in T$. The diameter between $\frac{\alpha}{2}$ and $\frac{\alpha+1}{2}$ divides $T$ into two open semi-circles $T_0^\alpha$ and $T_1^\alpha$, where the fixed point $0 = 1$ of $h$ shall belong to $T_1^\alpha$. The *itinerary* of a point $\beta \in T$ with respect to $\alpha$ is defined as

$$I^\alpha(\beta) = s_1 s_2 s_3 \ldots \quad \text{with} \quad s_i = \begin{cases} 0 & \text{for} \quad h^{i-1}(\beta) \in T_0^\alpha \\ 1 & \text{for} \quad h^{i-1}(\beta) \in T_1^\alpha \\ * & \text{for} \quad h^{i-1}(\beta) \in \{\frac{\alpha}{2}, \frac{\alpha+1}{2}\} \end{cases}$$

The itinerary of $\alpha$ itself, $\hat{\alpha} = I^\alpha(\alpha)$ , is called the *kneading sequence* of $\alpha$. In rough form, our main ideas can be stated as follows.

1. When the Julia set $J_c$ of $p_c(z) = z^2 + c$ is locally connected and $c$ has external angle $\alpha$ then $J_c$ is the quotient space of $T$ obtained by identification of points with equal itineraries.

2. If the boundary of the Mandelbrot set is locally connected, it is the quotient space of $T$ obtained by identification of points with equal kneading sequences.

3. If we confine ourselves to $\alpha$ with $\hat{\alpha} = I^\alpha(1-\alpha)$, we obtain all itineraries and kneading sequences of real unimodal maps [5, 12].

In the present paper, we shall work out the first point and its consequences: the $\overline{T}_i^\alpha$ form a Markov partition in the tree-like (non-hyperbolic) case, the branching points of the Julia set can be read from $\hat{\alpha}$ (sec. 6-8) and renormalization can be expressed by substitution of words (sec. 11).

We describe those $J_c$ which are locally connected but the abstract theory is more general, and in some respect more beautiful than the reality of complex polynomials. To *each* angle $\alpha$ on $T$ we construct an abstract Julia set as a quotient space of $T$. We shall distinguish three cases: the 'tree-like' case that $\hat{\alpha}$ is not periodic, the case where $\alpha$ is periodic under $h$, and the 'Siegel disk' case where $\alpha$ is not periodic but $\hat{\alpha}$ is.

We prove uniqueness of $J$ in all three cases. Beside the fact that a critical point in $J$ cannot be periodic, we only assume that $J$ is obtained as a quotient of $T$ by a homotopic process in the plane (external rays, cf. sec. 2). This topological condition is crucial for using laminations. In the tree-like case our assumption is a bit weaker (see below). Differentiability is not required in the present paper. Let us note that every continuous, orientation-preserving two-to-one map $h'$ on $T$ with a single fixed point is conjugate to $h$, so that our topological methods will work for $h'$ as well as for $h$.

There exist locally disconnected $J_c$ [6, 3, 11] but by the recent remarkable results of Yoccoz (cf. Hubbard [8]) these examples are rare exceptions. It seems that the question whether $J_c$ is locally connected belongs to conformal geometry rather than topology. Roughly speaking, certain topological spaces are too complicated to become realized by the conformal mapping $p_c$.

To give an impression of the technique, we state a few definitions and results. For fixed $\alpha$, two inverse branches of $h$ can be defined as $l_i^\alpha : T \setminus \{\alpha\} \to T_i^\alpha$, $i = 0, 1$. A closed equivalence relation $\sim$ on $T$ is said to be an *$\alpha$-equivalence* if

(a) $\frac{\alpha}{2} \sim \frac{\alpha+1}{2}$

(b) $\beta \sim \gamma$ implies $h(\beta) \sim h(\gamma)$

(c) $\beta \sim \gamma$, $\beta, \gamma \neq \alpha$ implies $l_0^\alpha(\beta) \sim l_0^\alpha(\gamma)$ and $l_1^\alpha(\beta) \sim l_1^\alpha(\gamma)$.

For each $\alpha$, there is a *minimal* $\alpha$-equivalence $\sim_\alpha$ which corresponds to Thurston's minimal lamination. The *dynamical* $\alpha$-equivalence $\approx_\alpha$ is given by the equality of $\alpha$-itineraries, with $*$ used as a joker for both 0 and 1. We show (sec. 3,4) that for all $\alpha$ in $T$, the space $T/\approx$ is the invariant factor [1] of the one-sided shift space $\{0,1\}^\infty$, given by the generating relation $0\hat{\alpha} \sim 1\hat{\alpha}$. Let us say that an $\alpha$-equivalence $\sim$ is *degenerate* if the equivalence class of $\frac{\alpha}{2}$ is periodic under the map $\tilde{h}$ induced by $h$ on $T/\sim$. In sec. 7 we prove

**Theorem 1.** (Uniqueness of $\alpha$-equivalences in the tree-like case)
For non-periodic $\hat{\alpha}$, there is only one non-degenerate $\alpha$-equivalence. Thus minimal and dynamical $\alpha$-equivalence coincide. If the point $c$ belongs to $J_c$ and has external angle $\alpha$, then $(J_c, p_c)$ is homeomorphic to $(T/\approx_\alpha, \tilde{h})$.

For the Siegel disk case (sec. 9), the dynamical equivalence will collaps certain Cantor sets and the minimal equivalence will turn them into circles. The latter yields the proper topology. In the periodic case (sec. 10), both equivalences coincide and collapse certain Cantor sets. These will turn into circles when instead of (a), $\alpha$ is identified with a well-defined "conjugate point" $\beta$.

To give an idea on how branching points of $J_c$ are connected with $\hat{\alpha}$, consider the fixed points of $p_c$ which correspond to itineraries $\overline{1} = 111\ldots$ and $\overline{0}$. The first one is always an endpoint of $J_c$ while the second is a branching point with $k+1$ branches if $\hat{\alpha}$ starts with $0^k1$. In fig. 1, $\hat{\alpha}$ begins with $(001)^3$ but not with $(001)^4$, which implies the existence of branching points with four branches, associated to the sequence $\overline{001}$.

**Remark** of the first author: in November 1987, six weeks before his death, I met Professor Horst Michel in a curriculum committee and showed him my first rather vague ideas on these questions. In his kind manner, he became interested, gave hints and said he would certainly like to read a written outline. Let me express my gratitude for his encouragement which has contributed to [1, 2] and the present paper.

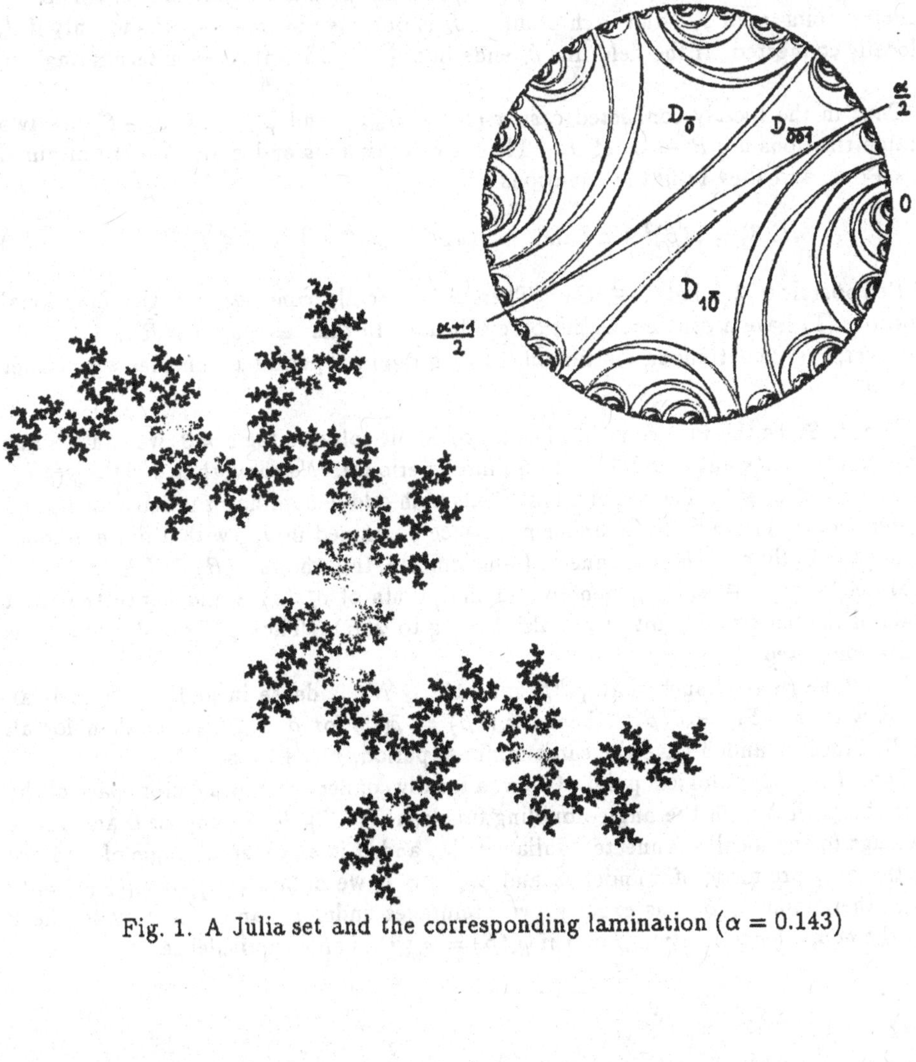


Fig. 1. A Julia set and the corresponding lamination ($\alpha = 0.143$)

## 2 Parametrization of locally connected Julia sets

We recall some well-known facts [3, 11] concerning Julia sets of $p_c(z) = z^2 + c$ in a topological version. For $c, z \in C$, let $O_c(z) = \{z, p_c(z), p_c^2(z), \ldots\}$ denote the forward orbit of $z$, and call $K_c = \{z \in C | O_c(z) \text{ is bounded}\}$ the filled-in Julia set. The boundary $J_c$ of $K_c$ is said to be the Julia set of $p_c$. $K_0 = D$ is the unit disk.

If $J_c$ is connected, there is a unique conformal isomorphism $\Phi_c : C \setminus K_c \to C \setminus D$ with $\lim_{z\to\infty} \Phi_c(z)/z = 1$ which conjugates $p_c$ and $p_0$, i.e. $\Phi_c p_c \Phi_c^{-1} = p_0$.

Let us define a potential $u_c(z) = \Phi_c(z)$ on $C \setminus K_c$ with field lines

$$\beta_c = \{z \in C \setminus K_c | \arg(\Phi_c(z) = 2\pi\beta\}$$

([6], p.65). Then the mapping $h(\beta) = 2\beta \bmod 1$ fulfils

$$p_c(\beta_c) = (h(\beta))_c \quad \text{and} \quad -\beta_c = ((\beta + \tfrac{1}{2}) \bmod 1)_c \quad \text{for all} \quad \beta \in [0, 1[.$$

According to Caratheodory's theorem, each field line $\beta_c$ has a continuous extension to a unique point $z_\beta$ of $J_c$, and each point of $J_c$ is obtained in this way, if and only if $J_c$ is locally connected. If the field line $\beta_c$ ends in $z \in J_c$, $\beta$ is called an external angle of $z$.

Thus in the locally connected case $\varphi_c(\beta) = z_{\beta \bmod 1}$ and $\varphi_c^-(\beta) = \varphi_c(-\beta)$ are two parametrizations $\varphi : R \to J_c$ of $J_c$. They are continuous and periodic with minimal period one, and they fulfil the equations

$$\varphi(2\beta) = \varphi(\beta)^2 + c \quad \text{and} \quad -\varphi(\beta) = \varphi(\beta + \tfrac{1}{2}), \ \beta \in R. \tag{1}$$

**Proposition 2.** Let $c \in C$. Then $J_c$ is locally connected iff the functional equations (1) have a continuous periodic solution. In this case, $J_c = \varphi(R)$. Moreover, every continuous solution of (1) with minimal period 1 coincides with either $\varphi_c$ or $\varphi_c^-$.

**Proof.** Since the first equation has no constant solution for $c \neq 0$, we can assume $\varphi$ is a continuous solution of (1) with minimal period q. We show that $\tilde{\varphi}(\beta) = \varphi(\beta/q)$ agrees with $\varphi_c$ or $\varphi_c^-$. The set of all rationals with odd denominator is dense in $R$, and all points of $\tilde{\varphi}(A)$ are periodic under $p_c$, hence contained in $J_c$, with a finite number of exceptions. Since $\tilde{\varphi}(R)$ is connected and infinite, this shows $\tilde{\varphi}(R) \subseteq J_c$.

Moreover, if $z \in \tilde{\varphi}(R)$, then by (1) the points of $p_c^{-1}(z)$ – and hence the limit points of the backward orbit of $z$ – also belong to $\tilde{\varphi}(R)$. Thus $\tilde{\varphi}(R) = J_c$, and $J_c$ is locally connected.

Now take $\beta_0 \in R$ such that $\{2^n\beta_0 \bmod 1 | n \in N\}$ is dense in $[0, 1[$. There is an $r \in R$ with $\varphi_c(r\beta_0) = \tilde{\varphi}(\beta_0)$, hence $\varphi_c(r\beta) = \tilde{\varphi}(\beta)$ for $\beta = 2^n\beta_0$ and then for all $\beta \in R$. Since $\varphi_c$ and $\tilde{\varphi}$ have the same minimal period, $r = \pm 1$. ■

Thus, from a topological point of view, a locally connected $J_c$ is a factor space of the circle compatible with the angle-doubling function $h$ [6, 7]. Now suppose 0 and hence $c$ belongs to the locally connected Julia set $J_c$, and $\alpha$ is an external angle of $c$. Since 0 is the only preimage of c under $p_c$ and $\varphi_c(\alpha) = c$, we obtain $\varphi_c(\frac{\alpha}{2}) = \varphi_c(\frac{\alpha+1}{2}) = 0$. Each other point $\varphi_c(\beta)$ has exactly two preimages under $p_c$, and it is easy to check that the equivalence relation $\beta \approx \gamma$ if $\varphi_c(\beta) = \varphi_c(\gamma)$ is an $\alpha$-equivalence.

On the other hand, each $\alpha$-equivalence $\sim$ defines a factor space $J$ of $T$ and a map $\tilde{h}: J \to J$ such that the projection $\varphi: T \to J$ is a semiconjugacy from $h$ to $\tilde{h}$ (i.e. $\varphi h = \tilde{h}\varphi$). Moreover, $\tilde{h}$ has two inverse branches $\tilde{l}_0^\alpha$ and $\tilde{l}_1^\alpha$ defined on the whole set $J$: just let $\tilde{l}_i^\alpha(\varphi(\alpha))$ be the equivalence class of $\frac{\alpha}{2}$ for $i = 0, 1$. Clearly, $J$ is Hausdorff iff $\sim$ is closed [9]. We say $\sim$ is non-degenerate if the equivalence class of $\frac{\alpha}{2}$ is not periodic under $\tilde{h}$. An $\alpha$-equivalence associated with the Julia set of some $p_c$ must be non-degenerate: if 0 has a periodic orbit, this orbit is superstable. For each $\alpha$, the smallest $\alpha$-equivalence (the intersection of all $\alpha$-equivalences in $T \times T$) will be called $\sim_\alpha$.

## 3 Invariant factors of shift spaces

In [1] we introduced a concept related to $\alpha$-equivalences: factor spaces $A$ of the one-sided shift space $\{0, 1, \ldots, m\}^\infty$ with mappings semiconjugate to shift maps. Here we shall only need a special case.

Let $\{0,1\}^* = \bigcup_{n=0}^\infty \{0,1\}^n$ be the set of 0-1-words $w = w_1 w_2 \ldots w_n$ , and $\{0,1\}^\infty$ the set of one-sided sequences $s = s_1 s_2 \ldots$ . Let $\lambda$ denote the empty word, $|w|$ the length of $w$, $ws$ and $w^k$ the concatenation, $\overline{w} = www\ldots$ the periodic sequence and $s|_n = s_1 \ldots s_n$ the initial subword of $s$ with length $n$ for $w \in \{0,1\}^*$, $s \in \{0,1\}^\infty$ . On $\{0,1\}^\infty$ we have the left shift $\sigma(s_1 s_2 \ldots) = s_2 s_3 \ldots$ and right shift maps $\tau_0, \tau_1$ defined by $\tau_i(s_1 s_2 \ldots) = i s_1 s_2 \ldots$ .

An equivalence relation $\sim$ on $\{0,1\}^\infty$ is said to be *invariant* (strongly invariant in [1]) if for all $s, t \in \{0,1\}^\infty$

(a) $s \sim t$ implies $\sigma(s) \sim \sigma(t)$

(b) $s \sim t$ implies $\tau_0(s) \sim \tau_0(t)$ and $\tau_1(s) \sim \tau_1(t)$ .

If $\sim$ is closed, the compact Hausdorff space $F = \{0,1\}^\infty / \sim$ is called an invariant factor. On $F$ there are continuous maps $\tilde{\sigma}([s]) = [\sigma(s)]$ and $\tilde{\tau}_i([s]) = [\tau_i(s)]$, $i = 0, 1$. Conversely, a given compact Hausdorff space $A$ is (homeomorphic to) an invariant factor iff there is a continuous $\tilde{\sigma}: A \to A$ with exactly two inverse branches $\tilde{\tau}_0, \tilde{\tau}_1$ (that is, $\tilde{\sigma}\cdot\tilde{\tau}_0 = \tilde{\sigma}\cdot\tilde{\tau}_1 = id_A$ and $A = \tilde{\tau}_0(A) \cup \tilde{\tau}_1(A)$ ), such that

$$\bigcap_{n=1}^{\infty} \tilde{\tau}_{s_1} \cdot \tilde{\tau}_{s_2} \cdot \ldots \cdot \tilde{\tau}_{s_n}(A) \quad \text{is a singleton} \quad \psi(s) \quad \text{for each} \quad s \in \{0,1\}^\infty. \tag{2}$$

If (2) is true, then $s \sim t$ iff $\psi(s) = \psi(t)$, and $\psi$ is the projection onto the factor space.

Example. Let $A = [0,1]$ and $\tilde{\sigma}$ the tent map, $\tilde{\sigma}(x) = 2x$ for $0 \le x \le \frac{1}{2}$ and $\tilde{\sigma}(x) = 2(1-x)$ for $\frac{1}{2} \le x \le 1$ . Condition (2) is fulfilled for $\tilde{\tau}_0(x) = \frac{x}{2}$, $\tilde{\tau}_1(x) = 1 - \frac{x}{2}$. The fixed point of $\tilde{\tau}_0$ is $\psi(\overline{0}) = 0$. So $01\overline{0}$ and $11\overline{0}$ are the two sequences assigned to the critical point $\frac{1}{2}$ since $\tilde{\sigma}^2(\frac{1}{2}) = 0$. In fact $\sim$ is the smallest invariant equivalence relation which identifies $01\overline{0}$ and $11\overline{0}$. Since $k(x) = 2\cos\pi x$ is a conjugacy from $\tilde{\sigma}$ to $p_{-2}$, this invariant factor can also be considered as Julia set $J_{-2} = [-2, 2]$.

If $A = \tilde{\tau}_0(A) \cup \tilde{\tau}_1(A)$ is an invariant factor, the points $x \in \tilde{\tau}_0(A) \cap \tilde{\tau}_1(A)$ will be called critical points since $\tilde{\sigma}(x)$ has no other preimage than $x$. We are interested in factors with a single critical point. These spaces are dendrites (simply connected Peano continua) and hence embeddable into the plane [9]. Note that all locally connected and connected $J_c$ with $J_c = K_c$ belong to this class.

We give a construction for such factors. For $s \in \{0,1\}^\infty$ , let $\sim_s$ be the smallest closed invariant equivalence relation containing the identification $0s \sim 1s$ and $F(s)$ the corresponding factor. If $s$ is not periodic, $\sim_s$ is algebraically generated by the invariance condition (b):

$$r \sim_s t \text{ iff } r = t \text{ or there is a word } w \text{ with } r = w0s, t = w1s \text{ or } r = w1s, t = w0s.$$

This relation is closed since the sets $U_n = \{(r,t)| \ s_i = t_i \ \text{ for } \ i = 1,\dots,n\}$ form a neighbourhood base of the diagonal in $\{0,1\}^\infty \times \{0,1\}^\infty$, and only finitely many non-trivial equivalence classes intersect the complement of $U_n$.

A closed invariant $\sim$ generated by a single equation $t \sim wt$ looks more complicated. Condition (a) implies $t \sim w^n t$ for $n = 1,2,\dots$ and $t \sim \overline{w}$. Moreover, if $t = \overline{v}$ is periodic, all $ut$ with $u \in \{v,w\}^*$ and hence $\{v,w\}^\infty$ belong to the class of $t$.

Thus for $s = \overline{u0}$ as well as for $s = \overline{u1}$, the equivalence class of $s$ with respect to $\sim_s$ contains $\bullet_s = \{0u, 1u\}^\infty$. If $|0u| = |1u|$ is the minimal period of $s$, then $\sigma^k(\bullet_s) \cap \bullet_s = \emptyset$ unless $k$ is a multiple of $|0u|$, and $\bullet_s$ is a full equivalence class. If $s$ has smaller period, the class is larger. For example, $u = \lambda, u = 000$ and $u = 1$ all yield the trivial relation $\bullet_s = \{0,1\}^\infty$.

**Theorem 3.** (Classification of topologically self-similar dendrites with two pieces)

(a) A compact space $A$ is an invariant factor with a single critical point iff $A = F(s)$ for some $s \in \{0,1\}^\infty$.

(b) The non-trivial equivalence classes of $\sim_s$ are the sets $w\bullet_s, w \in \{0,1\}^*$ , where $\bullet_s = \{0s, 1s\}$ for non-periodic $s$ and $\bullet_s = \{0u, 1u\}^\infty$ for periodic $s$ with minimal period $u0$ or $u1$.

(c) $\sim_s$ has finite equivalence classes iff $s$ is not periodic.

(d) For all $s,t \in \{0,1\}^* \cup \{0,1\}^\infty$ we have $\sim_s \subset \sim_t$ iff $\bullet_s \subset \bullet_t$ .

**Proof.** For the remaining part of (a), suppose $A$ is an invariant factor with one critical point $\{x\} = \tilde{r}_0(A) \cap \tilde{r}_1(A)$ , and $x = \varphi(0wr) = \varphi(1wt)$ with $w = w_1 \dots w_n$ and $r_1 \neq t_1$. The $\sigma$-invariance implies $r \sim t$ , so $\varphi(r) = \varphi(t) = x$ . As above, $0wr \sim r \sim t \sim 1wt$ yields an equivalence class $\bullet$ containing $\{0w, 1w\}^\infty$.

Now we claim that $\bullet \neq \{0w, 1w\}^\infty$ implies the existence of some $k < n$ with $\bullet \supseteq \{0w_1 \dots w_k, 1w_1 \dots w_k\}^\infty$. Indeed, a sequence in $\bullet \setminus \{0w, 1w\}^\infty$ can be written as $ujw_1 \dots w_k v$, where $u \in \{0w, 1w\}^*, j \in \{0,1\}, k < n$ and $v \in \{0,1\}^\infty, v_1 \neq w_{k+1}$. Since $ujw\overline{0w}$ also belongs to $\bullet$, we conclude $v \sim w_{k+1} \dots w_n \overline{0w}$ by $\sigma$-invariance. Since we have only one critical point, these two sequences belong to $\bullet$. Now $jw_1 \dots w_k \bullet \subseteq \bullet$ since $v$ and $jw_1 \dots w_k v$ are in $\bullet$, and $(1-j)w_1 \dots w_k \bullet \subseteq \bullet$ because $w_{k+1} \dots w_n \overline{0w}$ and $(1-j)w\overline{0w}$ are in $\bullet$. The claim is proved.

When we apply the above conclusion finitely often, we either end with $\bullet = \{0,1\}^\infty$, or with $\bullet = \{0w', 1w'\}^\infty$, where $w'$ is a subword of $w$. This proves (a), and the other assertions follow easily. ∎

# 4 Itineraries and kneading sequences

Let $X$ be a topological space, $f : X \to X$ a continuous map and $\mathcal{P} = \{P_0, P_1, \ldots\}$ a partition of $X$. The *symbolic dynamics* of a point $x$ in $X$ with respect to $f$ and $\mathcal{P}$ is the sequence $I(x) = s_1 s_2 \ldots$ with $s_i = k$ iff $f^{i-1}(x) \in P_k$. This is an old idea. The binary representation $b(\beta) = b_1 b_2 \ldots$ of $\beta \in T$, for instance (with $\ldots\overline{1}$ excluded), is the dynamics of $\beta$ with respect to $h$ and $\mathcal{P} = \{[0, \frac{1}{2}[, [\frac{1}{2}, 1[\}$. Our $I^0(\beta)$ (sec. 1) is obtained from $b(\beta)$ by writing $\overline{*}$ for $\overline{0}$ and $w\overline{*}$ for $w1\overline{0}$.

For the topologist it is somewhat disgusting that $I$ is not a continuous map unless $\mathcal{P}$ consists of open-and-closed sets. However, in our case $\mathcal{P}$ contains the open semi-circles $T_0^\alpha$ and $T_1^\alpha$ which both become closed when we add the rest, $P_*^\alpha = \{\frac{\alpha}{2}, \frac{\alpha+1}{2}\}$. Using $*$ as a joker for both 0 and 1, and replacing the shift space by an invariant factor, we shall succeed in making $I^\alpha$ continuous.

We start with some simple remarks. The $n$-th coordinates of $I^\alpha(\beta)$ and $I^0(\beta)$ are different iff $h^{n-1}(\beta)$ lies in $[0, \frac{\alpha}{2}]$ or $[\frac{1}{2}, \frac{\alpha+1}{2}]$. Thus we can calculate itineraries directly from the binary representation:

$$I^\alpha(\beta) = s_1 s_2 \ldots \quad \text{with} \quad s_i = \begin{cases} b_i(\beta) & for \quad \sigma^i(b(\alpha)) > b(\beta) \\ 1 - b_i(\beta) & for \quad \sigma^i(b(\alpha)) < b(\beta) \\ * & for \quad \sigma^i(b(\alpha)) = b(\beta) \end{cases} \tag{3}$$

The *kneading sequence* $\hat{\alpha} = I^\alpha(\alpha)$ always starts with 0. Since $I^\alpha(\beta) = I^{1-\alpha}(1-\beta)$ and in particular $\hat{\alpha} = \widehat{1-\alpha}$ for all $\alpha, \beta$, we *assume throughout that* $\alpha \leq \frac{1}{2}$.

A point $\beta \in T$ is periodic under $h$ iff it is $0 = 1$ or rational with odd denominator: $\beta$ has period $p$ if we can write $\beta = m/(2^p - 1)$. Periodic points have periodic itineraries. The converse is true, except for one case (see proposition 6.2). $\hat{\alpha}$ is periodic iff it contains $*$.

$\beta$ is called *preperiodic* under $h$ if some $h^n(\beta) = h^{n+p}(\beta)$ for some minimal $n, p$ but $\beta$ is not periodic. These are the rationals with even denominator, $\beta = m/2^n(2^p - 1)$. Preperiodic points have preperiodic itineraries, and a preperiodic $\alpha$ has preperiodic $\hat{\alpha}$. (To see that $\hat{\alpha}$ is not periodic, write $b(\alpha) = b_1 \ldots b_k \overline{b_{k+1} \ldots b_{k+p}}$ with $b_k \neq b_{k+p}$. By (3), $\hat{\alpha}(k) \neq \hat{\alpha}(k+p)$. )

For $\beta \neq \alpha$ let $l_i^\alpha(\beta)$ be the point in $T_i^\alpha \cap \{\beta/2, (\beta+1)/2\}$ , $i = 0, 1$. Now take a word $w \in \{0,1\}^n$. The mapping $l_w^\alpha = l_{w_1}^\alpha \cdot \ldots \cdot l_{w_n}^\alpha$ is defined and continuous on the set of $\beta \in T$ with $h^i(\alpha) \neq \beta, i = 0, \ldots, n-1$. It is easy to see that $T_w^\alpha = l_w^\alpha(T)$ is the set of all $\beta$ such that the itinerary $I^\alpha(\beta)$ starts with $w$. This is a finite union of open intervals, with total length $2^{-n}$. The itineraries of the endpoints of the intervals are obtained from $w$ by replacing one or more $w_i$ by $*$. If we define

$$C_w^\alpha = \{\beta \in T |\ I^\alpha(\beta)(i) \in \{w_i, *\} \text{ for } i = 1, \ldots, n\},$$

then $C_w^\alpha \supseteq \overline{T_w^\alpha}$. Moreover, equality holds unless $\alpha$ is periodic. (If a point $\beta \in C_w^\alpha$ is not in $T_w^\alpha$ and neither a right nor a left endpoint of some interval of $T_w^\alpha$, there must exist two different integers $i, j \leq n$ with $h^i(\beta) = h^j(\beta) = \alpha$ . )

Now let $t \in \{0,1\}^\infty$ , and write $t_{|n}$ for $t_1 \ldots t_n$ . By compactness,

$$C_t^\alpha := \{\beta |\ I^\alpha(\beta)(i) \in \{t_i, *\} \text{ for } i = 1, 2, \ldots\} = \bigcap_{n=1}^{\infty} C_{t_{|n}}^\alpha \supseteq \bigcap_{n=1}^{\infty} \overline{T_{t_{|n}}^\alpha} \neq \emptyset.$$

There are points with itinerary $t$, maybe with some $t_i$ replaced by $*$. However, the $*$ in an itinerary has to be followed by $\hat{\alpha}$ , so there are points with proper itinerary $t$ unless $t$ has the form $w\hat{\alpha}$.

Let us define what we call the dynamical $\alpha$-equivalence on $T$. If $\beta, \gamma$ are points such that for each $i$, either $I^\alpha(\beta)(i) = I^\alpha(\gamma)(i)$ or $I^\alpha(\beta)(i) = *$ or $I^\alpha(\gamma)(i) = *$ , then $\beta$ and $\gamma$ should be equivalent. Let $\approx_\alpha$ denote the smallest closed equivalence relation with this property.

If $\hat{\alpha}$ is non-periodic, then $\beta \approx_\alpha \gamma$ iff either $I^\alpha(\beta) = I^\alpha(\gamma)$ or $I^\alpha(\beta) = wu\hat{\alpha}$ and $I^\alpha(\gamma) = wv\hat{\alpha}$ for some $w \in \{0,1\}^*$ and $u, v \in \{0,1,*\}$. If $\hat{\alpha} = \overline{u*}$, $\overline{u0}$ or $\overline{u1}$ for some word $u$, all points with itineraries in $w\{0u, 1u, *u\}^\infty$ are identified for each $w$.

**Theorem 4.** (Invariant factors of the circle are invariant factors of shift space)

(a) For each $\alpha$ in $T$ , the spaces $T/\approx_\alpha$ and $F(\hat{\alpha})$ are homeomorphic, and the homeomorphism is a semiconjugacy from $\tilde{h}$ and $\tilde{l}_i^\alpha$ to $\tilde{\sigma}$ and $\tilde{\tau}_i$.

(b) $\approx_\alpha$ is degenerate iff $\hat{\alpha}$ is periodic.

**Proof.** We saw that $I^\alpha$ can be considered as a map from $T$ onto $F(\hat{\alpha})$, with $\beta \approx_\alpha \gamma$ iff the itineraries represent the same point of $F(\hat{\alpha})$. We show that $I^\alpha$ is continuous, hence a quotient map. Basic neighbourhoods of $I^\alpha(\gamma)$ are given by fixing a finite number of coordinates of $I^\alpha(\gamma)$ which are $\neq *$. For all $\beta$ sufficiently near to $\gamma$, the itineraries of $\beta$ and $\gamma$ will agree in these coordinates. Observing that $I^\alpha(h(\beta)) = \sigma(I^\alpha(\beta))$ and $I^\alpha(l_i^\alpha(\beta)) = \tau_i(I^\alpha(\beta))$ we finish the proof of (a). If $\hat{\alpha} = \overline{u*}$ then $\bullet$ is invariant under $\tilde{\sigma}^{|u*|}$ while for non-periodic $\hat{\alpha}$ , the image of $\bullet = \{0\hat{\alpha}, 1\hat{\alpha}, *\hat{\alpha}\}$ under $\sigma^n$ does not contain a point of $\bullet$ for any $n$. ∎

**Remark.** Since we proved that all 0-1-sequences are itineraries with respect to every $\alpha$ (provided $*$ is used as a joker) we should also mention that only few sequences are kneading sequences. Some 0-1-words, as 010011, cannot be initial subwords of any $\hat{\alpha}$. In fact, $\hat{\alpha}$ starts with 01 iff $2\alpha > \frac{\alpha+1}{2}$. Now since $2 > 4\alpha > \alpha + 1$, the next digits 001 imply $4\alpha < \frac{\alpha+3}{2}$ , $8\alpha > 3 + \frac{\alpha}{2}$ and $16\alpha - 6 \in ]\frac{\alpha+1}{2}, 2\alpha[$. This means $32\alpha$ mod 1 is in $]\alpha, \frac{\alpha+1}{2}[$ so that the sixth letter of $\hat{\alpha}$ must be zero.

## 5 Invariant laminations

Thurston [15] uses a more geometric approach. He considers $T$ as the boundary of the unit disc $D$. A *lamination of the disc* is a set $\mathcal{S}$ of chords of $T$ such that $\bigcup\mathcal{S}$ is closed in $D$, and that any two of these chords do not intersect except at their endpoints (cf. fig. 1). Points of $T$ are considered as degenerate chords. A *gap* of $\mathcal{S}$ is the closure of a component of $D \setminus \bigcup\mathcal{S}$. For any chord or gap $S \in \mathcal{S}$ let $h(S) = \text{conv}\ h(S \cap T)$, where conv means convex hull. The lamination $\mathcal{S}$ is called *invariant* with respect to $h$ if [14]

- For each chord $S$ in $\mathcal{S}$, the image chord $h(S)$ and the opposite chord $-S$ belong to $\mathcal{S}$, and there is a chord $S'$ with $h(S') = S$.

Obviously, every $S$ will have two preimage chords so that the conditions are almost the same as in the definition of $\alpha$-equivalence. Let us fix a *non-periodic* $\alpha$ and construct an invariant lamination containing the chord $S_* = \text{conv}\ \{\frac{\alpha}{2}, \frac{\alpha+1}{2}\}$. Let $w$ be a 0-1-word.

Besides $C_w = C_w^\alpha = \overline{T_w^\alpha}$ we now consider $D_w = \text{conv } C_w$, and we define $S_{w*}$ to be the chord in $D$ connecting $l_w(\frac{\alpha}{2})$ and $l_w(\frac{\alpha+1}{2})$. Then $S_{w*} = D_{w0} \cap D_{w1}$ and $D_w = D_{w0} \cup D_{w1}$, which can be proved by induction on $|w|$.

Let $S^\alpha$ consist of the $S_{w*}$, $w \in \{0,1\}^*$, and of their limit chords. It is not difficult to see that $S^\alpha$ is the smallest invariant lamination containing $S_*$ [15], prop. II.4.5. Moreover, $S^\alpha$ defines the minimal $\alpha$-equivalence: $\beta \sim_\alpha \gamma$ iff $\beta$ and $\gamma$ can be joined through a finite number of chords from $S^\alpha$. (An elementary argument shows that this defines a closed relation.)

This lamination is tightly connected to our itineraries. By construction, each gap is obtained as $D_s^\alpha := \text{conv } C_s = \bigcap_{n=1}^\infty D_{s|_n}$ for some $s \in \{0,1\}^\infty$. Moreover, each chord in $S^\alpha$ which does not bound a gap also coincides with some $D_s$. Thus the family of all gaps of $S^\alpha$, and of all chords not contained in gaps, coincides with the family conv $C_s$, $s \in \{0,1\}^\infty$.

Let $S$ be a chord, $a$ the length of the subtended arc. Note that $a \leq \frac{1}{2}$ since $T$ has perimeter 1. The length $a'$ of the arc subtended by $h(S)$ is given by the tent map: $a' = 2a$ for $a \leq \frac{1}{4}$ and $a' = 1 - 2a$ for $a \geq \frac{1}{4}$. This implies a simple but important fact ([15], II.5.1):

**Lemma 5.1** Among all chords $h^i(S)$, $i = 1, 2, \ldots$ the first chord longer than $S$ coincides with the first chord which lies between $S$ and $-S$. ■

The following result of Thurston [15] is a combinatorial analogue of Sullivan's celebrated theorem on the non-existence of wandering domains [14]. We give a new proof.

**Theorem 5.2** (Thurston's structure theorem for quadratic laminations)

(a) On $(T, h)$ there are no wandering triangles: If $\alpha, \beta, \gamma \in T$, then some of the sets $\Delta_k = \text{conv } \{h^k(\alpha), h^k(\beta), h^k(\gamma)\}$, $k = 0, 1, 2, \ldots$ will intersect each other, or there exist $n$ such that $\Delta_k$ collapses to a chord for $k > n$.

(b) If $G$ is a gap in an invariant lamination, there exists an integer $n \geq 0$ such that either

- $h^n(G)$ is periodic: $h^n(G) = h^{n+p}(G)$, or
- $h^n(G)$ is a triangle with a diameter of $T$ as side, or a rectangle with a diameter of $T$ as diagonal.

**Proof.** First we note that (b) follows from (a). Note that $h$ maps each gap onto a gap or a chord. Given a gap $G$, take $\alpha, \beta, \gamma \in G \cap T$. If $\Delta_n$ and $\Delta_{n+p}$ intersect then $h^n(G) = h^{n+p}(G)$. The same is true if $h^n(G)$ and $h^{n+p}(G)$ contain a diameter in their interior. However, if some $\Delta_k$ collapses, then $h^n(G)$ contains a diameter for some $n < k$, and if $h^n(G)$ is not a triangle or rectangle, $h^{n+1}(G)$ is still a gap.

To prove (a), let $c \geq b \geq a$ denote the arc lengths of the sides of a triangle on $T$. Either $c = b + a$ or $a + b + c = 1$, but any two triangles from the latter case will intersect. Thus we can assume $a + b = c < \frac{1}{2}$. Moreover, there are at most three disjoint triangles with $c \geq \frac{1}{3} > b$. So we find $n_0$ such that for $k > n_0$, either $a \leq b \leq c \leq \frac{1}{3}$ or $a < \frac{1}{3} \leq b \leq c$.

Let $c' \geq b' \geq a'$ denote the lengths of the sides of the image triangle. In the first case, $a' = 2a$, $b' = 2b$, $c' = 2c$. In the second case $1 - 2c \leq 1 - 2b$ and $a = c - b < \frac{1}{2} - b$

implies $c' = 1-2b$, and we have $b' = 1-2c$, $a' = 2a$ for $a+c \leq \frac{1}{2}$, and $b' = 2a$, $a' = 1-2c$ for $a \geq \frac{1}{2} - c$.

In words: the shortest side can only interchange its place with the longest. For some $n > n_0$, this must happen: $a' = 1 - 2c := \delta$. Let $S$ denote the chord which had length $c$ at this step. In all succeeding steps, $a' < \delta$ is only possible, if the longest side of the original triangle lies between $S$ and $-S$. This cannot happen, since a triangle with short side $\geq \delta$ will not fit into that region. Thus all $\Delta_k$, $k \geq n$, have area larger than the area of a triangle with $a = b = \delta$, $c = 2\delta$. They cannot be pairwise disjoint. ■

Another result of Thurston (cf. [15], proof of II.5.3) will be needed.

**Proposition 5.3** If $G$ is a periodic polygonal gap with period $p$ in an invariant lamination, which does not contain the origin, then the return map $h^p$ acts transitively on the vertices of $G$.

**Proof.** We can assume that $G$ contains the longest chord $S'$ of all chords of the $h^i(G)$. There is at most one other chord $S"$ of length greater than $\frac{1}{3}$ in $G$. If $S"$ exists, then $G$ and no other $h^i(G)$ lies in the region between $S"$ and $-S"$ so that $h^p(S") = S'$ by lemma 5.1. Similarly, if $S$ is a chord longer than $\frac{1}{3}$ in some $h^j(G)$ then by induction on those $h^i(G)$ which are between $S$ and $-S$ we see that $h^{p-j}(G)$ must be either $S'$ or $S"$. Now for any chord of $G$, some image chord under $h^i$, $i = 1, 2, \ldots$ will be longer than $\frac{1}{3}$ and so some other image chord will be $S'$. ■

## 6 The combinatorics of initial subwords

A point $x$ in a locally arcwise connected space $X$ is said to have $k$ branches ($k \in N$) if $k$ is the maximum number for which there are arcs $I_1, \ldots, I_k$ in $X$ with $I_i \cap I_j = \{x\}$ for $i \neq j$. For $k = 1$ we call $x$ an *endpoint*, for $k \geq 3$ a *branching point*. Since points in $T/\sim_\alpha$ with $k$ branches correspond to polygonal gaps $D_s$ with $k$ sides in $S^\alpha$, we are interested in estimates of card $C_s$ for arbitrary $s$.

Throughout sec. 6 to 9, let $\alpha$ be non-periodic. In sec. 4 we saw that $C_w = \overline{T_w^\alpha}$ is a union of intervals (arcs) on $T$. Let us determine the number of these intervals.

**Proposition 6.1** For any 0-1-word $w$, $D_w$ is the polygon bounded by the arcs of $T_w$ and by those chords $S_{v*}$ for which there is an integer $k \geq 0$ and $u \in \{0,1\}$ with $w = vu\hat{\alpha}_{|k}$.

**Proof.** For every $k \geq 0$, the chord $S_*$ belongs to the boundary of the two domains $D_{0\hat{\alpha}_{|k}}$ and $D_{1\hat{\alpha}_{|k}}$. By continuity, $S_{v*}$ bounds $D_{v0\hat{\alpha}_{|k}}$ and $D_{v1\hat{\alpha}_{|k}}$, and no others. ■

For $w = w_1 \ldots w_n$ let $\pi_0^\alpha(w) = \text{card}\,\{k |\ 0 < k \leq n,\ w_{k+1} \ldots w_n = \hat{\alpha}_{|n-k}\}$. Note that $\pi_0^\alpha(w) \geq 1$ since for $k = n$ we get the empty word $\lambda = \hat{\alpha}_{|0}$. The definition is motivated by our proposition: $C_w$ consists of $\pi_0^\alpha(w)$ arcs.

For $s = s_1 s_2 \ldots \in \{0,1\}^\infty$ let $\pi_0^\alpha(s) = \liminf_{n\to\infty} \pi_0^\alpha(s_{|n})$. If the number of components of $C_s$ is greater than some $k$, there is $n_0$ such that $C_{s_{|n}}$ has more than $k$ components for every $n \geq n_0$. This shows card $C_s \leq \pi_0^\alpha(s)$.

**Proposition 6.2** If $\alpha$ is not periodic, card $C_s \leq \pi_0^\alpha(s)$. Moreover, card $C_s < \infty$ for $s \in \{0,1\}^\infty$, unless $\hat{\alpha}$ is periodic and $\sigma^n(s) = \hat{\alpha}$ for some $n$.

**Proof.** To verify card $C_s < \infty$, it suffices to consider periodic $s = \overline{w}$ since by theorem 5.2 only a periodic or preperiodic gap can have infinitely many points on $T$. If $\hat{\alpha}$ is not periodic, there must be a maximal initial subword $v$ of $\hat{\alpha}$ which is contained in $\overline{w}$ as a connected subword. Thus $\pi_0^\alpha(s) \leq |v| < \infty$. On the other hand, if $\hat{\alpha}$ is periodic and $s$ contains arbitrary long initial subwords of $\hat{\alpha}$, take a subword longer than the period of both sequences to see that $\sigma^n(s) = \hat{\alpha}$ for some $n$. ■

As an application, we determine some $s$ with card $C_s = 1$. The first coordinate of $\hat{\alpha}$ is 0, so the fixed point $\beta = 1 \in T$ of $h$ has itinerary $s = \overline{1}$. Since $\pi_0^\alpha(s) = 1$, this $\beta$ is always an endpoint in $J = T/\sim_\alpha$. For $\alpha = \frac{1}{2}$ the lamination consists of all vertical segments, $J$ is an interval ($J = J_{-2}$, cf. sec. 3), and the two endpoints have addresses $\overline{1}$ and $\hat{\alpha} = 0\overline{1}$. This is an exceptional case:

**Proposition 6.3** If $\alpha$ is non-periodic and $\alpha \neq \frac{1}{2}$, the set of endpoints of $J = T/\sim_\alpha$ has the cardinality of the continuum.

**Proof.** If $\hat{\alpha} = 01^k0\ldots$ with $k \geq 0$, then any $s \in \{01^{k+1}, 1\}^\infty$ obviously yields $\pi_0^\alpha(s) = 1$. ■

It can happen that card $C_s < \pi_0^\alpha(s)$. To see this, note that the convex set $D_{s_{|n}}$ is obtained from $D_{s_{|n-1}}$ by cutting away some piece, drawing the chord $S_{s_{|n-1}*}$. By proposition 6.1, this chord which corresponds to $\hat{\alpha}_0$ is always the new bounding chord, and the old chords which have been cut away belong to those initial subwords of $\hat{\alpha}$ in $s$ which end with $s_{n-1}$ and cannot be extended. In particular, if $s_n = 1$, a proper subword of $\hat{\alpha}$ will not begin with $s_n$ and so the new chord of the $n$-th step has to be cut away in the ($n$+1)-th step.

For example, let $\hat{\alpha}$ start with 0011, and let $s = \overline{00001}$. Clearly, $\pi_0^\alpha(s) = 2$. Let $w = 00001$, $v = 0000$. Now $D_{w0}$ is bounded by two arcs $A_1, A_2$, by $S_{v*}$ and $S_{w*}$ (the latter chord just cut away $S_{0*}$). The new chord $S_{w0*}$ for $D_{w00}$ does not cut away another chord, so it has both endpoints on one arc, say $A_1$. Now since $S_{wv*}$ for $D_{ww}$ removes two chords, the two bounding arcs of $D_{ww}$ and $D_{ww0}$ are contained in $A_1$. Repeating the argument, we see that the bounding arcs of $D_{w^{k+1}}$ are both contained in one of the bounding arcs of $D_{w^k}$. So $D_s$ is a single point.

Given $\hat{\alpha}$ and $s$, we define a directed graph $\Gamma^\alpha(s)$ with vertex set $N = \{1, 2, \ldots\}$ as follows. Each point $n$ is the initial point of one edge. If $s_{n+1} = 1$, the endpoint of the edge is $n+1$. If $s_{n+1} = 0$, consider the maximal $k$ such that $\hat{\alpha}_{|k} = s_{n+1}\cdots s_{n+k}$, and let $n+k+1$ be the endpoint of the edge. If $\hat{\alpha} = s_{n+1}s_{n+2}\ldots$, connect $n$ with itself by a loop. (For non-periodic $\hat{\alpha}$, this can happen only once.)

Obviously, $\Gamma^\alpha(s)$ is a union of trees, where each tree consists of an "infinite chain with finite branches". Let $\pi^\alpha(s)$ denote the number of components of $\Gamma^\alpha(s)$. For any $n$ and $w$, $\pi^\alpha(\sigma^n(s)) = \pi^\alpha(s) = \pi^\alpha(ws)$. Moreover, $\pi^\alpha(s) \leq \pi_0^\alpha(s)$.

**Theorem 6.4** (The initial subword index)
card $C_s = \pi^\alpha(s)$ for non-periodic $\hat{\alpha}$ and arbitrary $s$.

**Proof.** By proposition 6.2, it remains to show card $C_s \geq \pi^\alpha(s)$. Let $\pi = \pi^\alpha(s)$. Choose $n$ such that each component of $\Gamma^\alpha(s)$ contains a number $k < n$. Then $\pi^\alpha(s_{|n}) \geq \pi$ since edges of the graph represent initial subwords of $\hat{\alpha}$ in $s$. According to proposition 6.1, $D_{s_{|n}}$ is bounded by $\pi^\alpha(s_{|n})$ chords $S_{v*}$ and by the same number of arcs. There are at least $\pi$ arcs $A_i$ which have two neighbouring chords corresponding (in the sense of proposition 6.1) to subwords of different components of $\Gamma^\alpha(s)$. Our proof will show that all chords associated to one component are arranged consecutively, so that the number of arcs $A_i$ is exactly $\pi$. However, it will be sufficient to prove that each $D_{s_{|m}}$, $m \geq n$, contains at least one point on each $A_i$.

We use induction on $m$. $D_{s_{|m+1}}$ is obtained from $D_{s_{|m}}$ by drawing the chord $S_{s_{|m}*}$ which will cut away all chords corresponding to subwords ending at $s_m$. The chords cut away belong to the edges ending at $m$, and the new chord is represented by the edge starting in $m$, and all these edges belong to one component. Consequently, the new chord cannot cut away an arc $A_i$ completely. ∎

Let us calculate some $\pi^\alpha(s)$ for periodic $s$. The second fixed point of $h$ is given by $s = \overline{0}$. If $\hat{\alpha}$ starts with $O^k1$, in $\Gamma^\alpha(s)$ each number $n$ is connected with $n+k+1$, so $\pi^\alpha(s) = k+1$.

For $s = s_1s_2\ldots = \overline{w}$ the situation is a bit more intricate. We consider the case $w = 010$. If $\hat{\alpha}$ starts with 01 then $\Gamma^\alpha(s)$ contains the chain $C = s_1s_2s_4s_5\ldots s_{3n+1}s_{3n+2}\ldots$ If $\hat{\alpha} = 011\ldots$, there is another chain $s_3s_6\ldots s_{3n+3}\ldots$, and if $\hat{\alpha}$ would begin with $(010)^k011, k \geq 1$, this chain would split into $k+1$ chains $\{s_{3(i+n(k+1))}|n \in N\}$, $i = 1,\ldots,k+1$ so that $\pi^\alpha(s) = k+2$. However, if $\hat{\alpha} = (010)^k1\ldots$, then all these 'long' subwords in $s$ will only create branches to the chain $C$, independently of $k$, so that $\pi^\alpha(s) = 1$. Our argument easily extends to the general case:

**Proposition 6.5** Let $s = \overline{w}$ with $w = w_1\ldots w_p$, let $\hat{\alpha}$ be non-periodic, and let $w^kw_1\ldots w_q$ be the longest initial subword of $\hat{\alpha}$ contained in $s$ ($k \geq 0$, $0 \leq q < p$ ). With $\pi$ we denote the value $\pi^\alpha(s)$ for $k = 0$.

If the chain in $\Gamma^\alpha(s)$ starting at $s_{q+1} = w_{q+1}$ does not meet $s_{np}, n = 1,2,\ldots$ (it suffices to consider $n = 1,2$) then $\pi^\alpha(s) = \pi$ for every $k$, otherwise $\pi^\alpha(s) = \pi + k$. ∎

## 7 The normal case: tree-like Julia sets

For us, the normal case is that $\hat{\alpha}$ is not periodic. Theorem 1 says that for such $\alpha$, minimal and dynamical equivalence coincide and so there is a unique non-degenerate $\alpha$-equivalence. Here is the proof.

**Proof of theorem 1.** For non-periodic $\hat{\alpha}$, consider the lamination $S^\alpha$. Its gaps and chords outside gaps coincide with the sets conv $C_s$, $s \in \{0,1\}^\infty$. Thus the gaps are finite polygons. According to the description of the minimal $\alpha$-equivalence in terms of $S^\alpha$, all points of $C_s$ are $\sim_\alpha$-equivalent for each $s$. Moreover, points of polygonal gaps with a common chord are equivalent, too. Such a common chord cannot be a limit chord of the $S_{w*}$, so it must be some $S_{w*}$ itself. Thus if adjacent gaps exist, they belong to $C_{w0\hat{\alpha}}$ and $C_{w1\hat{\alpha}}$ for some word $w$, and points of these sets are $\sim_\alpha$-equivalent. The minimal $\alpha$-equivalence contains the dynamical equivalence.

Now assume that some $\alpha$-equivalence $\sim$ strictly contains $\sim_\alpha$. By theorem 4 (a), $T/\sim$ is a factor space of $F(\hat{\alpha})$ — and in fact a factor $\{0,1\}^\infty/\simeq$, where $\simeq$ is an invariant equivalence relation on $\{0,1\}^\infty$ strictly containing $\sim_{\hat{\alpha}}$. By theorem 3, this means that $T/\sim = F(s)$, where $s = \overline{u0}$ and $0\hat{\alpha} \in \{0u, 1u\}^\infty$ for some 0-1-word $u$. Thus $\sim$ is degenerate by theorem 4 (b). ∎

Theorems 1 and 4 describe the topology of $J = T/\sim_\alpha$ in an abstract way as a factor of a shift space. This description includes a Markov partition of $J$, and it shows that $J$ is obtained from the shift space $\{0,1\}^\infty$ by pairwise identification of countably many points. Moreover, $J$ is a dendrite ('tree-like') [9]. There are no tree-like Julia sets for periodic $\hat{\alpha}$ (cf. sec. 9,10).

Let us classify the gaps of $S^\alpha$ which correspond to branching points of $J$. Gaps come in families $\{l_w(G) | w \in \{0,1\}^*\}$. By theorem 5.2, there are two types of families: families with a periodic gap and families with a central gap. Periodic gaps belong to certain initial subwords of $\hat{\alpha}$ as we show in theorem 8.2.

We now assume we have a central gap $G$. It contains the diameter $S_*$, and there is the gap $-G$ symmetric to $G$ at the origin. Since the vertices of $G$ and $-G$ are identified by $\sim_\alpha$, we delete $S_*$ from the lamination, and consider $G_* = G \cup -G$ as the unique central gap of $S^\alpha$. With this convention, there are two cases. If $\alpha$ is preperiodic, then $G_*$ is a $2n$-gon if it maps onto a periodic $n$-gon by some iterate of $h$, and $G_*$ is a rectangle if it maps onto some periodic chord.

If $\alpha$ is not preperiodic, then a central gap $G_*$ can only be a rectangle, by theorem 5.2. Such rectangle exists iff $\alpha$ belongs to a chord of $S^\alpha$, which means $\pi^\alpha(\hat{\alpha}) = 2$ by theorem 6.4. We give another necessary condition.

Assume the rectangle $R$ with vertices $\frac{\alpha}{2}, \frac{\beta}{2}, \frac{\alpha+1}{2}, \frac{\beta+1}{2}$ is a gap. In this case, we shall say that $\beta$ is the *conjugate point* of $\alpha$. For each $n \geq 0$, the chord $h^n(\alpha)h^n(\beta)$ does not intersect $R$, so its length is not larger than the long side of $R$. It follows that the chord $h^{n+1}(\alpha)h^{n+1}(\beta)$ is not shorter than $\alpha\beta$. Thus the points of the forward orbit of $\alpha$ do not come to the arc $\alpha\beta$.

In this case $\hat{\beta} = \hat{\alpha}$, and we can state more. If we construct the lamination $S^\alpha$ from rectangles $l_w(R)$ instead of $l_w(S_*^\alpha)$, we see that it coincides with $S^\beta$ (except for the choice of diagonals).

**Theorem 7.** (Central gap and conjugate point for non-periodic $\alpha$)
Let $\alpha < \frac{1}{2}$ be non-periodic.

(a) A central gap exists iff $\pi^\alpha(\hat{\alpha}) > 1$. The central gap has $2\pi^\alpha(\hat{\alpha})$ vertices.

(b) If $\alpha$ is not preperiodic, a central gap must be a rectangle. A central rectangle can only exist if for some $\epsilon$, either $]\alpha - \epsilon, \alpha[$ or $]\alpha, \alpha + \epsilon[$ does not contain points of the forward orbit of $\alpha$.

(c) Assume $\alpha$ is not preperiodic and has the conjugate point $\beta$. Then $\hat{\alpha} = \hat{\beta}$, and $\sim_\alpha = \sim_\beta$. Hence $\beta$ has the conjugate point $\alpha$. Moreover, there is no point $\gamma$ between $\alpha$ and $\beta$ with $\hat{\gamma} = \hat{\alpha}$.

(d) When the forward orbit of $\alpha$ does not hit $]\alpha, 1-\alpha[$ then $\beta$ exists and $\beta = 1 - \alpha$. If $\alpha$ has a conjugate point $\beta > \frac{1}{2}$ then $\beta = 1 - \alpha$.

**Proof.** In the last statement of (c), $\gamma$ cannot be eventually periodic, and the diameter corresponding to $\gamma$ must intersect the short sides of $R$. This implies $I^\gamma(\alpha) = I^\gamma(\beta) = \hat{\gamma}$. By theorem 1, $\sim_\gamma$ identifies all three points which contradicts (b). For (d) we note that all $h^i(\alpha)$ are in $]-\frac{\alpha}{2}, \frac{\alpha}{2}[$ or on the opposite arc. Thus $I^\alpha(1-\alpha) = \hat{\alpha}$, so that $\alpha$ and $1-\alpha$ are identified by $\sim_\alpha$ and their preimages form a central gap. The second assertion follows from (c). ∎

If $\alpha$ is an accumulation point of its own orbit under $h$, from the left and from the right, then $\alpha$ represents an endpoint in *any* abstract Julia set $T/\sim_\gamma$ (for non-periodic $\hat{\gamma}$, see sec. 9,10), and also in the abstract Mandelbrot set [2]. The $\alpha$ with property (d) will be studied in part 3 of our paper; they correspond to real values $c$ of $p_c$. Finally, let us note that central gaps are quite rare: the set of $\alpha \in T$ which satisfy (b) has Lebesgue measure zero (some word must not appear in the binary representation of $\alpha$).

# 8 The periodic branching points

Let $\beta, \gamma$ be rational points on $T$ with odd denominator. We want to determine all $\alpha$ for which $\beta\gamma$ is a chord in $S^\alpha$. Since the points are periodic under $h$, this chord must be a side of some $n$-gon $D_s = \text{conv } C_s$ ($D_s = \beta\gamma$ would be considered as 2-gon), and therefore $I^\alpha(\beta) = I^\alpha(\gamma) = s$. If $p$ denotes the period of $s$, then by proposition 5.3, $h^i(\beta)h^i(\gamma)$, $i = 0, 1, \ldots np-1$, are the different sides of $D_{\sigma^k(s)}$, $k = 0, 1, \ldots p-1$.

Thus a necessary condition for the existence of any $\alpha$ is that $\beta$ and $\gamma$ have the same period $np$ under $h$, and that the $h^i(\beta)h^i(\gamma)$ do not intersect except at their endpoints. Among all these chords ($i = 0, \ldots np-1$) there is a largest chord $\beta^+\gamma^+$, and the image under $h$ of this chord will be denoted by $\beta^-\gamma^-$ since it is the smallest one (cf. lemma 5.1).

Now a necessary condition for $\alpha$ is that the diameter $S_*$ from $\frac{\alpha}{2}$ does not intersect $\beta^+\gamma^+$. In other words, $\alpha$ must belong to the arc $\beta^-\gamma^-$ which is equivalent to the fact that $S_*$ is contained in the region $R$ between $S = \beta^+\gamma^+$ and $-S$. If this condition is fulfilled, then by lemma 5.1 $S_*$ cannot intersect any of the chords $h^i(\beta)h^i(\gamma)$. Endpoints of chords which form a polygon must have the same itinerary, for all $\alpha \in \beta^-\gamma^-$ : we just assign 0 and 1 to the parts of $T \setminus R$. Moreover, by the above property of $h^p$ all chords corresponding to some $s$ must form a polygon. We proved

**Theorem 8.1** (The $\alpha$ which belong to a fixed branching point)
Let $\beta, \gamma$ be periodic under $h$.

(a) In order that the chord $\beta\gamma$ belongs to some $S^\alpha$, it is necessary and sufficient that $\beta$ and $\gamma$ have the same period $q$, and that the chords $h^i(\beta)h^i(\gamma)$, $i = 0, 1, \ldots q$ do not cross each other.

(b) If this condition is fulfilled, then $I^\alpha(\beta) = I^\alpha(\gamma)$ holds exactly for those non-periodic $\alpha$ which belong to the shortest of the arcs $h^i(\beta)h^i(\gamma)$. Moreover, $s = I^\alpha(\beta)$ is constant for all these $\alpha$. (For $\beta < \frac{1}{2}, \beta < \gamma$ we obtain $s$ by assigning 0 to the points of the arc $\beta^+\gamma^+$ and 1 to the opposite points.) ∎

As a corollary, we show that the itinerary of a periodic branching point of $J$ has as period an initial subword of $\hat{\alpha}$, and the number of branches can be counted as for $s = \overline{0}$.

**Theorem 8.2** (The itineraries of periodic gaps in $S^\alpha$)
Let $\alpha$ be non-periodic, and $s$ a sequence not of the form $u\hat{\alpha}$. Let $D_s$ be a gap in $S^\alpha$ with $k \geq 3$ sides. Then $s$ is eventually periodic with some minimal period $p$. More precisely, $\sigma^n(s) = \overline{w}$ for some $n$ and $w = \hat{\alpha}_{|p}$, and $\hat{\alpha}$ begins with $w^{k-2}$.

**Proof.** $s$ belongs to a point $\beta \in T$ which is periodic under $h$. (For non-periodic $\hat{\alpha}$ this was shown, for periodic $\hat{\alpha}$ see below.) Replacing $s$ by some $\sigma^m(s)$, we can assume that $\beta = \beta^+$. The proof of proposition 5.3 implies that all $h^i(\beta^-)h^i(\gamma^-)$ with $i = 0, 1, \ldots p(k-2) - 1$ must be shorter than $\frac{1}{3}$. Thus the first $p(k-2)$ coordinates of $I^\alpha(\beta^-)$ and $I^\alpha(\alpha) = \hat{\alpha}$ must coincide. ∎

# 9 The Siegel disk case

Now we consider the case that $\hat{\alpha} = \overline{w}$ is periodic with minimal period $p = |w|$, but $\alpha$ is not periodic under $h$ — the forward orbit of $\alpha$ is infinite. Such $\alpha$ can be defined recursively. For instance, the binary number $\alpha = .01001\,01001001\,01001\ldots$ constructed as $\lim a_n$ where $a_1 = 01$, $b_1 = 001$, $a_{n+1} = a_n b_n$ and $b_{n+1} = a_n b_n b_n$ fulfils $\hat{\alpha} = \overline{0}$, by (3) in sec. 4 and an inductive argument.

We shall study the infinite gap $D_{\hat{\alpha}}$ and its preimages under $h$. For $i = 1, \ldots, p$, the gap $D_{\overline{u}}$ with $u = w_i \ldots w_p w_1 \ldots w_{i-1}$ is invariant under the action of $h^p$, by definition. Let $v = v_1 \ldots v_m$ denote a word which does not have $w$ as terminal subword, and let $w' = w_1 \ldots w_{p-1}$, $v' = v_1 \ldots v_{m-1}$.

**Theorem 9.1** (Abstract Siegel disks)
Let $\alpha$ be non-periodic and $\hat{\alpha} = \overline{w}$. The bounding chords of the gap $D_{v\overline{w}}$ in $S^\alpha$ are $S_{v'*}$ and $S_{vw^k w'*}$, $k = 0, 1, \ldots$, and the gap does not contain rational points of $T$. The minimal $\alpha$-equivalence identifies the endpoints of bounding chords, so that the Cantor set $C_{v\overline{w}}$ becomes a circle $C$. The action on $C$ induced by $h^p$ is conjugate to an irrational rotation.

**Proof.** The graph $\Gamma^\alpha(\hat{\alpha})$ will have loops at all vertices $kp$, $k = 1, 2, \ldots$ According to sec. 6, the region $D_{\overline{w}}$ is bounded by the chords $S^k := S_{w^k w'*}$, $k = 0, 1, \ldots$ There are less than $p$ other bounding chords $S'$ corresponding to chains of proper subwords of $w$ in $\Gamma^\alpha(\hat{\alpha})$. Moreover, such chords are limit chords of the lamination $S^\alpha$, so they are permuted among themselves under the action of $h^p$. Thus, if there were *any* bounding chords $S'$ different from the $S^k$ then $D_{\overline{w}}$ would contain a fixed point $\beta$ for some $h^{pm}$.

We derive a contradiction. $h^p$ is one-to-one on $D_{\overline{w}}$, except for the collapsing chord $S_{w'*}$. Thus $h^{pm}$ preserves the circular order of $D_{\overline{w}}$. Moreover, $h^{pm}(S^{k+m}) = S^k$, and the $S^k$ are disjoint, and dense in $D_{\overline{w}}$. This is impossible.

To derive the last statement from Denjoy's theorem, Thurston proved that $h^p$ does not possess a proper closed invariant subset $K$ in $D_{\overline{w}}$ ([15], II.5.3): Since the backward image of the critical chord is dense, $K$ could not contain any $S^k$. Generate a new lamination by adding the chords of the boundary of conv $K$. This lamination has a periodic gap, which leads to the previous contradiction. ∎

**Theorem 9.2** (Uniqueness of topology in the Siegel disk case)
Let $\alpha$ be non-periodic with $\hat{\alpha} = \overline{w}$. Then

(a) The Cantor sets $C_{v\overline{w}}$ turn into singletons for $\approx_\alpha$ and into circles for $\sim_\alpha$ . Two points which are not identified by $\sim_\alpha$ and not contained in one $C_{v\overline{w}}$ have different itineraries.

(b) $\sim_\alpha$ is the only non-degenerate $\alpha$-equivalence for which the quotient space can be obtained from $T$ by a homotopic process in the plane.

(c) If the point $c$ belongs to the Julia set $J_c$ and has external angle $\alpha$, then $(J_c, p_c)$ is conjugate to $(T/\sim_\alpha, \tilde{h})$.

**Proof.** (a): Two points which are not in the same gap are separated by some chord $S_{v*}$. Proposition 6.2 says that there are no infinite gaps beside the $D_{v\overline{w}}$.

(b): Let us try to identify two points $\beta, \gamma$ which are not identified by $\sim_\alpha$: we draw the chord $\beta\gamma$ in $S^\alpha$. If both points belong to the same gap $D_{v\overline{w}}$ then in order to get an invariant lamination, we have to draw an image chord $S$ in $D_{\overline{w}}$, and all $h^{kp}(S)$, $k = 1, 2, \ldots$ Since $h^p$ acts as an irrational rotation, this leads to the degenerate dynamical equivalence.

If $\beta$ and $\gamma$ do not belong to the same infinite gap, we have to identify them with all endpoints of chords crossing the chord $\beta\gamma$, in order to obtain the quotient space by a homotopic process. However, $I^\alpha(\beta) \neq I^\alpha(\gamma)$ by (a), so that $\beta\gamma$ crosses a gap $D_{v\overline{w}}$. Again, this leads us to an equivalence relation containing $\approx_\alpha$ which is degenerate.

(c): Yoccoz has shown that $J_c$ is locally connected if the critical value $c$ has an external angle [8] . The rest follows from (b). ∎

## 10 The periodic case

Now let $\alpha$ be periodic under $h$, that is, $\alpha = q/(2^p - 1)$ where $p > 1$ is the minimal period. Then $\hat{\alpha} = \overline{w*}$ with $w = 0w_2 \ldots w_{p-1}$. There will be some ambiguity concerning the lamination $S^\alpha$ but the minimal $\alpha$-equivalence is well-defined. We start with the simplest case $w = 0$, $\alpha = \frac{1}{3}$. Obviously, $S_* = [\frac{1}{6}, \frac{2}{3}]$, $S_{0*} = [\frac{1}{3}, \frac{5}{12}]$ but for $S_{00*}$ we know only one endpoint $\frac{7}{24}$ since $l_0(\alpha)$ is not defined. We can connect this point either to $\frac{1}{6}$ or to $\frac{2}{3}$, or to both. In any case, all three points are identified by $\sim_\alpha$ .

In the general case, let $\dot{\alpha} = h^{p-1}(\alpha)$ denote the periodic point among $\frac{\alpha}{2}$ and $\frac{\alpha+1}{2}$, and $\ddot{\alpha}$ the preperiodic one. That is, $\dot{\alpha} = \frac{\alpha}{2}$ for even $q$ and $\ddot{\alpha} = \frac{\alpha}{2}$ for odd $q$. For any word $v \in \{0w, 1w\}^*$, the point $\gamma_v = l_v(\ddot{\alpha})$ is unique. Now we can decide that $l_{0w}(\dot{\alpha}) = l_0(\alpha)$ is either $\frac{\alpha}{2}$ or $\frac{\alpha+1}{2}$, or draw both chords whenever this can be done without crossing previous chords. In each case we can construct $S^\alpha$ as in sec. 5, and we see that all points $\gamma_v$ will be connected to $\dot{\alpha}$ by a finite chain of chords.

Thus $\sim_\alpha$ identifies all $\gamma_v$, and all their accumulation points, with the critical points $\dot{\alpha}, \ddot{\alpha}$. Note that $I^\alpha(\gamma_v) = v\overline{*w}$, and $\gamma_v$ separates $C_{v0}$ and $C_{v1}$, in each construction. Thus (cf. below)

$$C_\bullet = \text{cl } \{\gamma_v | \ v \in \{0w, 1w\}^*\}$$

includes all points with itinerary in $\{0w, 1w, *w\}^\infty$. With theorem 3 (b) and 4 we get

**Proposition 10.1** (The strong equivalence for periodic $\alpha$ )
For periodic $\alpha$, minimal and dynamical $\alpha$-equivalence coincide. They are degenerate, however. ∎

For our purposes we define $S^\alpha$ as follows. The chord $S^0 := S_{w*} = l_w(S_*)$ has endpoints $\alpha = l_w(\dot\alpha)$ and $\gamma_w = l_w(\ddot\alpha)$. Define $l_0(\alpha)$ so that $l_0$ preserves the order of these two points: $l_0(\alpha) = \frac{\alpha}{2}$ , $l_1(\alpha) = \frac{\alpha+1}{2}$ if $\alpha < \gamma_w$, and conversely otherwise. This guarantees that $l_i$ is continuous on the arc subtended by $S_{w*}$ , and so the arc length of $S_{0w*}$ and $S_{1w*}$ is half the length of $S^0 = S_{w*}$, in particular less than $\frac{1}{3}$.

Another concept is needed. Let us define the *characteristic symbol* of $\alpha$ as

$$e = e^\alpha = \begin{cases} 1 & \text{if } \gamma_w \text{ is between } \alpha \text{ and } \dot\alpha \\ 0 & \text{if } \gamma_w \text{ is between } \alpha \text{ and } \ddot\alpha \end{cases} \qquad \text{and} \qquad e' = 1 - e.$$

It is easy to check that $l_e(\alpha) = \dot\alpha$ and $l_{e'}(\alpha) = \ddot\alpha$. Moreover, $S^0 = S_{w*}$ separates $C_{w0}$ and $C_{w1}$, by definition, and $e$ was chosen so that $C_{we}$ is the exterior part (we shall show that $C_{we}$ is the arc $\gamma_w\alpha$ ). In fact, the definition of $e$ says that for small positive $\epsilon$, the point $h^{p-1}(\alpha + \epsilon(\gamma_w - \alpha))$ lies in $T_e$.

We draw all chords $S_{v*}$ as in sec. 5 (cf. fig. 2a). Since $l_{we}(\dot\alpha) = \dot\alpha$ and $l_{we'}(\dot\alpha) = \ddot\alpha$, the chords $S_{uv*}$ and $S_{v*}$ will have a common endpoint for $u = (we)^k$ and $u = we'(we)^{k-1}$, $k = 1, 2, \ldots$ and for arbitrary words $v$. Thus the chords $S^k = S_{w(e'w)^k*}$, $k = 0, 1, \ldots$ form a chain starting at $\alpha$ : $\gamma_w$ is in $S^0$ and $S^1$, and $\gamma_{we'w}$ in $S^1$ and $S^2$ ... So all $S^k$ are bounding chords of one region $C_{\overline{we'}}$ in $C_w$, and there is a limit of the endpoints of the $S^k$ which we call *the conjugate point* $\beta$ *of* $\alpha$.

Our next statement gives a simple formula for calculation of $\beta$ from $\alpha$ and shows that the chord $\alpha\beta$ is a minor in the sense of Thurston. Since $h^p(S^k) = S^{k-1}$, the point $\beta$ is fixed by $h^p$. Let $h^{p-1}(\beta) = \dot\beta$, and let $\ddot\beta$ the opposite point.

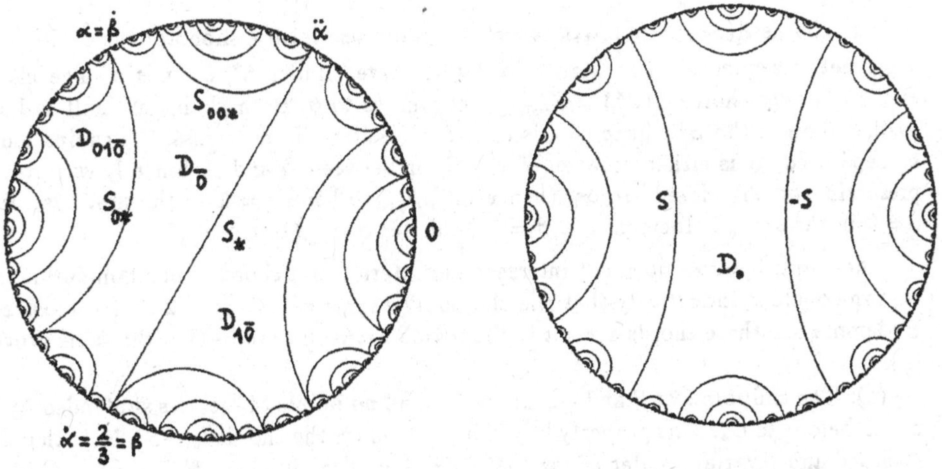


Fig. 2. The laminations $S^\alpha$ and $S_0^\alpha$ for $\alpha = \frac{1}{3}$ , $w = 0$ , $e = 1$

**Theorem 10.2** (The conjugate point of a periodic point)
Let $\alpha < \frac{1}{2}$ be periodic, $p > 1$ the minimal period, $\hat{\alpha} = \overline{w*}$ and $\beta$ the conjugate point.

(a) The open arcs $h^i(\alpha\beta)$, $i = 0, 1, \ldots, p-1$ are disjoint, and $\alpha\beta$ is the shortest among them. Moreover, $\hat{\alpha} = \hat{\beta}$.

(b) $\beta = \alpha + \frac{2^p}{2^p-1}(\gamma_w - \alpha)$

(c) $C_\bullet$ is a Cantor set on $T$ with cutout intervals $\dot{\alpha}\beta$, $\ddot{\alpha}\beta$ and $l_v(\ddot{\alpha}\ddot{\beta})$, $v \in \{0w, 1w\}^*$.

(d) The following conditions are equivalent:

- $\beta$ belongs to the orbit of $\alpha$
- $\overline{we'}$ (with $e = e^\alpha$) has period less than $p$
- $\dot{\alpha}\beta$ is not a limit chord of $S^\alpha$.

(e) There is no point $\gamma$ between $\alpha$ and $\beta$ with $\hat{\gamma} = \hat{\alpha}$. If $\beta > \frac{1}{2}$ then $\beta = 1 - \alpha$.

(f) The conjugate point of $\beta$ is $\alpha$, and $e^\alpha = e^\beta$.

**Proof.** The chords $h^i(S^k)$ with $k \geq 1$ and $0 \leq i \leq p-1$ are transformed into each other by $h$ as follows:

$$\ldots S^k, h(S^k), \ldots, h^p(S^k) = S^{k-1}, \ldots, S^1, h(S^1), \ldots, h^{p-1}(S^1) = l_e(S^0).$$

The last chord was by definition shorter than $\frac{1}{3}$, and this holds for all of them: If there were chords $h^i(S^k)$ of length greater $\frac{1}{3}$, choose $k$ minimal and then choose $i$ maximal. Since between this long chord and its opposite chord, there is the diameter $S_*$ and the adjoining chord $l_e(S^0)$, we obtain a contradiction to lemma 5.1.

It follows that the arc lengths of the $h^i(S^k)$, in the above order, form a geometric sequence with factor $\frac{1}{2}$. Thus the $S^k$, $k = 0, 1, \ldots$ are arranged monotonously, and (b) is proved.

The chords $l'_e(S^k) = S_{(e'w)^{k+1}}$, $k = 0, 1, \ldots$ do also form a monotonous chain, and a geometric sequence. The limit point is preserved under $h^p$, so it is $\dot{\beta}$. The chain of the opposite chords $l_e(S^k) = S_{ew(e'w)^k}$ converges to $\ddot{\beta}$. Both chains are half as long as the chain of the $S^k$. Since chords of our lamination do not cross, the chains must be separated: $\alpha$ is either between $\dot{\alpha}$ and $\dot{\beta}$, or between $\ddot{\alpha}$ and $\ddot{\beta}$. Since $l_e$ was order-preserving on the closed arc $\alpha\gamma_w$(hence on $\alpha\beta$), we have $e = 1$ in the first case and $e = 0$ in the second. Incidentally, $\beta = \alpha$ holds for $\alpha = \frac{1}{2^p-1}$, $p > 1$.

The chord $S = \dot{\alpha}\dot{\beta}$ does not intersect the interior of a chord of our lamination, so we can conclude inductively that the chords $h^i(\alpha\beta)$, $i = 0, \ldots, p-2$ do not cross $S_*$. By lemma 5.1, these chords are not in the region between $S$ and $-S = \ddot{\beta}\ddot{\alpha}$. This proves (a).

(c): The endpoints of $S$ and $-S$ are in $C_\bullet$, and no points of the arcs subtended by $S$ or $-S$ belong to $C_\bullet$. This property has to be verified for the chords $l_v(-S)$. Note that $C_\bullet$ is closed and invariant under $h^p$, and $h^p(\dot{\alpha}) = \dot{\alpha} = h^p(\ddot{\alpha})$ implies $h^p(S) = h^p(-S) = S$. Consider the chord $l_{e'w}(-S)$ with endpoints $\gamma_{e'w}$ and $l_{e'w}(\ddot{\beta})$. Since $\dot{\beta}$ is the limit point of chords $S_{(e'w)^k}$, $l_{e'w}(\ddot{\beta})$ must be the limit of the $S_{e'wew(e'w)^k}$. Our assertion holds for $l_{e'w}(-S)$ and the opposite chord $l_{ew}(-S)$, and extends to the other chords by an inductive argument.

(d): Suppose first $\beta$, or equivalently, $\dot{\beta}$ belongs to the forward orbit of $\alpha$. We prove the other two conditions. Since $\dot{\beta} = h^m(\ddot{\alpha})$ for some $m < p$, the chords $h^m(S_{(e'w)^k*})$, $k = 1, 2, \ldots$ form a chain of chords starting at $\dot{\beta}$, similar to the chain of chords running from $\ddot{\alpha}$ to $\dot{\beta}$. The supremum of this chain is $\dot{\gamma} = h^m(\dot{\beta})$. If $\dot{\gamma} \neq \dot{\alpha}$, there will be a succeeding chain, but after less than $p$ steps we shall arrive at $\dot{\alpha}$. The gap surrounded by the chains of chords is $D_{\overline{e'w}}$, and it is fixed by $h^m$. Thus $\overline{e'w} = \sigma^m(\overline{e'w})$, the period of $\overline{e'w}$ is $\gcd(m, p)$.

Now suppose $\overline{e'w}$ has minimal period $m < p$, so that $D_{\overline{e'w}}$ is invariant under $h^m$. Let $G$ be the convex hull of all points in $D_{\overline{e'w}}$ which have period $p$ (their itineraries may contain $*$). If $\dot{\alpha}$ and $\dot{\beta}$ are the only two vertices of $G$ then $m = \frac{p}{2}$ and $\dot{\beta} = h^m(\dot{\alpha})$. Otherwise, $G$ is a periodic gap to which we can apply proposition 5.3 to show that $\dot{\beta}$ is on the orbit of $\dot{\alpha}$.

Finally, assume $\dot{\beta}$ does not belong to the forward orbit of $\alpha$. We show that $S$ is a limit chord of our lamination. Since the minimal period of $\overline{e'w}$ is $p$, beside the chords $S_{(e'w)^k*}$ there are less than $p$ limit chords bounding $S_{\overline{e'w}}$, corresponding to the chains of initial subwords of $\hat{\alpha}$ of length less than $p$ in $\overline{e'w}$. Since the $S_{v*}$ are dense in $T$, the limit chords form a finite chain connecting $\dot{\beta}$ with $\dot{\alpha}$. If there was more than one chord in the chain, then together with the chord $\dot{\beta}\dot{\alpha}$ we would have a polygon $G$ fixed by $h^p$. By proposition 5.3, the return map would act transitively on vertices of $G$ which contradicts the assumption. (See fig. 3 for this case.)

(e): Assume there is a $\gamma$ strictly between $\alpha$ and $\beta$ with $\hat{\gamma} = \overline{w*}$. Then $I^\gamma(\alpha) = I^\gamma(\beta) = \overline{we'}$. In the lamination $S^\gamma$, the gap $D_{\overline{e'w}}$ has at least two vertices $\dot{\alpha}, \dot{\beta}$ different from $\dot{\gamma} = h^{p-1}(\gamma)$ which by the proof of (d) implies that these vertices are on the orbit of $\gamma$. This contradicts (a). For the second assertion, remember that $\hat{\alpha} = \widehat{1-\alpha}$.

(f): By (e) it suffices to verify that the point $\delta_w = l_w(\ddot{\beta})$ in the lamination $S^\beta$ lies between $\alpha$ and $\gamma_w$. In $S^\alpha$, the image chords of $\alpha\gamma_w$ do not intersect other chords, and $h^i(\alpha)$ is never between $S$ and $-S$. Thus $h^i(\gamma_w)$ is also outside the central region, and the same holds for $h^i(\delta_w)$ in $S^\beta$. So the chords $\alpha\beta$ and $\gamma_w\delta_w$ do not intersect. For the last statement, we proved already that $e = 1$ iff $S$ is "on the left of $-S$ ". ∎

**Remarks.** (1) Lavaurs ([10], cf. [2]) gave a method to determine conjugate points but our formula (b) is straightforward. For example, $\alpha = \frac{1}{5}$ with $\ddot{\alpha} = \frac{1}{10}$ and $\hat{\alpha} = \overline{001*}$ yields $\gamma_1 = l_1(\ddot{\alpha}) = \frac{1}{20}$ , $\gamma_{01} = \frac{21}{40}$ , $\gamma_{001} = \frac{21}{80}$ and $\beta = \frac{4}{15}$.
(2) We consider $\dot{\alpha}$ as a point of $T_e$ : $I(\dot{\alpha}) = \overline{ew}$, and $\ddot{\alpha}$ as a point of $T_{e'}$. Then the map $I^\alpha : C_\bullet \to \{0w, 1w\}^\infty$ is a homeomorphism. The endpoints of $S$ have itineraries $\overline{0w}$ and $\overline{1w}$, and those of $-S$ have $0w\overline{1w}$ and $1w\overline{0w}$.

Now let us improve the degenerate lamination $S^\alpha$. The idea is the same as for the Siegel disk: instead of collapsing a Cantor set, we identify endpoints of cutout intervals and turn the Cantor set into a circle. Instead of the diameter $S_*$ we use the chord $S = \dot{\alpha}\dot{\beta}$ as generator of the lamination. The forward images of $S$ do not meet the central region, and the backward images form the cutout intervals of $C_\bullet$ and its preimages. Adding the limit chords, we have a lamination $S_0^\alpha$ (cf. fig. 2b).

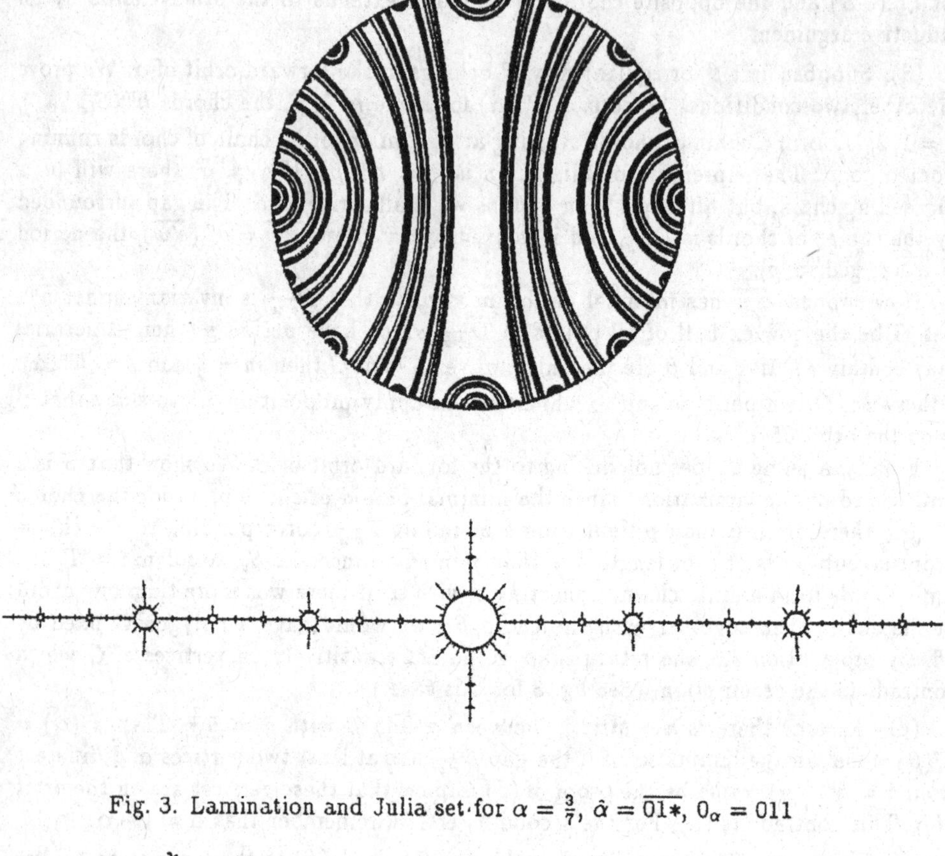

Fig. 3. Lamination and Julia set for $\alpha = \frac{3}{7}$, $\hat{\alpha} = \overline{01*}$, $0_\alpha = 011$

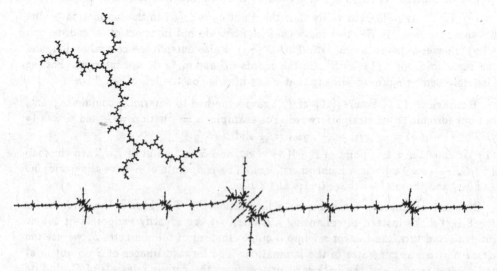

Fig. 4. Renormalization: the Julia set of $\delta = \frac{1}{4}$, $\hat{\delta} = 00\overline{1}$, inserted into the holes of the figure above, gives the Julia set of $\gamma = \frac{3}{7} + \frac{1}{64}$, $\hat{\gamma} = 011011\overline{010} = 0_\alpha 0_\alpha \overline{1_\alpha}$

**Theorem 10.3** (The topology of Julia sets for periodic $\alpha$ )
Let $\alpha$ be periodic and $\beta$ the conjugate point.

(a) Among all equivalence relations on $T$ which are forward invariant under $h$ and do not identify a point of $T_0^\alpha$ with a point of $T_1^\alpha$, there is a largest relation $\simeq_\alpha$. It corresponds to the lamination $S_0^\alpha$ generated by $S$.

(b) If $\frac{\alpha}{2}$ and $\frac{\alpha+1}{2}$ are assigned to $T_0, T_1$ as in remark (2) then all points of $T$ have 0-1-sequences as itineraries and $\beta \simeq_\alpha \gamma$ iff either $I(\beta) = I(\gamma)$ or $\{I(\beta), I(\gamma)\} = \{v1w\overline{0w}, v0w\overline{1w}\}$ or $\{I(\beta), I(\gamma)\} = \{v\overline{0w}, v\overline{1w}\}$ for some 0-1-word $v$, where in the last case $v$ must not end with $0w$ or $1w$.

(c) Suppose the critical value $c$ of $p_c$ belongs to the immediate basin of an attractive or indifferent periodic point $z$ of period $r$, and the image of the fixed point $d$ of $p_c^r$ on the boundary of the basin has external angle $\alpha$. Then $d$ has also the external angle $\beta$, and $(J_c, p_c)$ is conjugate to $(T/\simeq_\alpha, h)$.

(d) If $\beta = h^m(\alpha)$ for some $m$ then the external angles of $d$ are $h^{jm}(\alpha), j = 1, \ldots, k = r/\gcd(r, m)$, and $d$ is in the closure of the basins of all $h^{jm}(z)$. If $\beta$ does not belong to the forward orbit of $\alpha$ then there are no further external angles of $d$, and the immediate basins of the $p_c^i(z)$ , $i = 0, \ldots, r$ have disjoint closures.

**Proof.** (a): Let $\sim$ be invariant under $h$, and suppose the chords $\gamma\delta$ with $\gamma \sim \delta$ do not cross $S_*^\alpha$. Then they do not cross any chord of $S^\alpha$, and no chord $\alpha\gamma_v$ with $v \in \{0w, 1w\}^*$ since $\alpha$ and $\gamma_v$ are connected by a finite chain in $S^\alpha$. So $S$ and its preimages can be added to the lamination of $\sim$ to obtain an invariant lamination containing $S_0^\alpha$. However, since the $S_{u*}$, $u \in \{0,1\}^*$ are dense in $S^\alpha$, any lamination strictly containing $S_0^\alpha$ contains chords crossing some $S_{u*}$. For (b) see remark (2).

(c): It is known [6, 7, 3, 11] that in this case $J$ is locally connected, the boundary of the basin is a Jordan curve $K$, and the action of $p_c^r$ on $K$ is conjugate to angle-doubling on $T$. The external rays semiconjugacy between $h$ and $p_c$ shows that $\alpha$ has period $r$, and $\dot\alpha, \ddot\alpha$ are the external angles of $d, -d$. Since $d$ and $-d$ are connected by an arc through the basin, the equivalence relation on $T$ induced by $J$ cannot identify across $S_*^\alpha$ or its preimages, nor across $S$. This implies that the point $d'$ with external angle $\beta$ belongs to $K$. Now $d = d'$ since $p_c^r$ has a unique fixed point on $K$. The rest follows from (a), and (d) from 10.2, (d). ∎

## 11 Renormalization

For any periodic $\alpha$, we have constructed a circle $T' = C_\bullet / \simeq_\alpha$ in the corresponding Julia set, in the same way as we describe $T$ in terms of binary numbers: $v0\overline{1} = v1\overline{0}$ for arbitrary words $v$, and $\overline{0} = \overline{1}$. (Note that binary numbers are itineraries for $\alpha = 0$.) The correspondence between binary numbers in $T$ and itineraries of points in $T'$ given by the substitution $0 \mapsto ew$, $1 \mapsto e'w$ induces a one-to-one map $g^\alpha : T \to T'$ with $g(0) = S$ (more precisely $g(0) = \dot\alpha$ and $g(1) = \dot\beta$), $g(\frac{1}{2}) = -S$. This map conjugates $h$ on $T$ to $\tilde{h}^p$ on $T'$.

Another important circle in the Julia set is $T" = h(C_\bullet)/\simeq_\alpha$. The map $f^\alpha = l_w g^\alpha : T \to T"$ is one-to-one (since $l_w$ is one-to-one on $C_\bullet$), and it conjugates $h$ on $T$ to the return map $\tilde{h}^p$ of $T"$. Clearly, this map is given by the substitution $0 \mapsto we$, $1 \mapsto we'$.

If we construct a Julia set by identifying some $\frac{\gamma}{2}$ with its opposite point in $T$, and $\frac{\gamma}{2}$ belongs to $C^\alpha_\bullet$, we can try to make the identification within the Julia set of $\alpha$, in the circle $T'$, its image and preimage circles. Since the Julia set is unique, both methods give the same result. The external angle of $\gamma$ with respect to $T"$ can be any number between 0 and 1. So the identification in $T'$ gives small Julia sets of arbitrary shape while the topological structure of the whole Julia set outside the circles remains unchanged. See fig. 4.

In sec. 8 we proved that $\alpha \sim_\gamma \beta$ in all Julia sets associated to some $\gamma$ between $\alpha$ and the conjugate point $\beta$ (for rational $\gamma$, use $\simeq_\gamma$). Here we are only concerned with $\gamma \in h(C_\bullet)$. Everytime when $\gamma$ leaves this set, some preimages of $\alpha$ and $\beta$ will be identified in a different way.

Renormalization and substitution were already touched by theorem 3 (d). Let us call it *"tree of words"*: $s \prec w$ if $0s \in \{0w, 1w\}^\infty$ (or $s \in \{w0, w1\}^\infty$). (Morse-Thue numbers are those with infinitely many successors in this order.) $s \prec w$ means that the invariant factor $F(s)$ of $\{0,1\}^\infty$ has a subset $F'$ which is an invariant factor of $\{0w, 1w\}^\infty$. If $0s'$ is the image of $0s$ under the substitution $0w \mapsto 0$, $1w \mapsto 1$, then $F' = F(s')$. For any $w$, the partially ordered set of all predecessors together with $w$ is isomorphic to the whole tree. Similar statements hold for our kneading sequences:

**Theorem 12.** (Renormalization and substitution of kneading sequences)
Let $\alpha$ be periodic with period $p$, $\hat{\alpha} = \overline{w*}$ and $e = e^\alpha$.

(a) The substitution $0 \mapsto we$, $1 \mapsto we'$ induces a correspondence between binary numbers in $T$ and $\alpha$-itineraries of points in $T"$, and a conjugacy
$$f^\alpha : (T, h) \to (T" = h(C^\alpha_\bullet)/\simeq_\alpha, \tilde{h}^p).$$

(b) $f^\alpha(b) = \alpha + (\gamma_w - \alpha)\sum_{n=0}^\infty b_n 2^{-p(n-1)}$ for every binary number $\delta = .b_1 b_2 \ldots$ (numbers with two binary representations yield two endpoints of a chord)

(c) For each $\delta$ in $T$, the substitution $0 \mapsto 0_\alpha = we$, $1 \mapsto 1_\alpha = we'$, $* \mapsto *_\alpha = w*$ transforms $\hat{\delta}$ into the kneading sequence of $f^\alpha(\delta)$.

**Proof.** (b): $f^\alpha(\delta)$ has itinerary $we_1we_2\ldots$ with $e_k = e$ for $b_k = 0$ and $e_k = e'$ for $b_k = 1$. It is the limit of the monotone sequence $\beta_n = l_{we_1\ldots we_n}(\alpha)$ with $I^\alpha(\beta_n) = we_1\ldots we_n\overline{we}$. If $e_n = e$, then $\beta_n = \beta_{n-1}$. If $e_n = e'$, then the chord $\beta_{n-1}\beta_n = S_{we_1\ldots we_{n-1}w*}$ has the same length as $S^{n-1} = S_{(we')^{n-1}w*}$, namely $2^{-p(n-1)}$, which is easily verified by induction (cf. proof of 10.2).

(c): Take $\gamma$ in $h(C^\alpha_\bullet)$. By the results of 10.2, $\hat{\gamma} = wt_1wt_2w\ldots$ with $t_k \in \{0, 1, *\}$. If $I^\alpha(\gamma) = we_1we_2\ldots$, then $e_k \neq t_k$ holds for those $k$ for which $h^{pk}(\gamma)$ is between $\alpha$ and $\gamma$ (consider the diameter $S^\gamma_*$ in $T'$). Now let $\gamma = f^\alpha(\delta)$ with $\delta = .b_1b_2\ldots$ and $\hat{\delta} = s_1s_2\ldots$ Note that $b_k \neq s_k$ iff $h^k(\delta)$ is between 0 and $\delta$ (cf. sec. 4), and apply (a) to complete the proof. ■

**Remark.** When this paper was written, we found that C. Penrose (Queen Mary and Westfield College, London) derived similar ideas. We recommend his thesis [16] which in some way complements our results.

## References


[1] Bandt, C. and Keller, K., Self-similar sets 2. A simple approach to the topological structure of fractals. *Math. Nachr.*, to appear

[2] Bandt, C. and Keller, K., Symbolic dynamics for angle-doubling on the circle, II. Description of the abstract Mandelbrot set. Preprint, Greifswald 1991

[3] Blanchard, P., Complex analytic dynamics on the Riemann sphere. *Bull. Amer. Math. Soc.* 11 (1984), 85-141.

[4] Branner, B. and Hubbard, J.H., The iteration of cubic polynomials. Part 1: The global topology of parameter space. *Acta Math.* 160 (1988), 143-206.

[5] Collet, P. and Eckmann, J.-P., *Iterated maps on the interval as dynamical systems.* Birkhäuser 1980.

[6] Douady, A. and Hubbard, J. *Étude dynamique des polynômes complexes, Première partie.* Publications Mathématiques d'Orsay, 1984.

[7] Douady, A. and Hubbard, J. On the dynamics of polynomial-like mappings. *Ann. Sci. Ecole Norm. Sup. (4),* 18 (1985), 287-343.

[8] Hubbard, J.H., according to J.-C. Yoccoz. Puzzles and quadratic tableaux. Preprint, Paris 1990.

[9] Kuratowski, K. *Topology, Vol. 1, 2,* New York and Warszawa 1966,1968.

[10] Lavaurs, P. Une déscription combinatoire de l'involution définie par M sur les rationnels à dénominateur impair. *C. R. Acad. Sc. Paris Série I, t. 303* (1986), 143-146

[11] Lyubich, M. Yu. Dynamics of rational transformations: topological picture. *Uspekhi Mat. Nauk* 41 (1986) no. 4 (250), 35-95, 239.

[12] Milnor, J. and Thurston, W.P. On iterated maps of the interval,*Lecture Notes in Mathematics* 1342 (1988), 465-563.

[13] Rees, M. A partial description of parameter space of rational maps of degree two: Part 1. *Acta Math.*, to appear

[14] Sullivan, D. Quasiconformal homeomorphisms and dynamics I. Solution of the Fatou-Julia problem on wandering domains, *Annals Math.* 122 (1985), 401-418.

[15] Thurston, W.P. On the combinatorics and dynamics of iterated rational maps. Preprint, Princeton 1985.

[16] Penrose, C. *On quotients of the shift associated with dendrite Julia sets of quadratic polynomials.* Thesis, Warwick 1990.

# Spectral Decomposition, Periods of Cycles and a Conjecture of M. Misiurewicz for Graph Maps

A.M. Blokh

All-Union Hematological Scientific Centre, SU-125167 Moscow, Nowozykovski pr. 4a

**Abstract**

We describe a spectral decomposition of the set $\omega(f) = \bigcup_{x \in X} \omega(x)$ for a continuous map $f : X \to X$ of a one-dimensional branched manifold ("graph") into itself similar to that of Jonker-Rand [JR], Hofbauer [H] and Nitecki [N] (see also [B1-B3]; the analogous decomposition holds for the sets $\Omega(f)$, $\overline{Per(f)}$. Denoting by P(f) the set of all periods of cycles of a map f we then verify the following Misiurewicz conjecture: for a graph X there exists an interger $L = L(X)$ such that for a continuous map $f : X \to X$ the inclusion $P(f) \supset \{1, \ldots, L\}$ implies that $P(f) = \mathbf{N}$ (we prove also that such a map f has a positive entropy). It allows us to prove the following
**Theorem.** Let $f : X \to X$ be a continuous graph map. Then the following statements are equivalent.
1) The map f has positive entropy.
2) There exists such n that $P(f) \supset n\mathbf{N} = \{i \cdot n | i \in \mathbf{N}\}$.

## 1 The spectral decomposition

A.N. Sharkovsky constructed a decomposition of the set $\omega(f) = \bigcup_{x \in I} \omega(x)$ for continuous interval maps $f : I \to I$ in [S]. Then in [JR] Jonker and Rand constructed for unimodal maps a decomposition which is in fact close to that of Sharkovky; however they used completely different methods of symbolic dynamics. In [H] a decomposition for piecewise-monotone maps with discontinuities was constructed by Hofbauer and then Nitecky [N] considered the decomposition for piecewise-monotone continuous maps from a more general point of view. The author's papers [B1,B2] were devoted to the case of arbitrary continuous interval maps and contained a different approach to the problem in question; it allowed us to obtain some new corollaries (e.g. describing generic properties of invariant measures for interval maps). A similar approach was used in [B3] to construct the decomposition for graph maps. We describe it briefly in Section 1.

Let X be a graph, $f : X \to X$ be a continuous map. We use the terms *edge, vertex, endpoint* etc. in the usual sense; the numbers of edges and endpoints of X are denoted by Edg(x) and End(X). We construct a decomposition of the set $\omega(f)$. First we need some definitions. A connected closed set $Y \subset X$ is called *subgraph*. A subgraph Y is called *periodic* (of period k) if $Y, \dots, f^{k-1}Y$ are pairwise disjoint and $f^kY = Y$; the set $\bigcup_{i=0}^{k-1} f^iY \equiv orbY$ is called a *cycle of subgraphs*. Let $Y_0 \supset Y_1 \supset \dots$ be periodic subgraphs of periods $m_0, m_1, \dots$, then $m_{i+1}|m_i (\forall i)$. If $m_i \to \infty$ then the subgraphs $\{Y_j\}_{j=0}^{\infty}$ are said to be *generating*. We call any invariant closed set $S \subset Q = \bigcap_j orbY_j$ a *solenoidal set* and denote the solenoidal set $Q \cap \omega(f)$ by $S_\omega(Q)$ (note that $\omega(f)$ is closed for a graph map f [B3]).

One can use a transitive shift in an Abelian zero-dimensional infinite group as a model for the map on a solenoidal set. Namely, let $D = \{n_i\}_{i=0}^{\infty}$ be a sequence of integers, $n_{i+1}|n_i(\forall i)$ and $n_i \to \infty$. Consider the group $H(D) \subset \mathbf{Z}_{n_0} \times \mathbf{Z}_{n_1} \times \dots$ defined as follows: $H(D) \equiv \{(r_0, r_1, \dots) | r_{i+1} \equiv r_i \pmod{n_i} \forall i\}$. Denote by $\tau$ the minimal shift in H(D) by the element (1,1,...).

**Theorem 1** [B1-B3]. *Let $\{Y_j\}_{j=0}^{\infty}$ be generating subgraphs with periods $\{m_j\}_{j=0}^{\infty} = D, Q = \bigcap_{j \geq 0} orbY_j$. Then there exists a continuous surjection $\varphi : Q \to H(D)$ with the following properties:*

1. *$\tau \circ \varphi = \varphi \circ f$ (i.e.$\varphi$ semiconjugates $f|Q$ to $\tau$);*
2. *there exists a unique set $S \subset Q \cap \overline{Per\, f}$ such that $\omega(x) = S$ for any $x \in Q$ and if $\omega(z) \cap Q \neq \emptyset$ then $S \subset \omega(z) \subset S_\omega$;*
3. *for any $\overline{r} \in H(D)$ the set $J = \varphi^{-1}(\overline{r})$ is a connected component of $Q$ and $\varphi|S_\omega$ is at most 2-to-1;*
4. *$h(f|Q) = 0$.*

Let us turn to another type of an infinite limit sets. Let $\{Y_i\}_{i=1}^{l}$ be a collection of connected graphs, $K = \bigcup_{i=1}^{l} Y_i$. A continuous map $\psi : K \to K$ which permutes them cyclically is called *non-strictly periodic (or 1-periodic)*; for example if Y is a periodic subgraph then $f|orbY$ is non-strictly periodic. In what follows we will consider monotone semiconjugations between non-stricly periodic graph maps (a continuous map $g : X \to X$ is *monotone* provided $g^{-1}(y)$ is connected for any $y \in Y$ ). We need the following

**Lemma 1.** *Let X be a graph. Then there exists a number r=r(X) such that if $M \subset X$ is a cycle of subgraphs and $g : M \to Y$ is monotone then*

$$card(\partial(g^{-1}(y))) \leq r(X) \ (\forall y \in Y).$$

Lemma 1 suggests the following definition. If $\varphi : K \to M$ is monotone, semiconjugates the non-strictly periodic map $f : K \to K$ to the non-strictly periodic map

$g : M \to M$ and there is a closed f-invariant set $F \subset M$ such that $\varphi(F) = M$ and $\varphi^{-1}(y) \cap F \subset \partial(\varphi^{-1}))$ $(\forall y \in M)$ then we say that $\varphi$ *almost conjugates* $f|F$ *to* $g$.
Let Y be an n-periodic subgraph, orbY=M. Denote by E(M,f) the set $\{x \in M|$ for any neighbourhood U of x in M we have $\overline{orbU}$=M} provided it is finite. We call the set E(M,f) a *basic set* and denote it by B(M,f) provided Per(f|M)$\neq \emptyset$; otherwise (if Per(f|M)=$\emptyset$) we denote E(M,f) by C(M,f) and call it a *circle-like set.*

**Theorem 2** [B1-B3]. *Let Y be an n-periodic subgraph, M=orbY and E(M,f)$\neq \emptyset$. Then there exists a transitive non-strictly periodic map $g : K \to K$ and a monotone surjection $\varphi : M \to K$ which almost conjugates $f|B(M,F)$ to g.. Furthermore, the following properties hold:*

1. *E(M,f) is a perfect set;*
2. *$f|E(M,f)$ is transitive;*
3. *if $\omega(z) \supset E(M,f)$ then $\omega(z) = E(M,f)$;*
4. *If E(M,f)=C(M,f) is a circle-like set then K is the union of n circles, g permutes them, $g^n$ on any of them is an irrational rotation and $h(g|C(M,f)) = 0$;*
5. *if E(M,f)=B(M,f) is a basic set then $\overline{B(M,f)} \subset Perf$ and there exists a number k and a subset $D \subset B(M,f)$ such that the sets $f^iD \cap f^jD$, $0 \leq i < j < kn$, are finite, $f^{kn}D = D$, $\bigcup_{i=0}^{kn-1} f^iD = B(M,f)$ and $f^{kn}|D$ is topologically mixing.*

To formulate the decomposition theorem denote by $Z_f$ the set of all cycles maximal with respect to inclusion among all limit sets of f.

**Teorem 3** [B1-B3]. *Let $f : X \to X$. be a continuous graph map. Then there exist a finite number of circle-like sets $\{C(K_i, f)\}_{i=1}^k$, an at most countable family of of basic sets $\{B(L_j, f)\}$ and a family of solenoidal sets $\{S_\omega(Q_\alpha)\}_\alpha$ such that*

$$\omega(f) = Z_f \bigcup \left( \bigcup_{i=1}^{k} C(K_i) \right) \bigcup \left( \bigcup_j B(L_j) \right) \bigcup \left( \bigcup_\alpha S_\omega(Q_\alpha) \right).$$

*Moreover, there exist numbers $\gamma(X)$ and $\nu(X)$ such that $k \leq \gamma(X)$, the only possible intersections in the decomposition are amog basic sets and at most $\nu(X)$ basic sets can intersect.*
Theorem 2 shows that one can consider mixing graph maps as models for graph maps on basic sets. The following theorem seems to be important in this connection; to formulate it we need the definition of maps with specification (see for example [DGS]).

**Theorem 4** [B1-B3]. *Let $f : X \to X$ be a continuous mixing graph map. Then f has the specification.*
It is well known ([DGS]) that maps with the specification have nice properties concerning

the set of invariant measures. Using them and the Theorems 1-4 we can describe generic properties of invariant measures for graph maps. First we need some definitions. Let $T : X \to X$ be a map of a compact metric space into itself. The set of all T-invariant Borel normalized measures is denoted by $D_T$. A measure $\mu \in D_T$ with $\operatorname{supp} \mu$ contained in one cycle is said to be a *CO-measure*. The set of all CO-measures concentrated on cycles with minimal period p is denoted by $P_T(p)$. Let V(x) be the set of accumulation points of time-averages of iterations of the point x. A point $x \in X$ is said to have *maximal oscillation* if $V(x) = D_T$.

**Theorem 5** [B1-B3]. *Let B be a basic set. Then:*

1. *for any l the set* $\bigcup_{p \geq l} P_{f|B}(p)$ *is dense in* $D_{f|B}$*;*
2. *the set of all ergodic non-atomic invariant measures* $\mu$ *with* $\operatorname{supp} \mu = B$ *is a residual subset of* $D_{f|B}$*;*
3. *if* $V \subset D_{f|B}$ *is a non-empty closed connected set then the set of all points x with V(x)=V is dense in X (in particular every measure* $\mu \in D_{f|B}$ *has generic points);*
4. *points with maximal oscillation are resudual in B.*

**Theorem 6** [B1-B3]. *Let* $\mu$ *be an invariant measure. Then the following properties of* $\mu$ *are equivalent:*

1. *there exists a point x such that* $\operatorname{supp} \mu \subset \omega(x)$*;*
2. *the measure* $\mu$ *has generic points;*
3. *the measure* $\mu$ *is concentrated on a circle-like set or can be approximated by CO-measures.*

In particular, CO-measures are dense in the set of all ergodic measures which are not concentrated on circle-like set.
Properties of maps with specification and Teorem 2 imply the following

**Lemma 2.** *Let* $f : X \to X$ *be a graph map, B a basic set of f. Then there exists t such that* $P(f|B) \supset t\mathbf{N} = \{t \cdot i | i \in \mathbf{N}\}$.

# 2 Misiurewicz conjecture

During the problem session at Czecho-Slovak summer mathmatical school near Bratislava in 1990 M. Misiurewicz formulated the following
CONJECTURE. *Let X be a graph. Then there exists L=L(X) such that for any continuous map* $f : X \to X$ *the inclusion* $P(f) \supset \{1, \ldots, L\}$ *implies that P(f)=***N***.*
We show now that this conjecture is true and give a sketch of the proof. Some lemmata will be given without proofs. First let us formulate the following number-theoretic

**Lemma 3.** *Let $R$ be a positive integer. Then one can find $N = N(R) > R$ such that for any $M \geq N$ there exist positive integers $0 = a_0 < a_1 < \ldots < a_l = M$ with the following properties:*

1. $a_{i+1} - a_i \geq R$ $(0 \leq i \leq l-1)$;

2. *for any proper divisor $s$ of $M$ there exists $j$, $1 \leq j \leq l-1$, such that $a_j | s$.*

PROOF. Let $M = p_1^{b_1} \cdot \ldots \cdot p_k^{b_k}$, where $p_1 < p_2 < \ldots < p_k$ are prime integers. Let be $m_i = M/p_i$, $i \leq i \leq k$. Clearly the numbers $\{m_i\}$ have the required property 2. So it is sufficient to find numbers $a_0 = 1 < a_1 < \ldots < a_l = M$ such that $a_{j+1} - a_j \geq R$ $(\forall j)$ is true and for any i there exists j such that $a_j | m_i$. We suppose that $\{q_1 < q_2 \ldots < q_r\}$ is the set of all prime integers less than R+1. Let be $\alpha = \min\{1/q_{i+1} - 1/q_i\}_{i=1}^{r-1}$, $N = \max\{R/\alpha, 3q_r\}$. Now: if $\frac{M}{p_k p_{k-1}} \geq R$ then $\frac{M}{p_i} - \frac{M}{p_{i+1}} \geq \frac{M}{p_k p_{k-1}} \geq R$. If $\frac{M}{p_k p_{k-1}} < R$ then $p_1 < p_2 < \ldots < p_{k-2} \leq R$ and therefore $m_i - m_{i+1} = M/p_i - M/p_{i+1} \geq \alpha M \geq \alpha N \geq R$ $(1 \leq i \leq k-2)$. Thus it remains to consider the differences $(M/p_{k-1} - M/p_{k-2})$, $(M/p_k - M/p_{k-1})$ which is left to the reader. □

We define that a subset of a graph is an *intervall* if it is a homeomorphic image of the intervall [0,1]; we use for intervals standard notations. Let us fix until the end of the proof a graph X and a continuous map $f : X \to X$.

**Lemma 4.** *There exists a number $m = m(X) < 4Edg(X)$ such that if $A \in X$ and $[a, b_1], [a, b_2], \ldots, [a, b_{m+1}]$ are intervals then one of them contains some of the others.* Suppose that there exist an edge $I = [a, b] \subset X$ and two periodic points $P \in I$ of prime period $p > m(X)$ and $Q \in I$ of prime period $q > m(X)$, $p \neq q$ such that if $J = [P, Q] \subset I$ then $J \cap (orbQ \cup orbP) = \emptyset$; fix them for the Lemmata 5-9.

**Lemma 5.** *It holds $f^{p(q-1)m(X)}[P, Q] \supset orbQ$, $f^{q(p-1)m(X)}[P, Q] \supset orbP$ and so $f^t[P, Q] \supset orbQ \cup orbP$ for $t \geq pqm(X) - \min\{p, q\}m(X)$.*

PROOF. Consider all intervals of type $\{[P, c_i]\}_{i=1}^{k}$ which do not contain points of orbQ, but $c_i \in orbQ$. Then $k \leq m(X)$ and we may assume $Q = c_1$. On the other hand for any i there is j=j(i) such that $f^p[P, c_i] \supset [P, c_j]$. Hence there exist numbers l and n such that $l + n \leq k$ and, say $f^{pl}[P, c_1] \supset [P, c_2]$, $f^{pn}[P, c_2] \supset [P, c_2]$ which implies that $f^{pnj}[P, c_2] \supset \{f^{ipn}c_2\}_{i=0}^{j}$. But p,q are prime numbers and $n \leq m(X) < q$; thus $\{f^{ipn}c_2\}_{i=0}^{q-1} = orbQ$ and $f^{pn(q-1)+lp}[P, c_1] \supset orbQ$. It implies that $f^{p(q-1)m(X)}[P, Q] \supset orbQ$. Similarily it follows $f^{q(p-1)m(X)}[P, Q] \supset orbP$. □

We define subintervals of I with endpoints from orbQ or orbP to be *basical intervals* provided their interiors do not contain points from orbP or orbQ. A basical interval is called *P-interval, Q-interval* or *PQ-interval* depending on periodic orbits which contain its endpoints. Furthermore, suppose that there are two intervals $G \subset X$ and $H \subset X$ and a continuous map $\varphi : X \to X$ such that $\varphi(G) \supset H$ and there is a subinterval $K \subset G$ such that $\varphi(K) = H$; then say that *G $\varphi$-covers H*. Note the following property: If G $\varphi$-covers H and H $\psi$-covers M then G $\varphi \circ \psi$-covers M.

**Lemma 6.** *Let $Z \subset X$ be an interval, $Y = [\alpha, \beta] \subset X$ be an edge and $g : X \to X$ be a continuous map; suppose that $\alpha, \beta \in g(Z)$. Then there are points $\gamma, \delta \in Y$ such that Z* *g-covers* $[\alpha,\gamma]$ *and* $[\delta,\beta]$*; moreover,* $g(Z)\cap Y = [\alpha,\gamma]\cup[\delta,\beta]$.

**Lemma 7.** *Let A be a PQ-interval. Then for any* $i \geq pqm(X)$ *this interval* $f^i$*-covers all basical intervals except at most one.*

**Lemma 8.** *Suppose that* $card\{orbP \cup I\} \geq 4$, $card\{orbQ \cup I\} \geq 4$*. Then the following assertions are true:*

1. *Either for any P-interval M there exists* $i < p^2$ *such that* $f^iM$ *contains a PQ-interval or there exist two P-intervals Y and Z such that the interval Y* $f^i$*-covers Y,Z and the interval Z* $f^i$*-covers Y,Z for* $i \geq (p-1)^2$.

2. *Either for any Q-interval N there exists* $i < q^2$ *such that* $f^iN$ *contains a PQ-interval or there exist two Q-intervals Y' and Z' such that the interval Y'* $f^i$*-covers Y',Z' and the interval Z'* $f^i$*-covers Y',Z' for* $i \geq (q-1)^2$.

PROOF. We will prove assertion 1., only. Consider a P-interval $[c,d]$ which has a neighbouring PQ-interval, say $[d,e]$. Let c be closer to a than d. Divide the proof into four steps.
STEP 1. If $f^i[c,d]$ contains a PQ-interval then for any P-interval M there exists $j \leq p-1+i$ such that $f^iM$ contains a PQ-interval. Indeed, one can find for any P-interval M $m < p$ such that either $f^mM \supset [c,d]$ or $f^mM \supset [d,e]$ which implies the required.
STEP 2. Suppose there exists such an $i \leq (p-1)^2$ that $f^i[c,d]$ contains a PQ-interval. Then for any P-interval M there exists an integer $j < (p-1)^2 + p$ such that $f^jM$ contains a PQ-interval. Step 2 follows easily from Step 1.
If it exists, let $x \in orbP$ be the point that is closest to a and lies on the other side of e than d.
STEP 3. Suppose that $f^i[c,d]$does not contain PQ-intervals for $i < (p-1)^2$. Then for $i \geq (p-1)(p-2)$ the interval $[c,d]$ $f^i$-covers $[a,d]$ (and $[x,b]$ provided that x exists).
Let $l < p$ be such that $f^lc = d$. Then $f^l[c,d] \supset [c,d]$ and moreover $[c,d]$ $f^l$-covers $[c,d]$. But p is a prime integer which similarily to Lemma 5 implies that $f^i[c,d] \supset orbP$ for $i \geq l(p-2)$. Since $f^i[c,d]$ does not contain $[d,e]$ for $l(p-2) \leq i < l(p-1)$ we have by Lemma 6 that $[c,d]$ $f^i$-covers $[a,d]$ (and $[x,b]$ provided that x exists). But $[c,d]$ $f^l$-covers $[c,d]$ which easily implies that for any $i \geq l(p-2)$ $[c,d]$ $f^i$-covers $[a,d]$ (and $[x,b]$ provided that x exists).
STEP 4. Suppose that $f^i[c,d]$ does not contain PQ-intervals for $i < (p-1)^2 + p$. Then for any P-interval M and $i \geq (p-1)^2$ we have that M $f^i$-covers $[a,d]$ (and $[x,b]$ provided that x exists).
Clearly, there is $l < p$ such that either M $f^l$-covers $[c,d]$ or M $f^l$-covers $[d,e]$. Now by step 3 $f^{(p-1)(p-2)}[c,d] \supset M$; so if M $f^l$-covers $[d,e]$ then $f^{(p-1)(p-2)+l}[c,d] \supset [d,e]$ which is a contradiction. Thus m $f^l$-covers $[c,d]$ and by step 3 we get the required.
Now suppose that there exists a P-interval M such that $f^iM$ does not contain PQ-intervals for $i < p^2$. Then by step 1 $f^i[c,d]$ does not contain PQ-intervals for $i < p^2 - (p-1) = (p-1)^2 + p$. Now by step 4 and some simple geometrical arguments we may assert that there exist two P-intervals Y and Z such that $Y \cap Z = \emptyset$ and for any $i \geq (p-1)^2$ the interval Y $f^i$-covers Y,Z and the interval Z $f^i$-covers Y,Z. □

**Lemma 9.** *Suppose that* $card\{orbP \cap I\} \geq 4$, $card\{orbQ \cap I\} \geq 4$. *Let be* $T = T(p,q) = N(pqm(X) - \min(p,q)m(X) + [\max(p,q) - 1]^2)$. *Then* $P(f) \supset \{i | i \geq T\}$ *and* $h(f) > 0$.
PROOF. We use the Lammata 3 and 8 and consider several cases.
CASE A. There are P-intervals Y and Z such that each of them $f^i$-covers both of them for $i \geq (p-1)^2$.
Let $k \geq N((p-1)^2)$ be an integer. By Lemma 3 there exist integers $1 = a_0 < a_1 < \ldots < a_l = k$, $a_{i+1} - a_i \geq (p-1)^2$ such that for any proper divisor s of k there is $a_i$ with $a_i | s$. Properties of $f^i$-covering imply now that there exists an interval $K \subset Y$ such that $f^{a_i}K \subset Z$ for any $1 \leq i \leq l-1$ and $f^k K = Y$. Hence there exists a point $\xi \in Y$ such that $f^{a_i}\xi \in Z$ and $f^k\xi = \xi$; by the properties of the numbers $\{a_i\}$ it implies that k is the minimal period of the point $\xi$ and so $P(f) \supset \{i | i \geq N((p-1)^2)\} \supset \{i | i \geq T\}$. Standard one-dimensional arguments show also that $h(f) > 0$.
CASE B. There are Q-intervals Y' and Z' such that each of them $f^i$-covers both of them for $i \geq (q-1)^2$.
Similarily to case B we have $P(f) \supset \{i | i \geq N((q-1)^2)\}$ and $h(f) > 0$.
CASE C. For any basical interval M there exists a number $s = s(M) < < [\max(p,q) - 1]^2$ such that $f^s M$ contains a PQ-interval.
Let for definiteness $p > q$. Then similarily to Lemma 7 we can conclude by Lemma 5 that any basical interval M $f^i$-covers all basical intervals except at most one of them for $i \geq H = pqm(X) - qm(X) + (p-1)^2$. Choose four basical intervals $\{M_j\}_{j=1}^4$ which are pairwise disjoint and show that for any $k \geq N(H)$ there exists a periodic point $\xi$ of minimal period k.
Let be $k \geq N(H)$. As in case A choose integers $1 = a_0 < a_1 < \ldots < a_l = k$ with the properties from Lemma 3. Let $u = a_l - a_{l-1}$. Then it is easy to see that there is a basical interval , say $M_1$, that at least two other intervals, say $M_2$ and $M_3$, $f^u$-cover $M_1$. On the other hand one can easily show that there are two numbers $i, j \in \{2,3,4\}$ and two intervals $K_i \subset M_1$ and $K_j \subset M_1$ such that for any $1 \leq v \leq l-2$ we have $f^{a_v}(K_i) \subset M_{i(v)}$ and $f^{a_v}(K_j) \subset M_{j(v)}$ , where $i(v) \neq 1$, $j(v) \neq 1$ and $f^{a_{l-1}}K_i = M_i$, $f^{a_{l-1}}K_j = M_j$ . Clearly one of the numbers i,j belongs to the set{2,3}; let i be, say, equal to 2. Then choosing corresponding subintervals and using simple properties of f-coverings one can easily find an interval $K \subset M_1$ such that $f^{a_v}K \cap M_1 = \emptyset$, $1 \leq v \leq l-1$ and $f^k K = M_1$. Thus f has a periodic point of minimal period k. Moreover, it is clear that $h(f) > 0$ which completes the proof. □

**Theroem 7.** *Let X be a graph,* $s = Edg(X) + 1$ *and* $\{p_i\}_{i=1}^s$ *be ordered prime intergers greater than* $4Edg(x)$. *Let be* $L = L(X) = T(p_s, p_{s-1})$. *If a continuous map* $f : X \to X$ *is such that* $P(f) \supset \{1, 2, \ldots, L\}$ *then* $P(f) = \mathbf{N}$ *and* $h(f) >= 0$.
PROOF. Clearly in the situation of Theorem 7 on can find two periodic points with the properties from Lemma 9. It completes the proof. □

REMARK [B4]. If X is a tree then one may set $L(X) = 2(p-1)[End(X) - 1]$, where p is the least prime integer greater than End(X).

**Main Theorem.** *Let $f : X \to X$ be a continuous graph map. Then the following assertions are equivalent:*

1. *The map $f$ has positive entropy.*
2. *There exists $n$ such that $P(f) \supset n\mathbf{N} = \{i \cdot n | n \in \mathbf{N}\}$.*

PROOF. By the decomposition , if $h(f) > 0$ then f has a basic set. By Lemma 2 it implies assertion 2. On the other hand by Theorem 7 assertion 2. implies assertion 1., which completes the proof. □

This work was partly written when the author was visiting the Max-Planck-Institut für Mathematik in Bonn. It is a pleasure for him to express his gratitude to MPI for kind hospitality.

## REFERENCES

[*B*1] BLOKH, A.M.: On the limit behaviour of one-dimensional dynamical systems. 1,2. (in russ.) Preprints NN 1156-82,2704-82, Moscow, 1982.

[*B*2] BLOKH, A.M.: Decomposition of dynamical systems on an interval. Russ. Math. Surv., vol. 38, no. 5, (1983)133-134.

[*B*3] BLOKH, A.M.: On dynamical systems on one-dimensional branched manifolds, 1,2,3. (in russ.) Theory of functions, functional analysis and appl. 46(1986)8-18, 47(1987)67-77, 48(1987)32-46.

[*B*4] BLOKH, A.M.: On Misiurewicz conjecture for tree maps. (1990) to appear.

[*DGS*] DENKER, M., GRILLENBERGER, C., SIGMUND, K.: Ergodic theory on compact spaces. Lecture Notes in Math., vol. 527, Berlin, 1976.

[*H*] HOFBAUER, F.: The structure of piecewise-monotonic transformations. Erg. Theory & Dyn. Syst. 1(1981)159-178.

[*JR*] JONKER, L., RAND, D.: Bifurcations in one dimension. 1: The non-wandering set. Inv. Math. 62(1981)347-365.

[*N*] NITECKI, Z.: Topological dynamics on the interval, Erg. Theory and Dyn. Syst. 2, Progress in Math. vol. 21, Boston, (1982)1-73.

[*S*] SHARKOVSKY, A.N.: Partially ordered systems of attracting sets. (in russ.), DAN SSSR vol.170, no. 6, (1966)1276-1278.

# The Abramov-Rokhlin Formula

Thomas Bogenschütz* Hans Crauel**

**Abstract**

The Abramov-Rokhlin formula states that the entropy of a measure-preserving transformation $S$ equals the sum of the entropy of a factor $T$ of $S$ and the entropy of $S$ relative to $T$. We prove this formula for non-invertible transformations and apply it to skew-product transformations.

**Key words:** entropy relative to a factor; skew-product transformation.

## 1 Introduction

Consider the following type of skew-product. Let $\vartheta : \Omega \to \Omega$ be a measure-preserving transformation of a probability space $(\Omega, \mathcal{F}, P)$ and $\{\varphi(\omega) : \omega \in \Omega\}$ a family of measurable transformations of a measurable space $(X, \mathcal{B})$ such that the map $(\omega, x) \mapsto \varphi(\omega)x$ is measurable from $\Omega \times X$ to $X$. We can then define the skew-product transformation $\Theta$ on $\Omega \times X$ by $\Theta(\omega, x) = (\vartheta\omega, \varphi(\omega)x)$.

If there exists a fixed measure $\rho$ on $X$, i.e. $\rho$ is left invariant under (at least $P$-almost) all $\varphi(\omega)$, then $\mu = \rho \otimes P$ is $\Theta$-invariant. Abramov and Rokhlin [1] computed the entropy of $\Theta$ for this special case. They showed that for any finite partition $\mathcal{P}$ of $X$ the limit

$$h_\mu(\varphi; \mathcal{P}) = \lim_{n\to\infty} \frac{1}{n} \int H_\rho(\bigvee_{i=0}^{n-1} \varphi^{-1}(i,\omega)\mathcal{P})dP(\omega)$$

exists, where $\varphi(i,\omega) := \varphi(\vartheta^{i-1}\omega) \circ \ldots \circ \varphi(\omega)$, and proved the following formula:

$$h_\mu(\Theta) = h_P(\vartheta) + h_\mu(\varphi). \tag{A}$$

Here $h_\mu(\varphi) = \sup\{h_\mu(\varphi; \mathcal{P}) : \mathcal{P} \text{ finite partition of } X\}$ denotes the *fiber-entropy* of $\Theta$. A comprehensive exposition of the fixed measure case can be found in § 6.1.B, p. 254–257, of Petersen's book [8].

---

*Institut für Dynamische Systeme, Universität Bremen, Postfach 330440, 2800 Bremen 33, Federal Republic of Germany

**Fachbereich 9 Mathematik, Universität des Saarlandes, Im Stadtwald, 6600 Saarbrücken 11, Federal Republic of Germany

Existence of a fixed measure is a very restrictive assumption. In the present note we consider arbitrary $\Theta$-invariant measures $\mu$ on $\Omega \times X$ which have marginal $P$ on $\Omega$. Such measures can be shown always to exist if $X$ is a compact metric space and $\varphi(\omega)$ are continuous $P$-a. s.; see, e. g., Crauel [3], Theorem 1, p. 273. Suppose there is a family of conditional probabilities $\{\mu_\omega\}$ on $X$ such that $d\mu(\omega, x) = d\mu_\omega(x)dP(\omega)$. Then the limit

$$h_\mu(\varphi; \mathcal{P}) = \lim_{n\to\infty} \frac{1}{n} \int H_{\mu_\omega}(\bigvee_{i=0}^{n-1} \varphi^{-1}(i,\omega)\mathcal{P})dP(\omega)$$

exists; see Bogenschütz [2], Theorem 3.2, p. 5. The purpose of this paper is to prove that formula (A) still holds, where again $h_\mu(\varphi) = \sup\{h_\mu(\varphi; \mathcal{P}) : \mathcal{P} \text{ finite partition of } X\}$ .

To obtain our result we will prove a theorem on the entropy relative to a factor and then apply it to skew-product transformations. Such a 'relativizing' method has been used by other authors before (see [4], [6], [7], [9], [10]).

# 2 Main theorem

Let us first fix some notations. A dynamical system $(X, \mathcal{B}, \mu, S)$ is a probability space $(X, \mathcal{B}, \mu)$ together with a measure-preserving transformation $S$ of $X$. A dynamical system $(Y, \mathcal{C}, \nu, T)$ is said to be a factor of $(X, \mathcal{B}, \mu, S)$, if there exists a measure-preserving map $\pi : X \to Y$, such that $T \circ \pi = \pi \circ S$ $\mu$-a. s.

Let $h_\mu(S \mid \mathcal{A})$ denote the usual conditional entropy of $S$ with respect to a sub-$\sigma$-algebra $\mathcal{A} \subset \mathcal{B}$ satisfying $S^{-1}\mathcal{A} \subset \mathcal{A}$ (see Kifer [5], Definition 1.3, p. 41). Since $S^{-1}(\pi^{-1}\mathcal{C}) \subset \pi^{-1}\mathcal{C}$, the number

$$h_\mu(S \mid T) := h_\mu(S \mid \pi^{-1}\mathcal{C})$$

is well-defined and called the *entropy of $S$ relative to $T$*. Standard properties of conditional entropy can be found, e. g., in § 2.1, p. 33–47, of Kifer's book [5]. We will use them without explicit mentioning.

The usual metric entropies will be denoted by $h_\mu(S)$ and $h_\nu(T)$ respectively.

**Theorem (Abramov-Rokhlin formula)** *Suppose $(X, \mathcal{B}, \mu, S)$ is a dynamical system with a factor $(Y, \mathcal{C}, \nu, T)$, where $\mathcal{C}$ is countably generated (mod 0). Then*

$$h_\mu(S) = h_\nu(T) + h_\mu(S \mid T).$$

*(Both sides may be $\infty$.)*

*Remark:* This formula is due to Ledrappier and Walters [7]. The proof that they indicate rests on Pinsker's formula, so invertibility of $T$ is indispensible. The following proof uses ideas of Abramov and Rokhlin's original work [1] and does not need any invertibility assumptions.

*Proof:* Let $\mathcal{P}$ be a finite partition of $X$ and $\mathcal{Q}$ a finite partition of $Y$. We have, for $n \in \mathbb{N}$

$$\begin{aligned} H_\mu(\bigvee_{i=0}^{n-1} S^{-i}(\mathcal{P} \vee \pi^{-1}\mathcal{Q})) &= H_\mu(\bigvee_{i=0}^{n-1} S^{-i}\mathcal{P} \vee \pi^{-1}(\bigvee_{i=0}^{n-1} T^{-i}\mathcal{Q})) \\ &= H_\nu(\bigvee_{i=0}^{n-1} T^{-i}\mathcal{Q}) + H_\mu(\bigvee_{i=0}^{n-1} S^{-i}\mathcal{P} \mid \pi^{-1}(\bigvee_{i=0}^{n-1} T^{-i}\mathcal{Q})). \end{aligned}$$

The second term of this sum is $\geq H_\mu(\bigvee_{i=0}^{n-1} S^{-i}\mathcal{P} \mid \pi^{-1}\mathcal{C})$, so dividing by $n$ and letting $n$ tend to $\infty$ we obtain

$$h_\mu(S; \mathcal{P} \vee \pi^{-1}\mathcal{Q}) \geq h_\nu(T; \mathcal{Q}) + h_\mu(S \mid T; \mathcal{P}). \tag{B}$$

On the other hand we have, for $n, m \in \mathbb{N}$

$$\begin{aligned} H_\mu(\bigvee_{i=0}^{nm-1} S^{-i}\mathcal{P}) &\leq H_\mu(\bigvee_{i=0}^{nm-1} S^{-i}\mathcal{P} \vee \bigvee_{j=0}^{nm-1} S^{-j}(\pi^{-1}\mathcal{Q})) \\ &= H_\nu(\bigvee_{j=0}^{nm-1} T^{-j}\mathcal{Q}) + H_\mu(\bigvee_{i=0}^{nm-1} S^{-i}\mathcal{P} \mid \bigvee_{j=0}^{nm-1} S^{-j}(\pi^{-1}\mathcal{Q})). \end{aligned}$$

Since $\bigvee_{i=0}^{nm-1} S^{-i}\mathcal{P} = \bigvee_{i=0}^{n-1} S^{-im}(\bigvee_{k=0}^{m-1} S^{-k}\mathcal{P})$, the second term of this sum can be estimated from above by

$$\begin{aligned} \sum_{i=0}^{n-1} H_\mu(S^{-im}(\bigvee_{k=0}^{m-1} S^{-k}\mathcal{P}) \mid \bigvee_{j=0}^{nm-1} S^{-j}(\pi^{-1}\mathcal{Q})) &\leq \sum_{i=0}^{n-1} H_\mu(S^{-im}(\bigvee_{k=0}^{m-1} S^{-k}\mathcal{P}) \mid S^{-im}(\pi^{-1}\mathcal{Q})) \\ &= nH_\mu(\bigvee_{k=0}^{m-1} S^{-k}\mathcal{P} \mid \pi^{-1}\mathcal{Q}). \end{aligned}$$

Dividing by $nm$ and taking limits as $n \to \infty$ gives

$$h_\mu(S; \mathcal{P}) \leq h_\nu(T; \mathcal{Q}) + \frac{1}{m} H_\mu(\bigvee_{k=0}^{m-1} S^{-k}\mathcal{P} \mid \pi^{-1}\mathcal{Q}).$$

Suppose $\{\mathcal{Q}_i\}_{i \in \mathbb{N}}$ is an increasing sequence of finite partitions of $Y$ with $\sigma(\mathcal{Q}_i : i \in \mathbb{N}) = \mathcal{C}$ (here we use the assumption that $\mathcal{C}$ is countably generated). Then $h_\nu(T; \mathcal{Q}_i)$ increases to $h_\nu(T)$, and $H_\mu(\bigvee_{k=0}^{m-1} S^{-k}\mathcal{P} \mid \pi^{-1}\mathcal{Q}_i)$ decreases to $H_\mu(\bigvee_{k=0}^{m-1} S^{-k}\mathcal{P} \mid \pi^{-1}\mathcal{C})$, hence we may conclude

$$h_\mu(S; \mathcal{P}) \leq h_\nu(T) + \frac{1}{m} H_\mu(\bigvee_{k=0}^{m-1} S^{-k}\mathcal{P} \mid \pi^{-1}\mathcal{C}).$$

Letting $m$ tend to $\infty$ we get

$$h_\mu(S; \mathcal{P}) \leq h_\nu(T) + h_\mu(S \mid T; \mathcal{P}). \tag{C}$$

Taking suprema over $\mathcal{P}$ and $\mathcal{Q}$ in (B), and taking supremum over $\mathcal{P}$ in (C) completes the proof. □

Re-introducing skew-product transformations we have the following special case of the Abramov-Rokhlin formula.

**Corollary** *Let* $(\Omega, \mathcal{F}, P, \vartheta)$ *be a dynamical system and* $\{\varphi(\omega)\}$ *a family of measurable transformations of* $(X, \mathcal{B})$. *Furthermore, let* $\{\mu_\omega\}$ *be a family of conditional prababilities associated to a measure* $\mu$ *on* $\Omega \times X$, *which has marginal* $P$ *on* $\Omega$ *and is invariant under the skew-product transformation* $\Theta : \Omega \times X \to \Omega \times X, (\omega, x) \mapsto (\vartheta\omega, \varphi(\omega)x)$. *Let* $h_\mu(\varphi; \mathcal{P}) = \lim_{n\to\infty} \frac{1}{n} \int H_{\mu_\omega}(\bigvee_{i=0}^{n-1} \varphi^{-1}(i,\omega)\mathcal{P})dP(\omega)$ *and* $h_\mu(\varphi) = \sup\{h_\mu(\varphi; \mathcal{P})\}$, *where the supremum is taken over all finite partitions* $\mathcal{P}$ *of* $X$.

*If* $\mathcal{F}$ *is countably generated (mod 0), then*

$$h_\mu(\Theta) = h_P(\vartheta) + h_\mu(\varphi).$$

*Proof*: The dynamical system $(\Omega \times X, \mathcal{F} \otimes \mathcal{B}, \mu, \Theta)$ has $(\Omega, \mathcal{F}, P, \vartheta)$ as a factor, with $\pi : \Omega \times X \to \Omega$ the usual coordinate-projection. Thus $h_\mu(\Theta) = h_P(\vartheta) + h_\mu(\Theta \mid \mathcal{F} \times X)$ by our main theorem. But $h_\mu(\varphi) = h_\mu(\Theta \mid \mathcal{F} \times X)$, see Bogenschütz [2], Theorem 3.7, p. 8, so the corollary is proved. □

# References


[1] L. M. Abramov and V. A. Rokhlin, The entropy of a skew product of measure-preserving transformations, *Amer. Math. Soc. Transl. Ser. 2*, 48 (1966), 255–265

[2] T. Bogenschütz, Entropy for random dynamical systems, Report No. 235, Institut für Dynamische Systeme, Universität Bremen 1990

[3] H. Crauel, Lyapunov exponents and invariant measures of stochastic systems on manifolds, in: L. Arnold and V. Wihstutz (eds.), *Lyapunov Exponents, Proceedings, Bremen 1984*, Lecture Notes in Mathematics 1186, Springer 1986

[4] P. Hulse, Sequence entropy relative to an invariant σ-algebra, *J. London Math. Soc. (2)* 33 (1986), 59–72

[5] Y. Kifer, *Ergodic Theory of Random Transformations*, Birkhäuser 1986

[6] B. Kamiński, An axiomatic definition of the entropy of a $\mathbb{Z}^d$-action on a Lebesgue space, *Studia Mathematica*, T. XCVI (1990), 135—144

[7] F. Ledrappier and P. Walters, A relativised variational principle for continuous transformations, *J. London Math. Soc. (2)* 16 (1977), 568–576

[8] K. Petersen, *Ergodic Theory*, Cambridge University Press 1983

[9] J.-P. Thouvenot, Quelques propriétés des systèmes dynamiques qui se décomposent en un produit de deux systèmes dont l'un est un schéma de Bernoulli, *Israel J. Math.* 21 (1975), 177—207

[10] P. Walters, Relative pressure, relative equilibrium states, compensation functions and many-to-one codes between subshifts, *Trans. Amer. Math. Soc.* 296 (1986), 1–31

# Expanding Attractors with Stable Foliations of Class $C^0$

H.G. Bothe
Karl-Weierstraß-Institut für Mathematik
Mohrenstraße 39, 1086 Berlin

## 1 Introduction

For a compact manifold $M$ without boundary we consider the dynamics which is generated by a $C^r$ diffeomorphism $f : M \to M$ ($r \geq 1$), i.e. we try to describe how the points in $M$ behave under the action of the group $\mathbf{Z}$ of all integers on $M$ given by the iterates $f^k$ of $f$ ($k \in \mathbf{Z}$). In [5] S. Smale introduced the so-called Axiom-A-diffeomorphisms for which he could make an important step towards a satisfactory description of their dynamics. A diffeomorphism $f$ satisfies this Axiom A if the set $\Omega$ of all its non-wandering points is hyperbolic and the periodic points are dense in $\Omega$.
A compact invariant subset $\Lambda$ of $M$ is called hyperbolic if the restriction $T_\Lambda(M)$ of the tangent bundle $T(M)$ to $\Lambda$ splits into the direct sum $T^u \oplus T^s$ of two continuous subbundles which are invariant under the differential $df$ and where $f$ is expanding in the direction of $T^u$ and contracting in the direction of $T^s$, i.e. there is a Riemannian metric on $M$, called adapted metric, and a real $\gamma > 1$ such that for any $v \in T^u$ and any $w \in T^s$ we have $|df(v)| \geq \gamma\,|v|$, $|df(w)| \leq \gamma^{-1}\,|w|$.
Smale proved that for an Axiom-A-diffeomorphism $f$ the set $\Omega$ is the union of finitely many disjoint compact invariant subsets $\Lambda_1, ..., \Lambda_r$, called basic sets, each of which is topologically transitive in the sense that it contains a dense orbit. (This implies that the dimensions of the fibres of $T^u$ and $T^s$ are constant on each basic set.) Since a good deal of the structure of the whole dynamics on $M$ lies in the geometric structure of $\Omega$ and the restriction $f : \Omega \to \Omega$ we are led to ask for the structure of basic sets. An answer to this question has to include a description of all topological spaces $\tilde{\Lambda}$ with a homeomorphism $g : \tilde{\Lambda} \to \tilde{\Lambda}$ for which there is a basic set $\Lambda$ of a diffeomorphism $f$ and a homeomorphism $h : \tilde{\Lambda} \to \Lambda$ satisfying $hg = fh$. Though, using Markov partitions, it is easy to show that the class of all topological types of such spaces $\tilde{\Lambda}$ and mappings $g : \tilde{\Lambda} \to \tilde{\Lambda}$ is countable (see [2]), we do not know much about the geometry of these types. But there is one important exception: The expanding attractors for which, due to R. F. Williams [7], [8], [9], we have a satisfactory description. Here a basic set $\Lambda$ is called expanding attractor if it is attracting (i.e. if $\Lambda$ has a neighbourhood $U$ in $M$ such that $f(U) \subset U$ and $U \cap f(U) \cap f^2(U) ... = \Lambda$ ) and if the topological dimension of $\Lambda$ coincides with the dimension of the fibres $T^u_x$ of $T^u$ at points $x \in \Lambda$. To inform about the results on expanding attractors we need the following general facts concerning a compact invariant hyperbolic set $\Lambda$ of a $C^r$ diffeomorphism $f : M \to M$ (see [4]).
If $x \in \Lambda$ and $n$, $n'$ are the dimensions of the fibres $T^u_x$, $T^s_x$, then for the sets

$$W_x^u = \left\{ y \in M; \lim_{k \to \infty} d(f^{-k}(x), f^{-k}(y)) = 0 \right\},$$
$$W_x^s = \left\{ y \in M; \lim_{k \to \infty} d(f^{k}(x), f^{k}(y)) = 0 \right\}$$

($d$ the distance in $M$) there are one-to-one $C^r$ immersions $w_x^u : \mathbf{R}^n \to M$, $w_x^s : \mathbf{R}^{n'} \to M$ such that $W_x^u = w_x^u(\mathbf{R}^n)$, $W_x^s = w_x^s(\mathbf{R}^{n'})$. Therefore the sets $W_x^u$, $W_x^s$ are manifolds in $M$. They are called unstable or stable manifolds, respectively. The tangent space $T_x(W_x^u)$ of $W_x^u$ at $x$ is $T_x^u$, and $T_x(W_x^s) = T_x^s$. Therefore the intersection of $W_x^u$ and $W_x^s$ at $x$ is transverse. Moreover, for each $x \in \Lambda$ we can choose an $n$-ball $B_x^u$ which is a neighbourhood of $x$ in $W_x^u$ such that $B_x^u$ depends continuously on $x$ with respect to the $C^r$ topology. The analogous fact holds for the stable manifolds $W_x^s$.
If $\Lambda$ is an attracting basic set, then all unstable manifolds $W_x^u$ ($x \in \Lambda$) are contained in $\Lambda$, and the family $\mathcal{W}^s$ of all stable manifolds $W_x^s$ ($x \in \Lambda$) is a $C^0$ foliation of the open set $W^s$ consisting of all points $y \in M$ which are attracted by $\Lambda$. This set $W^s$ is called the region of attraction or the basin of $\Lambda$, and $\mathcal{W}^s$ will be called the stable foliation of $W^s$. If $n = \dim W_x^u > 1$ ($x \in \Lambda$) then $\mathcal{W}^s$ though consisting of $C^r$ leaves, is not always a $C^r$ foliation, even if $r = \infty$.
Now let $\Lambda$ be an $n$-dimensional expanding attractor ($n \geq 1$). The local structure of $\Lambda$ can be described as follows: If $x \in \Lambda$ then there is a homeomorphism $\beta : \mathbf{D}^n \times \mathbf{D}^{n'} \to V$ ($\mathbf{D}^n \subset \mathbf{R}^n$, $\mathbf{D}^{n'} \subset \mathbf{R}^{n'}$ the unit balls) onto a neighbourhood $V$ of $x$ in $M$ such that $V \cap \Lambda = \beta(\mathbf{D}^n \times C)$ where $C$ is a Cantor set in $\mathbf{D}^{n'}$. The disks $\beta(\mathbf{D}^n \times \{c\})$ ($c \in C$) are pieces of unstable manifolds while the disks $\beta(\{t\} \times \mathbf{D}^{n'})$ ($t \in \mathbf{D}^n$) are pieces of stable manifolds (see [3], [1]).

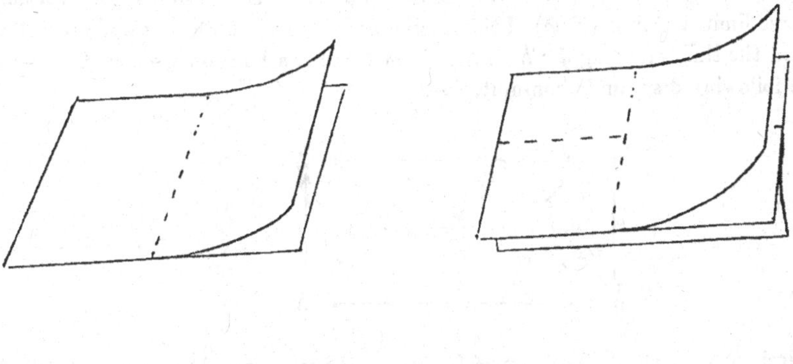

Fig. 1

An expanding attractor $\Lambda$ as defined so far is not necessarily connected (in the topological sense), but it always consists of a finite number $p$ of connected components which are cyclically permuted by $f$. Therefore, if $f$ is replaced by $f^p$, the attractor $\Lambda$ of $f$ splits into $p$ connected expanding attractors of $f^p$. Therefore it is sufficient to consider connected attractors, and we may include connectedness in the definition of expanding attractors.

To describe the global structure of an expanding attractor Williams introduced branched manifolds. A branched $n$-manifold $\Sigma$ may differ from an ordinary $n$-manifold by the fact that local branchings are allowed as indicated in Fig. 1 for $n = 2$ (for an exact definition see Section 2, below). Nevertheless a branched manifold $\Sigma$ has at each of its points $\tau$ a tangent space $T_\tau(\Sigma)$ and it is possible to define Riemannian metrics on $\Sigma$. Moreover a mapping $\varphi : \Sigma \to \Sigma'$ of branched manifolds may be differentiable, and, if so, for each $\tau \in \Sigma$ the differential $d_\tau\varphi : T_\tau(\Sigma) \to T_{\varphi(\tau)}(\Sigma')$ is defined. A $C^1$ mapping $\varphi : \Sigma \to \Sigma$ will be called expanding if there is a Riemannian metric on $\Sigma$ and a real $\gamma > 1$ such that $|d\varphi(v)| \geq \gamma\,|v|$ holds for each tangent vector $v$ of $T$. A continuous mapping $\varphi : \Sigma \to \Sigma'$ is called locally flattening if each $\tau \in \Sigma$ has a neighbourhood which is mapped to a subset of a smooth disk in $\Sigma'$.

If $\Lambda$ is an $n$-dimensional expanding attractor of a diffeomorphism $f$, then by a $W$-representation of $\Lambda$ we mean a commutative diagram

$$
\begin{array}{ccc}
\Lambda & \xrightarrow{\ f\ } & \Lambda \\
\pi \downarrow & & \downarrow \pi \\
\Sigma & \xrightarrow{\ \varphi\ } & \Sigma
\end{array}
$$

where $\varphi : \Sigma \to \Sigma$ is a locally flattening expanding $C^1$ mapping of a branched $n$-manifold $\Sigma$ and $\pi$ is a continuous projection onto $\Sigma$. It is assumed that the following conditions are satisfied:

• Each point $x \in \Lambda$ has a neighbourhood in $W_x^u$ which is mapped by $\pi$ homeomorphically onto a smooth open $n$-disk in $\Sigma$.

• If $\pi(x) = \pi(y)$ then $y \in W_x^s$, and $\pi f^k(x) = \pi f^k(y)$ for all $k \in \mathbf{Z}$ implies $x = y$.

The value of a $W$-representation is due to the fact that its lower part $\varphi : \Sigma \to \Sigma$ determines the structure of $\Lambda$ (i.e. the upper part $f : \Lambda \to \Lambda$) up to topological conjugacy. Moreover the topological type of $f : \Lambda \to \Lambda$ can be obtained from $\varphi : \Sigma \to \Sigma$ by a simple construction: Take the inverse limit $\bar\Lambda = \lim_{\leftarrow}(\Sigma, \varphi)$ of the sequence $\Sigma \xleftarrow{\varphi} \Sigma \xleftarrow{\varphi} ...$, the natural projection $\bar\pi : \bar\Lambda \to \Sigma$, and the shift mapping $\tilde\varphi : \bar\Lambda \to \bar\Lambda$. Then there is a homeomorphism $h : \bar\Lambda \to \Lambda$ such that the following diagram is commutative.

$$
\begin{array}{ccccc}
\Lambda & & \xrightarrow{\qquad f \qquad} & & \Lambda \\
\uparrow h & \pi \searrow & \Sigma \xrightarrow{\ \varphi\ } \Sigma & \swarrow \pi & \uparrow h \\
& \bar\pi \nearrow & & \searrow \bar\pi & \\
\bar\Lambda & & \xrightarrow[\qquad \tilde\varphi \qquad]{} & & \bar\Lambda
\end{array}
$$

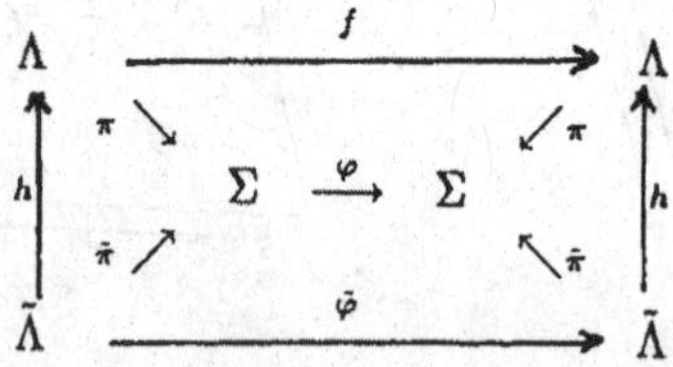

[The inverse limit $\bar\Lambda$ is the set of all sequences $(\tau_0, \tau_1, ...)$ of points from $\Sigma$ where $\tau_i = \varphi(\tau_{i+1})$, and $\bar\pi(\tau_0, \tau_1, ...) = \tau_0$, $\tilde\varphi(\tau_0, \tau_1, ...) = (\varphi(\tau_0), \varphi(\tau_1), ...) = (\varphi(\tau_0), \tau_0, \tau_1, ...)$. The topology in $\bar\Lambda$ is defined by the neighbourhoods

$$
U_k(\tau_0, \tau_1, ...) = \left\{ (\tau_0', \tau_1', ...) \in \bar\Lambda;\ \tau_i' = \tau_i \ for\ 0 \leq i \leq k \right\},
$$

where $(\tau_0, \tau_1, ...) \in \bar\Lambda$, $k = 0, 1, 2, ....$]

Due to their strange local structure expanding attractors seem to lie outside the domain of classical mathematics. On the other hand branched manifolds are so close to ordinary manifolds that they can be regarded as "honest" mathematical structures. So $W$-representations reduce the investigation of a strange but interesting mathematical object to a normal one. The main result of Williams' states that each expanding attractor has a $W$-representation provided its stable foliation is of class $C^1$. In this case a $W$-representation can be chosen so that $\pi$ maps the neighbourhood of $x$ mentioned in the first property of $W$-representations diffeomorphically onto a disk in $\Sigma$.

*It is the aim of this paper to construct W-representations for all expanding attractors without any further assumptions.*

Though our construction is influenced by this of Williams, for technical reasons it was convenient to make it selfcontained.

The main steps in our construction of $W$-representations can be sketched as follows:

0) We start with an $n$-dimensional expanding attractor $\Lambda$ of a $C^1$ diffeomorphism $f : M \to M$. Using elementary facts of differential topology we can embed $M$ in a euclidian space $\mathbf{R}^m$ and extend $f$ to a diffeomorphism $\bar{f} : \mathbf{R}^m \to \mathbf{R}^m$ whose differential $d\bar{f}$ maps each normal vector $v$ of $M$ in $\mathbf{R}^m$ to another normal vector which is shorter than $v$. Then $\Lambda$ is an expanding attractor of $\bar{f}$. For notational convenience we denote $\bar{f}$ by $f$ and forget $M$. This means that we assume $M = \mathbf{R}^m$. But the metric in $\mathbf{R}^m$ is still a Riemannian metric which is adapted to $\Lambda$ in the sense mentioned above.

1) We construct a $C^1$ foliation $\mathcal{F}$ of a neighbourhood $N$ of $\Lambda$ in the basin $W^s$ whose leaves are $n'$-dimensional and transverse to the unstable manifolds in $\Lambda$ $(n' = m - n)$. The leaf through a point $x \in N$ will be denoted by $F_x$. (This is done in Sect. 5.)

2) In Section 6 we define a branched $n$-manifold $\Sigma$ and a projection $\pi^F : \Lambda \to \Sigma$ where for each $\tau \in \Sigma$ the set $(\pi^F)^{-1}(\tau)$ is a Cantor set in a leaf $F_x$ of $\mathcal{F}$. The restrictions of $\pi^F$ to unstable manifolds are $C^1$ immersions in $\Sigma$.

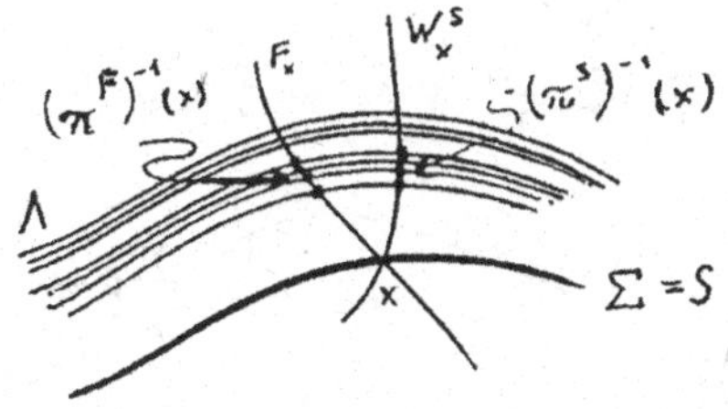


Fig. 2

3) In Section 7 we construct a $C^1$ immersion $\eta : \Sigma \to N$ such that $S = \eta(\Sigma)$ is transverse to $\mathcal{F}$ and to the stable foliation $\mathcal{W}^s$ of $\Lambda$. Moreover $\eta(\tau)$ will lie in the leaf $F$ of $\mathcal{F}$ containing $(\pi^F)^{-1}(\tau)$. Here in this sketch we shall assume that $\eta$ is an embedding. This allows to identify $\Sigma$ with $S$ so that $\eta = id$ and $\Sigma$ becomes a subset of $\mathbf{R}^m$. (In the detailed

construction below we do not make this identification, but this is not essential for the main idea). For each $x \in \Lambda$ the point $\pi^F(x)$ lies in $F_x$ not far from $x$. Then we define a mapping $\pi^s : \Lambda \to \Sigma$ such that $(\pi^s)^{-1}(x)$ is a Cantor set in the stable manifold $W^s_x$ containing $x$. This set $(\pi^s)^{-1}(x)$ is close to $(\pi^F)^{-1}(x)$ in $\Lambda$ in the sense that $(\pi^s)^{-1}(x)$ can be obtained from $(\pi^F)^{-1}(x)$ by a small move of the points of $(\pi^F)^{-1}(x)$ inside the unstable manifolds (see Fig. 2).

4) The mapping $\pi^s$ in 3) has the following property: For $x \in \Sigma$ there is a point $y \in \Sigma$ such that $f((\pi^s)^{-1}(x)) \subset (\pi^s)^{-1}(y)$. Then $x \mapsto y$ defines a locally flattening mapping $\varphi^s : \Sigma \to \Sigma$ for which the diagram

$$\begin{array}{ccc} \Lambda & \xrightarrow{f} & \Lambda \\ \pi^s \downarrow & & \downarrow \pi^s \\ \Sigma & \xrightarrow{\varphi^s} & \Sigma \end{array}$$

is commutative. If $\varphi^s$ happens to be an expanding $C^1$ mapping, then we have obtained a $W$-representation for $\Lambda$. But, if $W^s$ is not of class $C^1$, then $\varphi^s$ will merely be a $C^0$ immersion in the sense that small smooth $n$-disks are mapped homeomorphically onto smooth $n$-disks. (See Section 8).

5) In Section 9 we consider the sequence $\Sigma_k = f^k(\Sigma)$ $(k = 0, 1, \dots)$ of branched manifolds. For increasing $k$ the branched manifolds $\Sigma_k$ converge to $\Lambda$. The mappings $f^k \pi^s f^{-k} : \Lambda \to \Sigma_k$, $f^k \varphi^s f^{-k} : \Sigma_k \to \Sigma_k$ will be denoted by $\pi^s$, $\varphi^s$, respectively. Then the diagrams

$$\begin{array}{ccc} \Lambda & \xrightarrow{f} & \Lambda \\ \pi^s \downarrow & & \downarrow \pi^s \\ \Sigma_k & \xrightarrow{\varphi^s} & \Sigma_k \end{array}$$

are commutative. Now let $k$ be large. For each $x \in \Sigma_k$ we consider the point $y = f(x) \in \Sigma_{k+1}$ and the point $z \in \Sigma_k$ which lies in the intersection of a small disk in $F_y$ containing $y$ and an $n$-disk which is the image $\varphi^s(\Gamma)$ of a small neighbourhood $\Gamma$ of $x$ in $\Sigma_k$. (Here we use that $\varphi^s$ is locally flattening). By $x \mapsto z$ a mapping $\varphi_k : \Sigma_k \to \Sigma_k$ is defined (see Fig. 3).

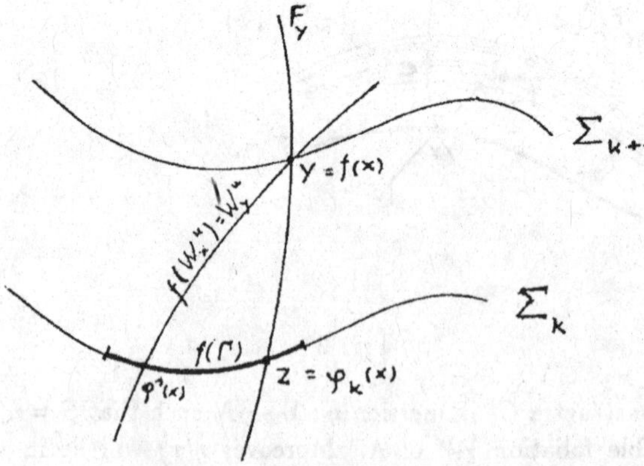


Fig. 3

Since $\mathcal{F}$ is a $C^1$ foliation $\varphi_k$ is a $C^1$ mapping, and since $\Sigma_k$ is close to $\Lambda$ and $f$ is expanding on the unstable manifolds, $\varphi_k$ is expanding too. Moreover for $k$ large the mapping $\varphi_k$ is close to $\varphi^s : \Sigma_k \to \Sigma_k$.
6) Let $\varphi, \varphi^* : \Sigma \to \Sigma$ be two mappings of a branched manifold $\Sigma$ which have the following properties: (a) $\varphi$ is a $C^0$ immersion. (b) $\varphi$ is locally flattening. (c) $\varphi^*$ is an expanding $C^1$ mapping. Then, if $\varphi^*$ is sufficiently close to $\varphi$, it is proved in Section 3 that there is a $C^0$ immersion $\chi : \Sigma \to \Sigma$ such that the following diagram is commutative:

$$\begin{array}{ccc} \Sigma & \xrightarrow{\varphi} & \Sigma \\ \chi \downarrow & & \downarrow \chi \\ \Sigma & \xrightarrow{\varphi^*} & \Sigma \end{array}$$

In the final step (Section 10) this is applied to the mappings $\varphi^s : \Sigma_k \to \Sigma_k$, $\varphi_k : \Sigma_k \to \Sigma_k$ ($k$ large). So we get a commutative diagram

$$\begin{array}{ccc} \Sigma_k & \xrightarrow{\varphi^s} & \Sigma_k \\ \chi \downarrow & & \downarrow \chi \\ \Sigma_k & \xrightarrow{\varphi_k} & \Sigma_k \end{array}$$

which, when combined with with the diagram in 5), yields our $W$-representation

$$\begin{array}{ccc} \Lambda & \xrightarrow{f} & \Lambda \\ \pi \downarrow & & \downarrow \pi \\ \Sigma & \xrightarrow{\varphi} & \Sigma \end{array}$$

where $\Sigma = \Sigma_k, \pi = \chi\pi^s, \varphi = \varphi_k$.
We shall use the following notations: $\{x \in X; H(x)\}$ means the set of elements in the set $X$ which have the property $H$. For a subset $X$ of a topological space $\mathrm{Cl}X$ and $\mathrm{Int}X$ denote the closure and the interior of $X$, respectively. If $M$ is a manifold, then $\partial M$ is the boundary and $\mathrm{Int}M = M \setminus \partial M$ the interior of $M$. By $\mathbf{R}^n$ we denote the $n$-dimensional real coordinate space, and $\mathbf{D}^n$ is the unit ball in $\mathbf{R}^n$. The word "smooth" means differentiable of class $C^1$.

# 2 Branched manifolds

In this section we define some concepts related to branched manifolds (especially those which appear in the main theorem). Then it is shown how certain identifications of points in a compact manifold can create a branched manifold. Later in Sect. 6 identifications of this kind will be a step on the way which leads from an expanding attractor to a branched manifold. Since some details in the concept of branched manifolds are not uniquely determined by our intuitive idea of these manifolds (see [9] and the remark below) we have to choose among several possible definitions. Our choice was suggested by the desire to avoid additional technical troubles in the constructions of this paper.
Let $\Delta_1, ..., \Delta_k$ be disjoint open $n$-disks ($n \geq 1, k \geq 1$) and let for each $\Delta_i$ a homeomorphism $\varphi_i : \Delta_i \to \mathbf{R}^n$ be given. Moreover we assume that for some pairs $(i, j)$, where $1 \leq i < j \leq k$, a subset $K(i, j)$ of $\mathbf{R}^n$ is fixed which is a positive half space in $\mathbf{R}^n$. (A positive half space in $\mathbf{R}^n$ is defined for some $i$ by $H_i = \{(x_1, ..., x_n) \in \mathbf{R}^n; x_i \geq 0\}$.) Then in the

union $\Delta_1 \cup ... \cup \Delta_k$ we perform the following identifications: If $K(i,j)$ is defined and $x \in \Delta_i$, $y \in \Delta_j$, $\varphi_i(x) = \varphi_j(y) \in K(i,j)$, then $x$ is identified with $y$. For the resulting space $\Gamma$ we have the continuous projection $\varphi : \Gamma \to \mathbf{R}^n$ which on each $\Delta_i$ (now regarded as a subset of $\Gamma$) coincides with $\varphi_i$. Obviously for $i \neq j$ the set $\varphi(\Delta_i \cap \Delta_j)$ can be obtained from the family of all positive half spaces by the operations $\cup, \cap$. It is assumed that $\Delta_i \cap \Delta_j$ is never empty. Then $\varphi(\Delta_i \cap \Delta_j)$ always contains the intersection of all positive half spaces, and $\Gamma$ is connected. Each space $\Gamma$ with a projection $\varphi : \Gamma \to \mathbf{R}^n$ which (up to topological equivalence) can be obtained in this way will be called a local branched manifold. By a branch of $\Gamma$ we mean an open disk in $\Gamma$ which is mapped by $\varphi$ homeomorphically onto $\mathbf{R}^n$ . As easily seen each branch of $\Gamma$ coinsides with one of the disks $\Delta_i$, and for any two branches $\Delta_i, \Delta_j$ of $\Gamma$ the set $\varphi(\Delta_i \cap \Delta_j)$ can be obtained from positive half spaces by the operations $\cup$ and $\cap$. Moreover, this set $\varphi(\Delta_i \cap \Delta_j)$ contains the intersection of all positive half spaces, and it is contained in the union of these half spaces. If $x$ lies in the intersection of all positive half spaces of $\mathbf{R}^n$ , then $\varphi^{-1}(x)$ consists of a single point, while for $x$ in the complement of the union of all positive half spaces the number of points in $\varphi^{-1}(x)$ is the same as the number of branches in $\Gamma$. The point in $\varphi^{-1}(o)$ will be called the centre of $\Gamma$.
Now let $(\Gamma, \varphi), (\Gamma', \varphi')$ be local branched $n$-manifolds for which $\Gamma, \Gamma'$ are open subsets of a space $\Sigma$. We say that $(\Gamma, \varphi), (\Gamma', \varphi')$ are $C^r$ compatible ($r \geq 1$) if either $\Gamma \cap \Gamma' = \emptyset$ or $\Gamma \cap \Gamma' \neq \emptyset$ and there is a $C^r$ diffeomorphism $\psi : \varphi(\Gamma \cap \Gamma') \to \varphi'(\Gamma \cap \Gamma')$ such that $\varphi' = \psi\varphi$ on $\Gamma \cap \Gamma'$. By a branched $n$-manifold $\Sigma$ of class $C^r$ ($r \geq 1$) we mean a compact metrizable space for which a family $\{(\Gamma_i, \varphi_i)\}$ of local $n$-manifolds is given such that any $\Gamma_i$ is an open subset of $\Sigma$, the sets $\Gamma_i$ cover $\Sigma$ and each pair $(\Gamma_i, \varphi_i), (\Gamma_j, \varphi_j)$ is $C^r$ compatible. The local branched manifolds $(\Gamma_i, \varphi_i)$ are called charts of $\Sigma$. In this paper all branched manifolds are assumed to be of class $C^1$ .
[For a more restrictive notion of branched $n$-manifold one can use local branched $n$-manifolds $\Gamma$ which are defined as follows: Start with at most $n+1$ disjoint open $n$-disks $\Delta_1, ..., \Delta_k$ and homeomorphisms $\varphi_i : \Delta_i \to \mathbf{R}^n$ ($i = 1, ..., k$). Then for $k-1$ pairs $(i,k)$ where $1 \leq i < j \leq k$ choose a positive half space $H(i,j)$ of $\mathbf{R}^n$ such that these spaces are different. Finally in $\Delta_1 \cup ... \cup \Delta_k$ we identify points $p \in \Delta_i$, $q \in \Delta_j$ if $H(i,j)$ is defined and $\varphi_i(p) = \varphi_j(q) \in H(i,j)$. If the result of these identifications is connected, then it is a local branched $n$-manifold $\Gamma$ in the restricted sense. By an additional argument of general position (after the slice refinement in Sect. 6) the construction below can be modified so that the resulting branched manifolds which appear in the $W$-representations are of this restricted kind.]
Points at which we can find charts with exactly one branch, i.e. near which $\Sigma$ looks like an ordinary manifold, are called regular points. The remaining points are the branch points of $\Sigma$ . For each point $\tau \in \Sigma$ we can define a chart $(\Gamma, \varphi)$ in $\Sigma$ with centre at $\tau$. These charts can be chosen so that the closures of its branches are closed $n$-disks. Then $\mathrm{Cl}\,\Gamma \cap (\Sigma \setminus \Gamma)$ is the union of the boundaries of these disks. A neighbourhood of this kind will be called a normal neighbourhood of $\tau$.
To make branched manifolds better accessible to our geometric intuition we remark that each branched $n$-manifold $\Sigma$ can be embedded in a space $\mathbf{R}^k$ ($k$ sufficiently large) so that the image of each branch $\Delta$ in a chart of $\Sigma$ is an $n$-dimensional $C^1$ subdisk of $\mathbf{R}^k$. We shall identify $\Sigma$ with its image in $\mathbf{R}^k$. Then for each $\tau \in \Sigma$ all smooth $n$-disks in $\Sigma$ which contain $\tau$ have the same tangent space at $\tau$ which will be regarded as the tangent space $T_\tau(\Sigma)$ of $\Sigma$

at $\tau$. Obviously $T_\tau(\Sigma)$ depends continuously on $\tau$. Concepts as Riemannian metrics on $\Sigma$, $C^1$ mappings between branched manifolds or branched manifolds and ordinary manifolds etc. have an obvious meaning. The differential at $\tau \in \Sigma$ of a $C^1$ mapping $\varphi : \Sigma \to \Sigma'$ will be denoted by $d_\tau\varphi : T_\tau(\Sigma) \to T_{\varphi(\tau)}(\Sigma')$.

By a smooth disk in a branched $n$-manifold $\Sigma$ we mean an $n$-dimensional open disk in $\Sigma$ which is of class $C^1$ in the obvious sense and whose closure in $\Sigma$ is a closed disk. A $C^1$ mapping $\varphi : \Sigma \to \Sigma'$ is called a $C^1$ immersion if for each $\tau \in \Sigma$ the linear mapping $d_\tau\varphi : T_\tau(\Sigma) \to T_{\varphi(\tau)}(\Sigma')$ is one-to-one. A continuous mapping $\varphi : \Sigma \to \Sigma'$ where $\Sigma$ and $\Sigma'$ have the same dimension, is called a $C^0$ immersion if each sufficiently small smooth disk in $\Sigma$ is mapped homeomorphically to a smooth disk in $\Sigma'$. As for ordinary manifold $C^1$ immersion are $C^0$ immersions.

A $C^1$ mapping $\varphi : \Sigma \to \Sigma$ of a branched $n$-manifold with a Riemannian metric is called expanding if there is a real $\gamma > 1$ such that for $\tau \in \Sigma$, $v \in T_\tau(\Sigma)$ we have $|d_\tau\varphi(v)| \geq \gamma\,|v|$. The number $\gamma$ will be called an expansion factor of $\varphi$. In this paper it is always assumed that an expanding mapping $\varphi : \Sigma \to \Sigma$ has the following additional property: If $\Gamma$ is a non empty open subset of $\Sigma$, then there is an exponent $k \geq 0$ such that $\varphi^k(\Gamma) = \Sigma$.

A $C^0$ immersion $\varphi : \Sigma \to \Sigma$ is called locally flattening if each $\tau \in \Sigma$ has a neighbourhood $\Gamma$ whose image $\varphi(\Gamma)$ lies in a smooth disk.

A Riemannian metric on a branched $n$-manifold $\Sigma$ defines a distance function $d$ in $\Sigma$: If $\tau, \tau'$ are points in $\Sigma$, then $d(\tau, \tau')$ is the infimum of the lengths of all $C^1$ curves running in $\Sigma$ from $\tau$ to $\tau'$. If $\kappa : \mathbf{R}^n \to \Sigma$ is a $C^0$ immersion, then we define a distance function $d_\kappa$ on $\mathbf{R}^n$: For points $t, t'$ in $\mathbf{R}^n$, we consider all arcs A running in $\mathbf{R}^n$ from $t$ to $t'$ for which $\kappa(A)$ is a rectifiable curve in $\Sigma$ and define $d_\kappa(t, t')$ to be the infimum of the lengths of all these curves $\kappa(A)$. If $\kappa$ is a closed mapping, i.e., if $\kappa$ maps closed sets to closed sets, then $d_\kappa$ is a complete metric on $\mathbf{R}^n$. A closed immersion $\kappa : \mathbf{R}^n \to \Sigma$ with $\kappa(\mathbf{R}^n) = \Sigma$ will be called a covering of $\Sigma$. A branched manifold which has a covering of this kind will be called strongly connected. All branched manifolds which appear in connection with expanding attractors will be strongly connected.

For a strongly connected branched $n$-manifold $\Sigma$ and a covering $\kappa$ we define a new distance $d^*$ by

$$d^*(\tau, \tau') = \{\inf_{t \in \mathbf{R}^n} d_\kappa(t, t');\ t \in \kappa^{-1}(\tau), t' \in \kappa^{-1}(\tau')\}$$

This distance $d^*(\tau, \tau')$ may be regarded as the minimal length of a curve running in $\Sigma$ from $\tau$ to $\tau'$ which can be lifted to $\mathbf{R}^n$. Obviously $d^*(\tau, \tau') = d^*(\tau', \tau)$, and $d^*(\tau, \tau') = 0$ holds if and only if $\tau = \tau'$. Moreover $d^*$ defines the original topology in $\Sigma$, i.e. a set $\Gamma$ in $\Sigma$ is open if and only if for any $\tau \in \Gamma$ there is a positive $\varepsilon$ such that $d^*(\tau, \tau') < \varepsilon$ implies $\tau' \in \Gamma$. Near branch points the triangle inequality will not hold for $d^*$. In spite of this $d^*$ is better adapted to the branch structure of $\Sigma$ than $d$ is.

For later use we describe the following construction which leads from a compact $n$-manifold $Q$ to a branched $n$-manifold $\Sigma$. Let $\sim$ be a binary relation in $Q$ which is reflexive ($x \sim x$) and symmetric ($x \sim y$ if and only if $y \sim x$) and has the following properties, where $\approx$ is the equivalence relation obtained from $\sim$ by transitive extension and $\bar{x}$ denotes the $\approx$-class of $x \in$Q.

(1) $\sim$ is closed, i.e. for sequences $x_1, x_2, \ldots, y_1, y_2, \ldots$ in $Q$ converging to $x$, $y$, respectively, $x_i \sim y_i$ implies $x \sim y$.

(2) For each $x \in \partial Q$ there is a point $y \in IntQ$ such that $x \sim y$.
(3) If $x \sim y$, $x \in IntQ$, then there are neighbourhoods $U_x$ of $x$, $U_y$ of $y$ in $Q$ and a $C^1$ embedding $\chi : U_y \to U_x$ such that $\chi(y) = x$ and $\chi(y') \sim y'$ for any $y' \in U_y$.
(4) If $x \in Q$ then for any $y \in \bar{x}$ we can find an open neighbourhood $U(y)$ in $Q$ and a $C^1$ embedding $\varphi_y : U(y) \to \mathbf{R}^n$ where $\varphi_y(y) = o$ and $\varphi_y(U(y))$ is either $\mathbf{R}^n$ (if $y \in IntQ$) or a positive half space of $\mathbf{R}^n$ (if $y \in \partial Q$) such that for $y_1, y_2 \in \bar{x}$, $y_1' \in U(y_1)$, $y_2' \in U(y_2)$ the relation $y_1' \approx y_2'$ implies $\varphi_{y_1}(y_1') = \varphi_{y_2}(y_2')$.
We shall show how under these conditions the factor space $\Sigma = Q/\approx = \{\bar{x};\ x \in Q\}$ can be equipped with the structure of a branched $n$-manifold so that the natural projection $\pi : Q \to \Sigma$ becomes a $C^1$ immersion.
By (4) each $x \in Q$ has a neighbourhood inside which $\approx$ is the identity $=$. Since $Q$ is compact this implies that the classes $\bar{x}$ ($x \in Q$) are finite and that the numbers of their elements are bounded. Therefore by (1) the relation $\approx$ is closed. Then it is well known from general topology that $\Sigma$ is metrizable and compact. Moreover it is not hard to see that for $x \in Q$ the neighbourhoods $U(y)$ in (4) can be chosen so small that for each $z$ in the neighbourhood

$$U(\bar{x}) = \bigcup_{y \in \bar{x}} U(y)$$

of $\bar{x}$ the whole class $\bar{z}$ lies in $U(\bar{x})$ and that $\Gamma = \pi(U(\bar{x}))$ is an open neighbourhood of $\pi(x) = \bar{x}$ in $\Sigma$. If $U(\bar{x})$ is sufficiently small, then by (3) for any $y, z \in \bar{x}$, $y' \in U(y), z' \in U(z)$ the relations $y \sim z$, $\varphi_y(y') = \varphi_z(z')$ imply $y' \sim z'$. By (2) for each $y \in \bar{x}$ there is a point $z \in \bar{x}$ such that $y \sim z$ and $\varphi_z(U(z)) = \mathbf{R}^n$. Let $\bar{x}_1$ be the set of all those $y \in \bar{x}$ for which $\varphi_y(U(y)) = \mathbf{R}^n$. Then for $y \in \bar{x}_1$ the projection $\pi : Q \to \Sigma$ maps $U(y)$ homeomorphically onto an open $n$-disk in $\Gamma$ and each of the sets $U(z)$ ($z \in \bar{x} \setminus \bar{x}_1$) is mapped by $\pi$ onto a subset of one of these disks. Therefore $\Gamma$ is the union of open $n$-disks $\Delta_1, ..., \Delta_k$. If $\bar{\varphi} : U(\bar{x}) \to \mathbf{R}^n$ is the mapping which on each set $U(y)$ ($y \in \bar{x}$) coincides with $\varphi_y$, then by (4) the mapping $\varphi = \bar{\varphi}\pi^{-1} : \Gamma \to \mathbf{R}^n$ is well defined, and the pair $(\Gamma, \varphi)$ has all properties of a local branched $n$-manifold. It is not hard to see that charts constructed in this way are $C^1$ compatible. (Use the fact that for $y \in U(\bar{x})$ we have $\bar{y} \subset U(\bar{x})$.) This shows that $\Sigma$ has the structure of a branched $n$-manifold and that $\pi : Q \to \Sigma$ is a $C^1$ immersion.

# 3 Conjugating Immersion of Branched Manifolds

In this section we shall show that for two immersions $\varphi, \varphi^* : \Sigma \to \Sigma$ of a branched manifold $\Sigma$ ($\varphi$ of class $C^0$, $\varphi^*$ of class $C^1$) there is another $C^0$ immersion $\chi : \Sigma \to \Sigma$ such that $\chi\varphi = \varphi^*\chi$ provided $\varphi^*$ is sufficiently close to $\varphi$ and $\varphi^*$ is expanding in the sense that $|d\varphi^*(v)| \geq \gamma|v|$ holds for some $\gamma > 1$ and all tangent vectors v of $\Sigma$. (It is assumed that a Riemannian metric has been fixed on $\Sigma$.) Since more details concerning the existence of $\chi$ will be needed the following definitions are necessary for the definitive formulation.
Let $\Sigma$ be a branched $n$-manifold with a Riemannian metric, and let $\Sigma'$ be the same branched manifold with another Riemannian metric such that for any tangent vector v of $\Sigma$ we have $|v|' \geq |v|$, where $|\,|, |\,|'$ are the metrics in $\Sigma$ and $\Sigma'$, respectively. Then we say that $\Sigma'$ is obtained from $\Sigma$ by a dilatation or, briefly, that $\Sigma'$ is a dilatation of $\Sigma$.
In this section we assume that a branched $n$-manifold $\Sigma$ with a covering $\kappa : \mathbf{R}^n \to \Sigma$ is fixed. Then we have the distance $d_\kappa$ in $\mathbf{R}^n$ (see the preceding section), and the distance

$d^*$ in $\Sigma$. If $\varepsilon > 0$ is sufficiently small, then for any $\tau \in \Sigma$ and any $t \in \kappa^{-1}(\tau)$ the set $\{t' \in \mathbf{R}^n;\, d_\kappa(t',t) < \varepsilon\}$ is mapped by $\kappa$ homeomorphically onto its image which will be called a flat $\varepsilon$-neighbourhood of $\tau$. If $\tau$ is a branch point, then no flat neighbourhood of $\tau$ can be an ordinary neighbourhood of $\tau$ in $\Sigma$, but each point has finitely many flat $\varepsilon$-neighbourhoods whose union is an ordinary neighbourhood.

Now we assume that a locally flattening $C^0$ immersion $\varphi : \Sigma \to \Sigma$ is given which satisfies $\varphi(\Sigma) = \Sigma$ and which has an extension number $\vartheta > 0$ with respect to $d^*$ (i.e. for $\tau \neq \tau'$ there is a $k \geq 0$ such that $d^*(\varphi^k(\tau), \varphi^k(\tau')) > \vartheta$).

It is the aim of this section to prove the following lemma.

**Lemma.** Let $\varphi : \Sigma \to \Sigma$ be the given $C^0$ immersion. Then for each $\gamma > 1$ there is a $\delta > 0$ such that for any $C^1$ immersion $\varphi^* : \Sigma' \to \Sigma'$ of a dilatation $\Sigma'$ of $\Sigma$ which has the properties (1) - (3) below we can find a $C^0$ immersion $\chi : \Sigma' \to \Sigma'$ such that the following diagram is commutative

$$\begin{array}{ccc} \Sigma' & \xrightarrow{\varphi} & \Sigma' \\ \chi\downarrow & & \downarrow\chi \\ \Sigma' & \xrightarrow{\varphi^*} & \Sigma' \end{array}$$

where $\varphi : \Sigma' \to \Sigma'$ is the old mapping $\varphi$ now regarded as a mapping of $\Sigma'$.

(1) $\varphi^*$ is expanding with the factor $\gamma$.

(2) $\varphi^*(\Sigma) = \Sigma$.

(3) $d'^*(\varphi(\tau), \varphi^*(\tau)) < \delta$ for all $\tau \in \Sigma$.

(Here $d'^*$ denotes the distance in $\Sigma'$ which is defined via the covering $\kappa : \mathbf{R}^n \to \Sigma'$ as $d^*$ is defined in $\Sigma$.)

If $\Sigma$ has no branch points, i.e. if $\Sigma$ is an ordinary manifold, then the proof of the lemma is simple. In the general case there are some difficulties caused by the possible branchings in $\Sigma$. We shall resolve theese branchings by the covering $\kappa : \mathbf{R}^n \to \Sigma$. Then the following sublemma can be applied in the resolved situation.

**Sublemma.** For each branched $n$-manifold $\Sigma$ with a Riemannian metric there is a positive $\delta_0$ such that any dilatation $\Sigma'$ of $\Sigma$ has the following property: If $0 < \delta \leq \delta_0$ and $\kappa_1, \kappa_2 : \mathbf{R}^n \to \Sigma'$ are $C^0$ coverings satisfying

$$\sup_{t \in \mathbf{R}^n} d'(\kappa_1(t), \kappa_2(t)) < \delta,$$

then there is a homeomorphism $h : \mathbf{R}^n \to \mathbf{R}^n$ such that $\kappa_2 = \kappa_1 h$ and

$$\sup_{t \in \mathbf{R}^n} d'_{\kappa_i}(h(t), t) \leq \delta \quad (i = 1, 2).$$

(Here $d'$ denotes the distance in $\Sigma'$ and $d'_{\kappa_i}$ are the corresponding distances in $\mathbf{R}^n$ as defined in the preceding section.)

**Proof of the Sublemma.** Looking at the local structure of $\Sigma$ and using that $\Sigma$ is compact it is not hard to prove that there is a positive $\delta_1$ such that for any dilatation $\Sigma'$ of $\Sigma$, any $C^0$ immersion $\kappa : \mathbf{R}^n \to \Sigma'$, any $t \in \mathbf{R}^n$ and any $\delta \in (0, \delta_1)$ the set

$$U(t, \delta, \kappa) = \{t' \in \mathbf{R}^n;\, d'_\kappa(t', t) < \delta\}$$

is an open neighbourhood of $t$ in $\mathbf{R}^n$ which is mapped by $\kappa$ homeomorphically onto an open subset of a smooth disk in $\Sigma'$. Once more using the compactness of $\Sigma$ we find a positive $\delta_0 < \delta_1$ such that for any dilatation $\Sigma'$ of $\Sigma$, any two immersions $\kappa_1, \kappa_2 : \mathbf{R}^n \to \Sigma'$ satisfying

$$\sup_{t\in\mathbf{R}^n} d'(\kappa_1(t),\kappa_2(t)) \leq \delta_0$$

and any $t \in \mathbf{R}^n$ the set $\kappa_2(U(t,\delta_0,\kappa_2))$ lies in $\kappa_1(U(t,\delta_1,\kappa_1))$. Then, if $t$ is any point in $\mathbf{R}^n$, we have the embedding

$$h_t = \kappa_1^{-1}\kappa_2 : U(t,\delta_0,\kappa_2) \to U(t,\delta_1,\kappa_1)\,,$$

and for each $t' \in U(t,\delta_0,\kappa_2)$ the mappings $h_t, h_{t'}$, coincide in a neighbourhood of $t'$. This implies that by $h(t) = h_t(t)$ a $C^0$ immersion $h : \mathbf{R}^n \to \mathbf{R}^n$ is defined which has the following property:

$$\sup_{t\in\mathbf{R}^n} d'(\kappa_i(h(t),t) \leq \sup_{t\in\mathbf{R}^n} d'(\kappa_1(t),\kappa_2(t)) \quad (i = 1,2).$$

Then it is an elementary fact in topology that $h : \mathbf{R}^n \to \mathbf{R}^n$ is a homeomorphism, and the sublemma is proved.

**Proof of the Lemma.** Since $\varphi$ is locally flattening and $\Sigma$ is compact we can find a positive $\rho$ such that for each $\tau \in \Sigma$ there is a flat $\rho$-neighbourhood $\Gamma(\tau)$ of $\varphi(\tau)$ as defined above which has the following property: If $t \in \kappa^{-1}(\tau)$ then there is an open $n$-disk $D$ in $\mathbf{R}^n$ containing $t$ such that $\varphi\kappa$ maps $D$ homeomorphically onto a disk which contains $\Gamma(\tau)$. We assume that $\rho$ and the sets $\Gamma(\tau)$ are fixed.

Let $\delta_0$ be chosen according to the sublemma. Then we fix $\delta > 0$ so that

$$\delta < \delta_0,\ \delta < \rho$$

$$\delta < \rho(\gamma - 1),\ \delta < (\gamma - 1)/2$$

For a moment we consider a dilatation $\Sigma'$ of $\Sigma$ with a distance function $d'$ and the given mappings $\varphi : \Sigma' \to \Sigma'$, $\kappa : \mathbf{R}^n \to \Sigma'$. If we replace in the definition of the distance functions $d_\kappa, d^*$ the metric $d$ by $d'$, we get distances $d'_\kappa$ in $\mathbf{R}^n$ and $d'^*$ in $\Sigma'$. Obviously $\vartheta$ is an extension number of $\varphi$ with respect to $d'^*$, and the number $\rho$ defined above does not lose its properties if $\Sigma$, $d$ are replaced by $\Sigma'$, $d'$. Moreover the sublemma remains true for $\Sigma'$ with the same $\delta_0$. Therefore it is sufficient to prove the lemma for $\Sigma' = \Sigma$.

Let $\varphi^* : \Sigma \to \Sigma$ be an immersion which satisfies the assumptions of the lemma. Then $\varphi\kappa, \varphi^*\kappa : \mathbf{R}^n \to \Sigma$ are coverings to which the sublemma applies, and we get a homeomorphism $h : \mathbf{R}^n \to \mathbf{R}^n$ such that

$$\varphi^*\kappa h(t) = \varphi\kappa(t),\ d_{\varphi\kappa}(h(t),t) \leq \delta$$

holds for all $t \in \mathbf{R}^n$.

In the next step of the proof we show that for each $\tau \in \Sigma$ we can choose a subset $\Gamma^*(\tau)$ of $\Gamma(\tau)$ which is mapped by $\varphi^*$ homeomorphically onto $\Gamma(\varphi(\tau))$.

Since $\Gamma(\tau)$ is a flat $\rho$-neighbourhood of $\varphi(\tau)$ we can find a point $t \in \kappa^{-1}(\varphi(\tau))$ such that

$$U = \{t' \in \mathbf{R}^n; d_\kappa(t',t) < \rho\}$$

is mapped by $\kappa$ homeomorphically onto $\Gamma(\tau)$. By the definition of the sets $\Gamma(\tau)$

$$U' = \{t' \in \mathbf{R}^n; d_{\varphi\kappa}(t',t) < \rho\}$$

is mapped by $\varphi\kappa$ homeomorphically onto $\Gamma(\varphi(\tau))$. Then we have

$$\begin{aligned} h(U') &= \{t' \in \mathbf{R}^n; d_{\varphi\kappa}(h^{-1}(t'), t) < \rho\} \\ &= \{t' \in \mathbf{R}^n; d_{\varphi^*\kappa}(t', h(t)) < \rho\} \\ &\subset \{t' \in \mathbf{R}^n; d_{\varphi^*\kappa}(t', t) < \rho + \rho(\gamma - 1)\} \\ &\subset \{t' \in \mathbf{R}^n; \gamma d_{\kappa}(t', t) < \rho\gamma\} \\ &= U, \end{aligned}$$

and $\Gamma^*(\tau) = \kappa h(U')$ has the properties required above.
Now for each $\tau \in \Sigma$ we define a sequence $\tau_0, \tau_1, \ldots$ of points in $\Gamma(\tau)$ by

$$\varphi^{*j}(\tau_i) \in \Gamma(\varphi^j(\tau)) \ (j = 0, \ldots, i - 1)$$

$$\varphi^{*i}(\tau_i) = \varphi^{i+1}(\tau).$$

Then, since $\varphi^*$ is expanding with the factor $\gamma$, $d^*(\tau_1, \varphi(\tau)) = d^*(\tau_1, \tau_0) < \delta\gamma^{-1}$, and $\varphi^{*i}(\tau_{i+1}) = (\varphi^i(\tau))_1$, $\varphi^{*i}(\tau_i) = \varphi(\varphi^i(\tau))$ implies $d(\tau_{i+1}, \tau_i) \leq \delta\gamma^{-i-1}$.
Therefore the sequence $\tau_0, \tau_1, \ldots$ converges to a point $\chi(\tau)$ in $\Sigma$. If $\tau'$ is close to $\tau$, then the points $\tau_i'$ are close to $\tau_i$ for all $i \leq k$ where $k$ is large. This implies that $\chi$ is continuous. The relation $\varphi^*\chi = \chi\varphi$ is an immediate consequence of the equalities $(\varphi(\tau))_i = \varphi^*(\tau_{i+1})$ $(i = 0, 1, \ldots)$. Moreover we have

$$d^*(\chi(\tau), \varphi(\tau)) \leq \delta \sum_{i=1}^{\infty} \gamma^{-i} = \delta/(\gamma - 1).$$

It remains to prove that $\chi$ is a $C^0$ immersion. Let $\tau \neq \tau' \in \Sigma$ be given. It is sufficient to prove that the assumptions $\chi(\tau) = \chi(\tau')$ and $d^*(\tau, \tau') < \vartheta$ lead to a contradiction. Since $\tau \neq \tau'$ and $d^*(\tau, \tau') < \vartheta$ we find an exponent $k \geq 1$ such that $d^*(\varphi^k(\tau), \varphi^k(\tau') > \vartheta$. On the other hand we have

$$d^*(\varphi^k(\tau), \chi\varphi^{k-1}(\tau)) \leq \delta/(\gamma - 1)$$

$$d^*(\varphi^k(\tau'), \chi\varphi^{k-1}(\tau') \leq \delta/(\gamma - 1)$$

$$\chi\varphi^{k-1}(\tau) = \varphi^{*k-1}\chi(\tau) = \varphi^{*k-1}\chi(\tau') = \chi\varphi^{k-1}(\tau')$$

and therefore

$$d^*(\varphi^k(\tau), \varphi^k(\tau')) \leq 2\delta/(\gamma - 1).$$

This contradicts $\delta < (\gamma - 1)\vartheta/2$.

# 4 Rectangles

Let $\Lambda$ be our $n$-dimensional expanding attractor in $\mathbf{R}^n$. For some $r \geq 0$ and $n' = m - n$ we consider a $C^r$ foliation $\mathcal{F}$ of an open neighbourhood $N$ of $\Lambda$ whose leaves are $n'$-dimensional. For $x \in N$ the leaf through $x$ will be denoted by $F_x$. If $r = 0$ it is assumed that the leaves are $C^1$ manifolds and that the tangent spaces $T_x F_x$ depend continuously on $x \in N$. We say that $\mathcal{F}$ is transverse to $\Lambda$ if for each $x \in \Lambda$ the manifolds $W_x^u$ and $F_x$ are transverse at $x$. The foliation $\mathcal{W}^s$ consisting of all stable manifolds $W_x^s$ ($x \in \Lambda$) is an example of a $C^0$ foliation transverse to $\Lambda$.

In $N$ we define a distance function $d^F$ by $d^F(x, y) =$ distance in $F_x$ if $y \in F_x$ and $d^F(x, y) = \infty$ if $F_x \neq F_y$. Moreover for each foliation $\mathcal{F}$ transverse to $\Lambda$ we fix two positive numbers $\rho = \rho(\mathcal{F})$ and $\varepsilon^* = \varepsilon^*(\mathcal{F})$ : $\rho$ is chosen so small that for each $x \in \Lambda$ the set $\{y \in F_x;\ d^F(x, y) \leq \rho\}$ is a closed $n'$-disk, and $\varepsilon^*$ is so small that for each finite sequence $x_1, ..., x_k$ in $\Lambda$ satisfying $d^F(x_i, x_{i+1}) < \varepsilon$ ($i = 1, ..., k-1$) we have $d^F(x_1, x_k) < \frac{1}{2}\rho$. (Since $\Lambda$ is compact it is not hard to find numbers $\rho, \varepsilon^*$ with these properties. If $x, y \in \Lambda$ and $d^F(x, y) \leq \frac{1}{2}\rho$, then we can find neighbourhoods $U_x$ of $x$ in $W_x^u$ and $U_y$ of $y$ in $W_y^u$ and a homeomorphism $h_x^y : U_y \to U_x$ such that $d^F(h_x^y(y'), y') < \rho$ holds for all points $y' \in U_y$. This mapping $h_x^y$ is uniquely determined and will be called the holonomy mapping along $\mathcal{F}$. If $r > 0$, then $h_x^y$ is a $C^1$ diffeomorphism.

By an $\mathcal{F}$-rectangle we mean the image $R = r(\mathbf{D}^n \times C)$ of a $C^0$ embedding $r : \mathbf{D}^n \times C \to \Lambda$ ($\mathbf{D}^n$ the $n$-dimensional closed unit disk, $C$ a Cantor set) which has the following properties:

(1) For $t \in \mathbf{D}^n$, $c \in C$, $x = r(t, c)$ the set

$$R_x^F = r(\{t\} \times C)$$

lies in $F_x$, is open in $F_x \cap \Lambda$ with respect to the metric $d^F$ and has $d^F$-diameter less than $\varepsilon^*(\mathcal{F})$. Moreover

$$R \cap \{y \in F_x; d^F(x, y) \leq \rho(\mathcal{F})\} = r(\{t\} \times C)$$

(2) If $\mathcal{F}$ is of class $C^1$, then each disk $r(\mathbf{D}^n \times \{c\})$ ($c \in C$) is a $C^1$ disk.

If $\mathcal{F} = \mathcal{W}^s$ then $\mathcal{F}$-rectangles will also be called $s$-rectangles, and we shall write $F_x = W_x^s$, $d^F = d^s$, $R_x^F = R_x^s$. The components of an $\mathcal{F}$-rectangle $R = r(\mathbf{D}^n \times C)$ are the disks $r(\mathbf{D}^n \times \{c\})$ ($c \in C$), and the interior of $R$ with respect to the topology of $\Lambda$ is given by

$$\mathrm{Int} R = r(\mathrm{Int}\mathbf{D}^n \times C).$$

The boundary $R \backslash \mathrm{Int} R$ of $R$ will be denoted by $\partial R$.

By a slice of an $\mathcal{F}$-rectangle $R$ we mean an $\mathcal{F}$-rectangle $R'$ each of whose components is a component of $R$. A finite family of slices of $R$ whose union is $R$ will be called a slice decomposition of $R$.

A finite family $\mathcal{R}$ of $\mathcal{F}$-rectangles whose interiors cover $\Lambda$ will be called a covering by $\mathcal{F}$-rectangles. If in a covering $\mathcal{R}$ by $\mathcal{F}$-rectangles each rectangle is replaced by one of its slice decompositions we get a new covering by $\mathcal{F}$-rectanlges which will be called a slice refinement of $\mathcal{R}$.

Looking at the local structure of $\Lambda$ as described in Section 1 we see that each point $x \in \Lambda$ has arbitrarily small neighbourhoods in $\Lambda$ which are $\mathcal{F}$-rectangles, and we can find coverings by $\mathcal{F}$-rectanlges consisting of arbitrarily small rectangles.

Now we describe how a covering $\mathcal{R}$ by $\mathcal{F}$-rectanlges defines a decomposition of $\Lambda$ in disjoint Cantor sets. For $x \in \Lambda$ we consider the set $V_x(\mathcal{R})$ of all y for which there are points $x = x_0, x_1, ..., x_k = y$ in $\Lambda$ and rectangles $R_1, ..., R_k$ in $\mathcal{R}$ such that $x_{i-1}, x_i \in R_i$ and $x_{i-1} \in (R_i)^F_{x_i}$ $(i = 1, ..., k)$. Obviously, for any two points $x, y \in \Lambda$ we have $V_x(\mathcal{R}) = V_y(\mathcal{R})$ or $V_x(\mathcal{R}) \cap V_y(\mathcal{R}) = \emptyset$, and each set $V_x(\mathcal{R})$ is a Cantor set. Since the sets $R^F_x$ have $d^F$-diameters less than $\varepsilon^*(\mathcal{F})$, the $d^F$-diameters of the sets $V_x(\mathcal{R})$ are less than $\frac{1}{2}\rho(\mathcal{F})$, and for any $R \in \mathcal{R}$ containing $x$ we have $V_x(\mathcal{R}) \cap R = R^F_x$. The family of all sets $V_x(\mathcal{R})$ $(x \in \Lambda)$ is upper semicontinuous, i.e. for each set $V_x(\mathcal{R})$ and each $\varepsilon > 0$ there is a $\delta > 0$ such that $d(V_x(\mathcal{R}), V_y(\mathcal{R})) < \delta$ implies that $V_y(\mathcal{R})$ is contained in the $\varepsilon$-neighbourhood of $V_x(\mathcal{R})$. Since the diameter of each set $V_x(\mathcal{R})$ is smaller than $\frac{1}{2}\rho(\mathcal{F})$ we can choose for each $y \in V_x(\mathcal{R})$ an open $n$-disk $U_y$ in $W^u_y$ such that for $y, z$ belonging to the same set $V_x(\mathcal{R})$ the holonomy map $h^y_z : U_y \to U_z$ is defined. The set

$$\tilde{U}_x = \bigcup_{y \in V_x(\mathcal{R})} U_y$$

is an open neighbourhood of $V_x(\mathcal{R})$ whose components are the disks $U_y$. It will be called a holonomy neighbourhood of $V_x(\mathcal{R})$. The mappings $h^y_x$ $(y \in V_x(\mathcal{R}))$ define a projection $\tilde{h}_x : \tilde{U}_x \to U_x$.

# 5 A $C^1$ foliation $\mathcal{F}$ transverse to $\Lambda$.

The aim of this section is the construction of a neighbourhood $N$ of $\Lambda$ and a $C^1$ foliation $\mathcal{F}$ of $N$ with $n'$-dimensional leaves $(n' = m - n)$ which is transverse to $\Lambda$ as defined in the preceding section. This foliation will be fixed for the rest of the paper. If $n' = 1$ then $\mathcal{F}$ with $N = W^s$ can be obtained by smoothening and integrating the line field on $W^s$ consisting of the tangents to the stable manifolds $W^s_x$. For $n' > 1$ this way to $\mathcal{F}$ seems to be blocked by the integrability conditions for plane fields. Therefore we could not avoid a construction like this which is described below. A family $R_1, ..., R_k$ of $s$-rectangles will be called *directed* if $x \in R_i \cap R_j$, $i < j$ implies $(R_j)^s_x \subset (R_i)^s_x$. If $R_1, R_2, ..., R_k$ is a directed family of $s$-rectanlges and for some fixed $i$ we have a slice decomposition $R'_1, ..., R'_\ell$ of $R_i$, then for each $j > i$ there is a slice decomposition $R'_{j,1}, R'_{j,2} ...$ of $R_j$ such that $R_1, ..., R_{i-1}, R'_1, ..., R'_\ell, R'_{i+1,1}, ..., R'_{i+2,1}, ..., R'_{k,1}, ...$ is a directed family of $s$-rectanlges. Let $X$ be a subset of $\mathbf{R}^m$ and $\rho > 0$. We consider a mapping which associates with each point $x \in X$ a euclidian $n'$-dimensional disk $\zeta(x)$ in $\mathbf{R}^m$ with centre $x$ and radius $\rho$. If the disks $\zeta(x)$ are disjoint we say that $\zeta$ is a disk bundle or a $\rho$-disk bundle over $X$. A disk bundle $\zeta$ defines a mapping $\bar{\zeta} : X \to G(m, n')$, where $G(m, n')$ denotes the Grassmann manifold consisting of all $n'$-dimensional linear subspaces of $\mathbf{R}^m$. We say that $\zeta$ is continuous or of class $C^r$, if $\bar{\zeta}$ has this property. The union of all disks $\zeta(x)$ $(x \in X)$ will be denoted by $|\zeta|$. If $R$ is an $s$-rectangle and $Q$ is a compact connected $n$-manifold lying in a component of $R$, then a $C^1$ disk bundle over $Q$ will be called a disk bundle for $R$ provided for each $x \in Q$ and each component $P$ of $R$ the intersection $\zeta(x) \cap P$ consists of exactly one point which lies in Int$P \cap$Int$\zeta(x)$, and the intersection is transverse. If $\zeta$ is a disk bundle for $R$, $x \in Q$, $y \in \zeta(x)$ we shall write $|\zeta|^* = |\zeta| \cap R$, $|\zeta|_y = \zeta(x)$, $|\zeta|^*_y = |\zeta| \cap R$. Later the following remarks will be used:

**A)** Let $X$ be a compact set lying in an $n$-dimensional $C^1$ manifold $W$ in $\mathbf{R}^m$, and let $\gamma : X \to G(m,n')$ be of class $C^1$ and transverse to $W$ (i.e. $\gamma$ can be extended to a $C^1$ mapping which is defined on an open neighbourhood of $X$ in $W$, and for each $x \in X$ the spaces $\gamma(x)$ and $T_x(W)$ are transverse). Then there is a disk bundle $\zeta$ over $X$ such that $\bar{\zeta} = \gamma$.
**B)** If $P$ is a compact $n$-dimensional $C^1$ manifold in $\mathbf{R}^m$ and $X$ is a closed subset of $P$, then for any continuous mapping $\gamma : X \to G(m,n')$ which is transverse to $P$ there is a continuous extension $\tilde{\gamma} : P \to G(m,n')$ which is still transverse to $P$. If $X'$ is a compact set lying in the interior of $X$ (with respect to the topology of $P$), then, if $\gamma$ is $C^1$ on Int$X$, we can find a $C^1$ mapping $\gamma' : P \to G(m,n')$ which is transverse to $P$ and which coincides with $\gamma$ on $X'$.
The existence of $\zeta$ in A) is a simple consequence of the assumptions that $\gamma$ is of class $C^1$ and that $X$ is compact. To prove B) we remark that for $x \in P$ the set $H_x$ of all those elements of $G(m,n')$ which are transverse to $T_x(P)$ is an open subset of $G(m,n')$ which topologically is an open disk. (There is a contracting isotopy which maps $H_x$ to a small disk in $G(m,n')$ whose centre is the linear space perpendicular to $T_x(P)$.) Therefore, $x \mapsto H_x$ is a disk bundle over $X$ and $\gamma$ may be regarded as a section in this bundle. To get $\tilde{\gamma}$ we merely have to apply a well known theorem about disk bundless (see [6],p.55), and $\gamma'$ can be obtained by ordinary smoothening techniques.
Let $R_1, ..., R_k$ be a directed family of $s$-rectangles with disk bundles $\zeta_1, ..., \zeta_k$, respectivley. We say that $\zeta_1, ..., \zeta_k$ are *well matched* if $x \in |\zeta_i|^* \cap |\zeta_j|^*$ , $i < j$ implies $|\zeta_j|_x \subset |\zeta_i|_x$.
**Lemma.**There is a directed covering by $s$-rectangles $R_1, ..., R_r$ with well matched disk bundles $\zeta_1, ..., \zeta_r$ such that $\Lambda = Int\,|\zeta_1|^* \cup ... \cup Int\,|\zeta_r|^*$ where $Int\,|\zeta_i|^*$ is the interior of $|\zeta_i|^*$ with respect to the topology of $\Lambda$ (i.e. if $P$ is a component of $|\zeta_i|^*$, then $IntP$ is a component of $Int\,|\zeta_i|^*$).
Before proving this lemma let us show how it leads to a neighbourhood $N$ of $\Lambda$ with a $C^1$ foliation $\mathcal{F}$ whose leaves are $n'$-dimensional and transverse to $\Lambda$. For $x \in \Lambda$ and the minimal index $i$ with $x \in |\zeta_i|^*$ we denote the disk $|\zeta_i|_x$ by $D_x$ and the set $|\zeta_i|_x^*$ by $D_x^*$. Then for any $x \in \Lambda$ belonging to a set $|\zeta_j|^*$ the disk $|\zeta_j|_x$ lies in $D_x$. There is a positive $\varepsilon_0$ such that for $1 \leq i \leq r$

$$d(|\zeta_i|^*, (|\zeta_i| \cap \Lambda) \setminus |\zeta_i|^*) > \varepsilon_0.$$

Since $\Lambda$ is compact this implies the existence of a positive $\varepsilon_1$ such that for $x, y \in \Lambda$, $z \in D_x \cap D_y$, $D_x \neq D_y$ we have $d(z,x) > \varepsilon_1$ or $d(z,y) > \varepsilon_1$. Using again that $\Lambda$ is compact and that the disks $D_x$ are transverse to $\Lambda$ we can find a positive $\varepsilon_2$ such that for any $z \in \mathbf{R}^m$ with $d(z,\Lambda) < \varepsilon_2$ there is exactly one disk $D_x$ for which $z \in D_x$, $d(z, D_x^*) < \varepsilon_1$. Let $N$ be the open $\varepsilon_2$-neighbourhood of $\Lambda$ in $\mathbf{R}^m$. For $z \in N$ we consider the disk $D_x$ just mentioned above and the set $F_z$ of all points $u \in D_x \cap N$ with $d(u, D_x^*) < \varepsilon_1$. Obviously for $z, z' \in N$ we have either $F_z = F_{z'}$ , or $F_z \cap F_{z'} = 0$, and the family of all sets $F_z$ $(z \in N)$ is a $C^1$ foliation of $N$ by $n'$-dimensional leaves which are transverse to $\Lambda$.
One step in the proof of the lemma will be the proof of the following sublemma. If $\zeta$ is a disk bundle for an $s$-rectangle $R$ which is defined over the submanifold $Q$ of a component of $R$, then by a restriction of $\zeta$ we mean a disk bundle $\zeta'$ for $R$ which is the restriction of $\zeta$ to a manifold $Q'$ in Int$Q$.
**Sublemma.** Let $R_1, ..., R_k, S_1, ..., S_\ell$ be a directed family of $s$- rectangles, where $S_1, ..., S_\ell$ are disjoint, and let $\zeta_1, ..., \zeta_k$ be a family of well matched bundes for $R_1, ..., R_k$. Moreover

we assume that $\zeta_1', ..., \zeta_k'$ are restrictions of $\zeta_1, ..., \zeta_k$, respectively , and that $S_1', ..., S_\ell'$ are $s$-rectangles such that $S_i' \subset Int S_i$ and each component of $S_i$ contains exactly one component of $S_i'$ $(i = 1, ..., \ell)$. Then there are $s$-rectangles $R_{k+1}, ..., R_{k'}$, with disk bundles $\zeta_{k+1}, ..., \zeta_{k'}$ which have the following properties:
(1) $R_1, ..., R_{k'}$ is a directed family.
(2) There are integers $k_0 = k < k_1 < ... < k_\ell = k'$ such that $R_{k_{i-1}+1}, ..., R_{k_i}$ is a slice decomposition of $S_i$ $(i = 1, ..., \ell)$.
(3) $\zeta_1', ..., \zeta_k', \zeta_{k+1}, ..., \zeta_{k'}$ are well matched.
(4) If $k_{i-1} + 1 \leq j \leq k_i$ then $R_j \cap S_i' \subset Int\,|\zeta_j|^*$ $(i = 1, ..., \ell)$.
**Proof of the Sublemma.** Since $S_1, ..., S_\ell$ are disjoint it is sufficient to consider the case $\ell = 1$. Accordingly we denote the rectangles $S_1$, $S_1'$ by $S$, $S'$. The bundles $\zeta_i$ $(i = 1, ..., k)$ are well matched. Therefore the mappings $\bar{\zeta}_i : |\zeta_i|^* \to G(m, n')$ define a mapping $\gamma$ : $|\zeta_1|^* \cup ... \cup |\zeta_k|^* \to G(m, n')$. For each $x \in |\zeta|^* \cup ... \cup |\zeta_k|^*$ the restriction of $\gamma$ to the interior of $W_x^u \cap (|\zeta_1|^* \cup ... \cup |\zeta_k|^*)$ is of class $C^1$ , and $\gamma(x)$ is transverse to $W_x^u$. If $P$ is a component of $S$ and

$$X = P \cap (|\zeta_1|^* \cup ... \cup |\zeta_k|^*),\ X' = P \cap (|\zeta_1'|^* \cup ... \cup |\zeta_k'|^*)$$

then we can apply the remark (B) and get a $C^1$ mapping $\gamma_P : P \to G(m, n')$ which is transverse to $P$ and which coincides with $\gamma$ on $X'$. Using the remark (A) we can find a $C^1$ disk bundle $\zeta_P^*$ over $P$ for which $\bar{\zeta}_P^* = \gamma_P$. Then it is easy to find a slice $S_P$ of $S$ which contains $P$ and a subbundle $\zeta_P$ of $\zeta_P^*$ which is a disk bundle for $S_P$ satisfying $S_P' \subset Int\,|\zeta_P|^*$ where $S_P' = S_P \cap S'$.
To finish the proof of the sublemma we merely have to choose a finite set of components $P_1, ..., P_j$ of $S$ for which the slices $S_{P_i}$ $(1 \leq i \leq j)$ cover $S$ and to choose in each slice $S_{P_i}$ a slice $R_{k+i}$ such that the new slices $R_{k+1}, ..., R_{k'}$ $(k' = k + j)$ are disjoint and still cover $S$. If these slices are sufficiently thin and the sets $|\zeta_{k+i}|_x$ are small, then $R_{k+1}, ..., R_{k'}$ together with the bundles $\zeta_{k+i} = \zeta_{P_i}$ $(i = 1, ..., j)$ have the properties which are required in the sublemma.
**Proof of the Lemma.** We start with two directed coverings $S_1, ..., S_p$ and $S_1', ..., S_p'$ by $s$-rectangles where $S_i' \subset Int S_i$ and each component of $S_i$ contains one component of $S_i'$ $(i = 1, ..., p)$. Then the construction of $R_1, ..., R_r, \zeta_1, ..., \zeta_r$ is carried out in $p$ steps. After the $q$-th step we shall have a directed slice refinement $R_1^q, ..., R_{k'}^q$ of $S_1, ..., S_q$ and corresponding well matched bundles $\zeta_1^q, ..., \zeta_{k'}^q$ such that, if $R_i^q$ $(1 \leq i \leq k')$ belongs to the slice decomposition of $S_{j(i)}$ in $R_1^q, ..., R_{k'}^q$, then $R_i^q \cap S_{j(i)}' \subset Int\,|\zeta_i^q|^*$.
The first step is a simple application of the sublemma for $k = 0$, $\ell = 1$ and $S_1$, $S_1'$ as chosen above. If the $q$-th step is done $(1 \leq q < p)$ and we have obtained $R_1^q, ..., R_{k'}^q, \zeta_1^q, ..., \zeta_{k'}^q$, then we proceed as follows: We choose a slice decomposition with disjoint slices $S_1^*, ..., S_\ell^*$ of $S_{q+1}$ and restrictions $\zeta_1', ..., \zeta_{k'}'$ of $\zeta_1^q, ..., \zeta_{k'}^q$ such that $R_1^q, ..., R_{k'}^q, S_1^*, ..., S_\ell^*$ is directed and $R_i^q \cap S_{j(i)}' \subset Int\,|\zeta_i'|^*$ holds for $1 \leq i \leq k'$ and $j(i)$ as defined above. Now we apply the sublemma to $R_1^q, ..., R_{k'}^q, S_1^*, ..., S_\ell^*, S_1'^*, ..., S_\ell'^*$ $(S_i'^* = S_i^* \cap S_{q+1}')$ and $\zeta_1^q, ..., \zeta_{k'}^q, \zeta_1', ..., \zeta_{k'}'$. What we obtain are the $s$-rectangles $R_i^{q+1}$ and the bundles $\zeta_i^{q+1}$ which are expected of the $(q + 1)$-st step. If the $p$-th step is done the $s$-rectangles $R_i = R_i^p$ and the bundles $\zeta_i = \zeta_i^p$ have the properties which are required in the lemma.

# 6 The Branched Manifold $\Sigma(\mathcal{R})$ and the Mapping $\pi^F$

Starting with a $C^1$ foliation $\mathcal{F}$ of a neighbourhood $N$ of $\Lambda$ with $n'$-dimensional leaves ($n' = m - n$) which are transverse to $\Lambda$ (see Section 5) we use a special kind of coverings $\mathcal{R}$ by $\mathcal{F}$-rectangles, called boundary transverse, to define a branched $n$-manifold $\Sigma(\mathcal{R})$ together with a projection $\pi^F : \Lambda \to \Sigma(\mathcal{R})$. This branched manifold $\Sigma(\mathcal{R})$ will be strongly connected and the restriction of $\pi^F$ to any unstable manifold $W_x^u$ in $\Lambda$ will be a $C^1$ immersion with $\pi^F(W_x^u) = \Sigma(\mathcal{R})$.

Let $\mathcal{R}$ be a covering by $\mathcal{F}$-rectangles. By the definitions given in Section 4 we have a decomposition of $\Lambda$ into the sets $V_x(\mathcal{R})$ ($x \in \Lambda$), and for each $V_x(\mathcal{R})$ we may choose an open neighbourhood $\tilde{U}_x$ in $\Lambda$ each of whose components is an open $n$-disk which contains exactly one point of $V_x(\mathcal{R})$ such that for $x, z \in V_x(\mathcal{R})$ the holonomy mapping $h_z^y : U_y \to U_z$ is a $C^1$ diffeomorphism mapping the component $U_y$ of $y$ in $\tilde{U}_x$ onto the component $U_z$ of $z$. The union of all the mappings $h_x^y : U_y \to U_x$ ($y \in V_x(\mathcal{R})$) will be denoted by $\tilde{h}_x : \tilde{U}_x \to U_x$. Now we choose a compact $n$-dimensional $C^1$ manifold $Q$ in $\Lambda$ whose components $Q_1, ..., Q_r$ are in one-to-one correspondence with the $\mathcal{F}$-rectangles $R_1, ..., R_r$ in $\mathcal{R}$, and $Q_i$ is a component of $R_i$. We say that $\mathcal{R}$ is boundary transverse if for each $x \in \Lambda$ there is a $C^1$ diffeomorphism $\psi_x : U_x \to \mathbf{R}^n$ such that $\psi_x(x) = o$ and any $y \in V_x(\mathcal{R}) \cap \partial Q$ has a neighbourhood $V$ in $Q \cap U_y$ which is mapped by $\psi_x h_x^y$ to a neighbourhood of $o$ in a positive half space in $\mathbf{R}^n$ (see Section 2).

If $\mathcal{R}_0$ is a covering by $\mathcal{F}$-rectangles then by a small push of the rectangles in $\mathcal{R}_0$ to general position we get a new covering $\mathcal{R}$ by $\mathcal{F}$-rectangles such that for each $x \in \Lambda$ there are at most $n$ points $y$ in $V_x(\mathcal{R}) \cap \partial Q$ and the corresponding $(n-1)$-manifolds $h_x^y(U_y \cap \partial Q)$ are transverse at $x$ in $U_x$. This shows that boundary transverse coverings by $\mathcal{F}$-rectangles exist. It may seem surprising that in the definition of boundary transverseness we did not exclude the case where there are points $x \in \Lambda, y \neq z \in V_x(\mathcal{R}) \cap \partial Q$ for which $h_x^y(U_y \cap \partial Q) = h_x^z(U_z \cap \partial Q)$. The reason to choose this definition was that by it slice refinements of boundary transverse coverings are boundary transverse.

Now let $\mathcal{F}, \mathcal{R} = \{R_1, ..., R_r\}$, $Q = Q_1 \cup ... \cup Q_r$ be as above, where $\mathcal{R}$ is boundary transverse. We consider the decomposition space $\Sigma(\mathcal{R})$ of $\Lambda$ whose elements are the sets $V_x(\mathcal{R})$ ($x \in \Lambda$) and the natural projection $\pi^F : \Lambda \to \Sigma(\mathcal{R})$. Since for each $x$ in the component $Q_i$ of $R_i$ and of $Q$ the set $(R_i)_x^F$ as defined in Section 4 lies in $V_x(\mathcal{R})$, the space $\Sigma(\mathcal{R})$ can also be obtained as the factor space of $Q$ with respect to the equivalence relation $\approx$ in $Q$ which is given by $x \approx y$ if and only if $V_x(\mathcal{R}) = V_y(\mathcal{R})$. This relation $\approx$ is the transitive extension of the relation $\sim$ in $Q$ which holds between elements $x, y \in Q$ if and only if $x \in Q_i$, $y \in Q_j$ and $(R_i)_x^F \cap (R_j)_y^F \neq \emptyset$. It is easily checked that the relations $\sim, \approx$ have the properties (1)-(4) which were needed in Section 2 to show that $Q/\!\approx\, = \Sigma$ is a branched $n$-manifold and that the natural projection $\pi^F$ of $Q$ to $Q/\approx$ is a $C^1$ immersion. So we see that $\Sigma(\mathcal{R})$ is a branched $n$-manifold and that the restriction of $\pi^F$ to an unstable manifold $W_x^u$ in $\Lambda$ is a $C^1$ immersion in $\Sigma(\mathcal{R})$. Moreover, the fact that each manifold $W_x^u$ is dense in $\Lambda$ implies $\pi^F(W_x^u) = \Sigma(\mathcal{R})$. Since $W_x^u$ is diffeomorphic to $\mathbf{R}^n$ we get a $C^1$ covering $\kappa : \mathbf{R}^n \to \Sigma(\mathcal{R})$ as defined in Section 2, and $\Sigma(\mathcal{R})$ is strongly connected.

# 7 The Mappings $\eta$, $\pi^s$, $\sigma^F$, $\sigma^s$.

The symbols $\mathcal{F}$, $N$, $\Sigma(\mathcal{R})$, $\pi^F$ will have the same meaning as in the preceding section. The distance functions corresponding to the foliations $\mathcal{F}$ and $\mathcal{W}^s$ will be denoted by $d^F$, $d^s$, respectively. The distance of points $x, y \in \Lambda$ inside an unstable manifold will be denoted by $d^u(x,y)$ (if $W^u_x \neq W^u_y$, then $d^u(x,y) = \infty$). We shall describe a construction which proves the following fact:
If $\mathcal{R}_0$ is a boundary transverse covering by $\mathcal{F}$-rectangles and $\varepsilon > 0$, then for any sufficiently fine slice refinement $\mathcal{R}$ of $\mathcal{R}_0$ there is a $C^1$ immersion $\eta : \Sigma(\mathcal{R}) \to N$ and a $C^0$ mapping $\pi^s : \Lambda \to \Sigma(\mathcal{R})$ such that the following conditions are satisfied, where $\sigma^F = \eta\pi^F : \Lambda \to N$, $\sigma^s = \eta\pi^s : \Lambda \to N$. (Since slice refinements of boundary transverse coverings are boundary transvers, $\Sigma(\mathcal{R})$ and $\pi^F$ are defined.)
(1) $d^F(x, \sigma^F(x)) < \varepsilon$ $(x \in \Lambda)$.
(2) The restriction of $\sigma^F$ to the unstable manifolds $W^u_x$ in $\Lambda$ are $C^1$ immersions in $N$.
(3) $(T_x W^u_x, d_x\sigma^F(T_x W^u_x)) < \varepsilon$ $(x \in \Lambda)$, where denotes the angle in $\mathbf{R}^n$.
(4) $d^s(x, \sigma^s(x)) < \varepsilon$ $(x \in \Lambda)$.
(5) For each $x \in \Lambda$ there is a unique point $y \in \Lambda$ such that $d^u(x,y) < \varepsilon$ and $\sigma^s(x) = \sigma^F(y)$.
(6) $\sigma^F(W^u_x)$ is transverse to $\mathcal{F}$ and to $\mathcal{W}^s$ $(x \in \Lambda)$.
The main step of the constructions in this section will be the definition of the mapping $\sigma^F : \Lambda \to N$ as expressed in the following lemma.
**Lemma.** If $\mathcal{R}_0$ is a boundary transverse covering by $\mathcal{F}$-rectangles and $\varepsilon > 0$, then for each slice decomposition $\mathcal{R}$ of $\mathcal{R}_0$ whose slices are sufficiently thin there is a continuous mapping $\sigma^F : \Lambda \to N$ which has the following properties:
($1_\sigma$) $\sigma^F(x) \in F_x$, and $d^F(x, \sigma^F(x)) < \varepsilon$ $(x \in \Lambda)$.
($2_\sigma$) The restrictions of $\sigma^F$ to the unstable manifolds are $C^1$ immersions.
($3_\sigma$) $(T_x(W^u_x), d_x\sigma^F(T_x(W^u_x))) < \varepsilon$ $(x \in \Lambda)$.
($4_\sigma$) If $x \in \Lambda$, $y \in V_x(\mathcal{R})$, then $\sigma^F(x) = \sigma^F(y)$.
Before proving the lemma we define $\eta$ and $\pi^s$. Since the counterimages of points in $\Sigma(\mathcal{R})$ under the projection $\pi^F : \Lambda \to \Sigma(\mathcal{R})$ (defined in Section 6) are the sets $V_x(\mathcal{R})$, ($4_\sigma$) implies that

$$\eta = \sigma^F(\pi^F)^{-1} : \Sigma(\mathcal{R}) \to N$$

is well defined, and ($2_\sigma$) together with the fact that $\pi^F$ is a $C^1$ immersion shows that $\eta$ is a $C^1$ immersion. If $\varepsilon$ in the lemma is sufficiently small then by ($1_\sigma$), ($3_\sigma$) the images $\sigma^F(W^u_x)$ of the unstable manifolds will be transverse to $\mathcal{F}$ and to $\mathcal{W}^s$. Moreover, if $\varepsilon$ is chosen small, there is a positive $\varepsilon'$ such that for each $x \in \Lambda$ there is a unique $y \in \Lambda$ for which $d^u(x,y) < \varepsilon'$, $d^s(x, \sigma^F(y)) < \varepsilon'$, and by $\pi^s(x) = \pi^F(y)$ we get a continuous mapping $\pi^s : \Lambda \to \Sigma(\mathcal{R})$ whose restrictions to unstable manifolds are $C^0$ immersions. Since for $\varepsilon$ small $\varepsilon'$ can be chosen small too, it is easy to see that $\eta$, $\pi^s$ have the properties (1)-(6).
**Proof of the Lemma.** Let $\mathcal{R} = \{R_1, ..., R_r\}$ be a slice refinement of the given boundary transverse covering $\mathcal{R}_0$ by $\mathcal{F}$-rectangles, and let $Q = Q_1 \cup ... \cup Q_r$ be a compact $n$-dimensional $C^1$ manifold where $Q_i$ is a component of $R_i$ and the $n$-disks $Q_i$ are disjoint $(i = 1, ..., r)$. For $0 \le k < n$ we consider the set $K^k$ of all points $x \in Q$ for which $V_x(\mathcal{R}) \cap \partial Q \neq \emptyset$ and the intersection of all linear subspace $d_y h^y_x(T_y(\partial Q))$ of $T_x(Q) = T_x(W^u_x)$ $(y \in V_x(\mathcal{R} \cap \partial Q))$ has the dimension $k$. (Here $h^y_x$ is the holonomy mapping along $\mathcal{F}$ as used in the preceding

section.) The set of all those $x \in Q$ for which $V_x(\mathcal{R}) \cap \partial Q = \emptyset$ will bedenoted by $K^n$. Since $\mathcal{R}_0$ is boundary transverse so is $\mathcal{R}$ (see Section 6), and it is easy to see that the sets $K^k$ have the following properties: Each $K^k$ $(0 \le k \le n)$ is a $k$-dimensional $C^1$ submanifold of $Q$ with empty boundary. Since $Q$ is compact, the family $\mathcal{K}^k$ of all components of $K^k$ is finite.
The family $\mathcal{K} = \mathcal{K}^0 \cup ... \cup \mathcal{K}^n$ is a decomposition (stratification) of $Q$ which has the following properties: The closure $\bar{K}$ of a manifold $K \in \mathcal{K}^k$ is a $k$-dimensional compact $C^0$ manifold whose boundary $\partial \bar{K}$ is the union of manifolds $K'$ belonging to $\mathcal{K}^0 \cup ... \cup \mathcal{K}^{k-1}$ ($\bar{K}$ is a $C^1$ manifold which may have corners). For $K_1, K_2 \in \mathcal{K}$ the intersection $K_1 \cap K_2$ is the union of manifolds belonging to $\mathcal{K}$.
For each $K \in \mathcal{K}$ we consider the set

$$V(K) = \bigcup_{x \in K} V_x(\mathcal{R})$$

Then $K$ is a component of $V(K)$. The family of all sets $V(K)$ where $K \in \mathcal{K}^k$ will be denoted by $\mathcal{V}^k = \mathcal{V}^k(\mathcal{R})$, and $\mathcal{V} = \mathcal{V}(\mathcal{R}) = \mathcal{V}^0 \cup ... \cup \mathcal{V}^n$. If $x \in V \in \mathcal{V}$ and $P$ is a component of $V$, then there is a homeomorphism $h : P \times V_x(\mathcal{R}) \to V$ which maps each set $\{y\} \times V_x(\mathcal{R})$ to $V_y(\mathcal{R})$ $(y \in P)$ and each set $P \times \{z\}$ to the component of $z$ in $V$ $(z \in V_x(\mathcal{R}))$. The homeomorphism $h$ can be extended to a homeomorphism $h : \bar{P} \times V_x(\mathcal{R}) \to \bar{V}$ where $\bar{P} = Cl\,P$, $\bar{V} = Cl\,V$. If $z \in \partial \bar{P}$, then $h(\{z\} \times V_x(\mathcal{R})) \subset V_z(\mathcal{R})$, and this inclusion can be proper.
If $\mathcal{R}^*$ is a slice refinement of $\mathcal{R}$ and $V \in \mathcal{V}(\mathcal{R})$, $V^* \in \mathcal{V}(\mathcal{R}^*)$, then $V \cap V^*$ is either empty or a slice in $V$ (i.e. the union of components of $V$) and, if the slices in $\mathcal{R}^*$ are sufficiently thin, then the sets $V^* \in \mathcal{V}(\mathcal{R}^*)$ become thin in the sense that the sets $V_x(\mathcal{R}^*)$ become small (see Section 4). On the other hand the components of the sets $V^*$ do not become smaller: If $x \in V^* \in \mathcal{V}(\mathcal{R}^*)$, $x \in V \in \mathcal{V}(\mathcal{R})$, then the component of $V$ containing $x$ lies in the component of $V^*$ containing $x$. Therefore, if the slices in $\mathcal{R}^*$ are thin, the components of a set $V^* \in \mathcal{V}^k(\mathcal{R}^*)$ $(k \ge 1)$ are large when compared with the sets $V_x(\mathcal{R}^*)$. This allows to assume that the rectangles in our slice refinement $\mathcal{R}$ of $\mathcal{R}_0$ are so thin that the sets $V \in \mathcal{V}^k(\mathcal{R})$ $(k \ge 1)$ are thin in the direction of the leaves of $\mathcal{F}$ but comparably large in the direction of the unstable manifolds $W_x^u$.
Let in each set $V \in \mathcal{V}$ a component $P_V$ be fixed. If the slices in $\mathcal{R}$ are sufficiently thin it is not hard to choose successively - first for all $V \in \mathcal{V}^0$, then for all $V \in \mathcal{V}^1$, then for all $V \in \mathcal{V}^2$ etc. - $C^1$ embeddings $\sigma_V : \bar{P}_V \to N$ which have the following properties:
(1) $\sigma_V(x) \in F_x$ $(x \in \bar{P}_V)$.
(2) If $V, V' \in \mathcal{V}$, $V \cap \partial \bar{V}' \neq \emptyset$, $x \in P_V$, $x' \in \partial \bar{P}'_V \cap V_x(\mathcal{R})$, then $\sigma_V(x) = \sigma_{V'}(x')$.
Moreover, if $\varepsilon_0 > 0$ is given, then the slice decomposition of $\mathcal{R}$ can be chosen so fine that instead of (1) we can fulfill the following condition
(1') $d^F(x, \sigma_V(x)) < \varepsilon$, $\ (T_x(W_x^u), d_x\sigma_V(T_x(W_x^u)) \le \varepsilon_0$ $(x \in \bar{P}_V)$.
Now for each $V \in \mathcal{V}^n$ we consider the projection $\pi_V : \bar{V} \to \bar{P}_V$ which is determined by $\pi_V(x) \in V_x(\mathcal{R}) \cap \bar{P}_V$. Then $\sigma_V \pi_V : V \to N$ is a mapping whose restrictions to the components of $V$ are $C^1$ embeddings, and by (2) all these mappings $\sigma_V \pi_V$ $(V \in \mathcal{V}^n)$ together define a $C^0$ mapping $\sigma^F : \Lambda \to N$ satisfying $d^F(x, \sigma^F(x)) < \varepsilon_0 + \varepsilon_1$ where $\varepsilon_1$ is the maximal diameter of a set $V_x(\mathcal{R})$ with respect to the metric $d^F$. (If the rectangles $\mathcal{R}$ in are thin, then $\varepsilon_1$ is small.) By an ordinary smoothening process at the boundaries $\partial \bar{P}_V$ of the sets $\bar{P}_V$ $(V \in \mathcal{V}^n)$ we can modify $\sigma^F$ so that it becomes a $C^1$ immersion on each unstable

manifold in $\Lambda$. If $\varepsilon_0$, $\varepsilon_1$ were chosen sufficiently small and the smoothening is taken $C^1$ close to $\sigma_V$ on each $\bar{P}_V$ ($V \in \mathcal{V}^n$), then the resulting mapping $\sigma^F : \Lambda \to N$ has the properties which are required in the lemma.

# 8 The mapping $\varphi^s : \Sigma(\mathcal{R}) \to \Sigma(\mathcal{R})$.

In this section we show that there is a boundary transverse covering $\mathcal{R}$ by $\mathcal{F}$-rectanlges for which in addition to $\Sigma(\mathcal{R})$, $\pi^F$, $\pi^s$, $\eta$, $\sigma^F$, $\sigma^s$ a locally flattening $C^0$ immersion $\varphi^s : \Sigma(\mathcal{R}) \to \Sigma(\mathcal{R})$ can be defined such that $\varphi^s(\Sigma(\mathcal{R})) = \Sigma(\mathcal{R})$ and

$$\begin{array}{ccc} \bullet\Lambda & \xrightarrow{f} & \Lambda \\ \pi^s \downarrow & & \downarrow \pi^s \\ \Sigma(\mathcal{R}) & \xrightarrow{\varphi^s} & \Sigma(\mathcal{R}) \end{array}$$

is commutative. To describe the construction of $\varphi^s$ some definitions are necessary.

By a triplet $T$ of rectangles we mean a triplet $(\tilde{R}, R, \tilde{R}')$ where $R$ is an $\mathcal{F}$-rectangle and $\tilde{R}$, $\tilde{R}'$ are $s$-rectangles such that $\tilde{R} \subset Int\, R$, $R \subset Int\, \tilde{R}'$ and each component of $\tilde{R}'$ contains a component of $R$ and a component of $\tilde{R}$. A family of triplets $\{(\tilde{R}_i, R_i, \tilde{R}_i')\}$ $(i = 1, ..., j)$ will be called a slice decomposition of the triplet $(\tilde{R}, R, \tilde{R}')$ if the families $\{\tilde{R}_i\}$, $\{R_i\}$, $\{\tilde{R}_i'\}$ are slice decompositions of $\tilde{R}$, $R$, $\tilde{R}'$, respectively. Then a triplet $\mathcal{T}$ of coverings is a triplet $(\tilde{\mathcal{R}}, \mathcal{R}, \tilde{\mathcal{R}}')$ where $\mathcal{R} = \{R_1, ..., R_k\}$ is a covering by $\mathcal{F}$-rectangles and $\tilde{\mathcal{R}} = \{\tilde{R}_1, ..., \tilde{R}_k\}$, $\tilde{\mathcal{R}}' = \{\tilde{R}_1', ..., \tilde{R}_k'\}$ are coverings by $s$-rectangles and each triplet $(\tilde{R}_i, R_i, \tilde{R}_i')$ $(1 \le i \le k)$ is a triplet of rectangles. Moreover it is assumed that $\mathcal{R}$ is boundary transverse and that for $x \in \tilde{R}_i' \in \tilde{\mathcal{R}}'$, $f(x) \in \tilde{R}_j'$ we have

$$f((\tilde{R}_i')^s_x) \subset (\tilde{R}_j')^s_{f(x)}.$$

(A covering $\tilde{\mathcal{R}}'$ by $s$-rectangles with this property will be called *f-adapted*). Then it is easy to see that for $x \in \Lambda$, $f(x) \in \tilde{R}_j$ the following inclusion holds

$$f(V_x(\tilde{\mathcal{R}}')) \subset (\tilde{R}_j)^s_{f(x)} \subset V_{f(x)}(\tilde{\mathcal{R}}).$$

By a slice refinement of the triplet $\mathcal{T}$ we means a triplet $\mathcal{T}^* = (\tilde{\mathcal{R}}^*, \mathcal{R}^*, \tilde{\mathcal{R}}'^*)$ of coverings which is obtained by replacing in $\mathcal{T}$ triplets of rectangles by slice decompositions of these triplets.

**Lemma 1.** There is a triplet of coverings.

**Lemma 2.** Each triplet $\mathcal{T}$ of coverings has arbitrarily fine slice refinements, i.e. for each $\varepsilon > 0$ there is a slice refinement $\mathcal{T}^* = (\tilde{\mathcal{R}}^*, \mathcal{R}^*, \tilde{\mathcal{R}}'^*)$ of $\mathcal{T}$ such that for $x \in R \in \tilde{\mathcal{R}}'^*$ the diameter of $R^s_x$ with respect to the metric $d^s$ is less than $\varepsilon$.

We show how these lemmas allow to define the mapping $\varphi^s$ mentioned at the beginning of this section. Let $\mathcal{T}_1 = (\tilde{\mathcal{R}}_1, \mathcal{R}_1, \tilde{\mathcal{R}}_1')$ be a triplet of coverings. It is sufficient to show that for any sufficiently fine slice refinement $\mathcal{T} = (\tilde{\mathcal{R}}, \mathcal{R}, \tilde{\mathcal{R}}')$ , for which the mappings $\pi^F$, $\pi^s$ : $\Lambda \to \Sigma(\mathcal{R})$ $\sigma^F$, $\sigma^s : \Lambda \to \mathbf{R}^m$, are defined as in the preceding section, we can find a locally flattening $C^0$ immersion $\varphi^s : \Sigma(\mathcal{R}) \to \Sigma(\mathcal{R})$ such that $\varphi^s\pi^s = \pi^s f$, $\varphi^s(\Sigma(\mathcal{R})) = \Sigma(\mathcal{R})$.

We assume that for each sufficiently fine refinement $\mathcal{T} = (\tilde{\mathcal{R}}, \mathcal{R}, \tilde{\mathcal{R}}')$ of $\mathcal{T}_1$ mappings $\sigma^F$, $\sigma^s$ are chosen and fixed and that this choice was made so that

$$\max(\sup_{x\in\Lambda} d^F(x,\sigma^F(x)), \sup_{x\in\Lambda} (T_x(W^u_x), d_x\sigma^F(T_x(W^u_x))))$$

tends to 0 if $\mathcal{T}$ becomes finer and finer. Then $\sup_{x\in\Lambda} d^s(x,\sigma^s(x))$ tends to 0, too. If $\mathcal{T}$ is fine, then the sets $\tilde{R}'_i \setminus R_i, R_i \setminus \tilde{R}_i$ ($\tilde{\mathcal{R}} = \{\tilde{R}_i\}$, $\mathcal{R} = \{R_i\}$, $\tilde{\mathcal{R}}' = \{\tilde{R}'_i\}$) are large when compared with the sets $V_x(\tilde{\mathcal{R}})$, $V_x(\mathcal{R})$, $V_x(\tilde{\mathcal{R}}')$ ($x \in \Lambda$). (This means that for any $C > 0$ all slice refinements $\mathcal{T} = (\tilde{\mathcal{R}}, \mathcal{R}, \tilde{\mathcal{R}}')$ which are sufficiently fine have the following property: If $\tilde{R}'_i$, $R_i$ are rectangles in $\tilde{\mathcal{R}}'$ or $\mathcal{R}$, respectivley, which belong to a triplet and if $P$ is a component of $\tilde{R}'_i \setminus R_i$, then for the $d^u$-distance of the two boundary components $\partial_1 P$, $\partial_2 P$ we have

$$d^u(\partial_1 P, \partial_2 P)/\sup_{x\in\Lambda} diam\, V_x(\tilde{\mathcal{R}}') > C,$$

and the same holds for the pairs $R_i$, $\tilde{R}_i$ of rectangles in $\mathcal{R}$, $\tilde{\mathcal{R}}$, respectively.) Then, looking at the definitions of $\sigma^s$ and $\pi^s$ we see that for $\mathcal{T}$ sufficiently fine $\pi^s(x) = \pi^s(y)$ implies $V_x(\tilde{\mathcal{R}}') = V_y(\tilde{\mathcal{R}}')$, and $V_x(\tilde{\mathcal{R}}) = V_y(\tilde{\mathcal{R}})$ implies $\pi^s(x) = \pi^s(y)$. Since $f(V_x(\tilde{\mathcal{R}}')) \subset V_{f(x)}(\tilde{\mathcal{R}})$ holds for all $x \in \Lambda$, the mapping

$$\varphi^s = \pi^s f(\pi^s)^{-1} : \Sigma(\mathcal{R}) \to \Sigma(\mathcal{R})$$

is well defined for these triplets $\mathcal{T}$. It remains to prove that $\varphi^s$ is a locally flattening $C^0$ immersion satisfying $\varphi^s(\Sigma(\mathcal{R})) = \Sigma(\mathcal{R})$; but this is a simple consequence of the definitions above.

Now we prove the two lemmas. Both of them will be immediate consequences of the following sublemmas.

**Sublemma 1.** There is an $f$-adapted covering by $s$-rectangles.

**Sublemma 2.** Each $f$-adapted covering by $s$-rectangles has arbitrarily fine slice refinements which are $f$-adapted, too.

To prove Lemma 1 we start with an $f$-adapted covering $\tilde{\mathcal{R}}'_0$ by $s$-rectangles. If $\tilde{\mathcal{R}}'$ is a sufficiently fine $f$-adapted slice refinement of $\tilde{\mathcal{R}}'_0$, it is easy to find coverings $\tilde{\mathcal{R}}$, $\mathcal{R}$ so that $\mathcal{T} = (\tilde{\mathcal{R}}, \mathcal{R}, \tilde{\mathcal{R}}')$ is a triplet of coverings. Lemma 2 is immediately implied by the second sublemma.

**Proof of Sublemma 1.** For $x \in \Lambda$, $\varepsilon > 0$ let $U^s_x(\varepsilon)$ be the set of all $y \in W^s_x \cap \Lambda$ for which there are points $x_0 = x, x_1, ..., x_k = y$ in $W^s_x \cap \Lambda$ such that $d^s(x_{i-1}, x_i) < \varepsilon$ ($i = 1, ..., k$), where $d^s$ denotes the distance in $W^s_x$. Since $W^s_x \cap \Lambda$ locally looks like a Cantor set, the intersection of all sets $U^s_x(\varepsilon)$ ($\varepsilon > 0$) is $\{x\}$. If we apply that the metric in $\mathbf{R}^m$ is adapted, we find for each sufficiently small positive $\varepsilon$ a positive $\varepsilon' < \varepsilon$ such that

$$f(U^s_x(\varepsilon)) \subset U^s_{f(x)}(\varepsilon')$$

holds for all $x \in \Lambda$. Then a covering $\tilde{\mathcal{R}}$ by $s$-rectangles is $f$-adapted provided for $y \in R \in \tilde{\mathcal{R}}$ we have

$$U^s_y(\varepsilon') \subset R^s_y \subset U^s_y(\varepsilon).$$

To find a covering $\tilde{\mathcal{R}}$ with this property we remark that for $\varepsilon'' = (\varepsilon + \varepsilon')/2$ and each $x \in \Lambda$ there is an $n$-disk $Q$ in $W_x^u$ which contains $x$ in its interior and for which

$$\bigcup_{y \in Q} U_y^s(\varepsilon'')$$

is an $s$-rectangle.

**Proof of Sublemma 2.** Let $\tilde{\mathcal{R}} = \{R_1, ..., R_k\}$ be an $f$-adapted covering by $s$-rectangles. We choose another covering $\tilde{\mathcal{R}}' = \{R'_1, ..., R'_k\}$ by $s$-rectangles such that $R'_i \subset Int\, R_i$ and each component of $R_i$ contains a component of $R'_i$. Then we may choose an integer $k_0$ which is so large that for $k \geq k_0$ the following holds true: If $x \in R'_i$, $f^k(x) \in R_j$ and $P$ is the component of $x$ in $R_i$, then $f^k(P)$ contains the component of $f^k(x)$ in $R_j$. We fix $k \geq k_0$ and denote for $x \in R'_i$, $f^k(x) \in R_j$ the union of all components of $R_j$ which intersect $f^k((R_i)_x^s)$ by $S(i,j,x)$. Then $S(i,j,x)$ is a slice of $R_j$ and the family $\tilde{\mathcal{R}}^*$ of all these rectangles $S(i,j,x)$ is a slice refinement of $\tilde{\mathcal{R}}$. To show that $\tilde{\mathcal{R}}^*$ is $f$-adapted we consider the covering $f^{-k}(\tilde{\mathcal{R}}^*)$ by all rectangles $f^{-k}(S(i,j,x))$. Since $f^{-k}(S(i,j,x))$ lies in $R_i$ and each component of $R_i$ contains a component of $f^{-k}(S(i,j,x))$, our assumption that $\tilde{\mathcal{R}}$ is $f$-adapted implies that $f^{-k}(\tilde{\mathcal{R}}^*)$ is $f$-adapted. Therefore, $\tilde{\mathcal{R}}^*$ is $f$-adapted, and if we choose $k$ large, then $\tilde{\mathcal{R}}^*$ can be made arbitrarily fine.

# 9 The Mappings $\varphi_k$

In this section we start with a boundary transverse covering $\mathcal{R}$ by $\mathcal{F}$-rectangles for which $\Sigma = \Sigma(\mathcal{R})$ and the mappings $\sigma^F$, $\pi^F$, $\sigma^s$, $\pi^s$, $\eta$, $\varphi^s$ are defined as in the preceding sections and construct a sequence $\Sigma_0 = \Sigma(\mathcal{R}), \Sigma_1, \Sigma_2, ...$ of branched $n$-manifolds with Riemannian metrics each of which coincides with $\Sigma(\mathcal{R})$ with the exception of the metric. For $k$ sufficiently large $\Sigma_{k+1}$ will be a dilatation of $\Sigma_k$ (see Section 3), and it will be possible to define a mapping $\varphi_k : \Sigma_k \to \Sigma_k$ such that the following conditions are satisfied:

(1) $\varphi_k$ is a locally flattening $C^1$ immersion, and $\varphi_k(\Sigma_k) = \Sigma_k$.

(2) The mappings $\varphi_k$ are expanding with a factor $\gamma > 1$ which is independent of $k$, i.e. for each $\tau \in \Sigma_k$, $v \in T_\tau(\Sigma_k)$ we have $|d_\tau \varphi_k(v)| \geq \gamma |v|$.

(3) $\lim_{k\to\infty} \sup_{\tau \in \Sigma_k} d_k^*(\varphi^s(\tau), \varphi_k(\tau)) = 0$,

where $\varphi^s$ is the old mapping $\varphi^s : \Sigma(\mathcal{R}) \to \Sigma(\mathcal{R})$ now regarded as mapping of $\Sigma_k$, and the distance $d_k^*$ is defined as follows: Choose an unstable manifold $W_x^u$ in $\Lambda$ and a homeomorphism $h : \mathrm{R}^n \to W_x^u$. Then $\kappa = \pi^s h : \mathrm{R}^n \to \Sigma_k$ is a covering of $\Sigma_k$, and $d_k^*$ is defined for this covering (see Section 2). Therefore, $d_k^*(\tau, \tau') < \varepsilon$ ($\varepsilon > 0$ small) holds if for some $n$-disk $D$ in $W_x^u$ which is mapped by $\pi^s$ homeomorphically onto a disk $\Delta$ in $\Sigma_k$ containing $\tau$, $\tau'$ the distance from $\tau$ to $\tau'$ inside $\Delta$ is less than $\varepsilon$.

The set $\eta(\Sigma)$ will be denoted by $S$. Then for $k = 0, 1, 2, ...$ we consider the $C^1$ immersions $\eta_k = f^k \eta : \Sigma \to \mathrm{R}^m$ and the corresponding mappings

$$\sigma_k^s = \eta_k \pi^s f^{-k} = f^k \sigma^s f^{-k} : \Lambda \to \mathrm{R}^m.$$

The sets

$$S_k = \eta_k(\Sigma) = \sigma_k^s(\Lambda) = f^k(S)$$

in general have no uniquely determined tangent spaces ($\eta$ is not necessarily an embedding), but for $\tau \in \Sigma$

$$T_\tau(S_k) = d_\tau \eta_k(T_\tau(\Sigma)) = d_{\eta(\tau)} f^k(T_\tau(S))$$

may be regarded as one of the tangent spaces of $S_k$ at $\eta_k(\tau)$, and, though $\sigma_k^s$ is not necessarily $C^1$ on the unstable manifolds of $\Lambda$, it is reasonable to write for $x \in \Lambda$

$$T_{\pi^s f^{-k}(x)} S_k = d_x \sigma_k^s(T_x(W_x^u)).$$

The branched manifold $\Sigma$ with the Riemannian metric for which $\eta_k$ is an isometry will be denoted by $\Sigma_k$. (The situation is simplified, and this will help our geometric intuition, if we assume that the immersions $\eta_k$ are $C^1$ embeddings. Then $\Sigma_k$ can be identified with $S_k$ and $\pi^s$ with $\sigma_k^s f^k$. This assumption is justified since with somewhat more care (and for $m$ large) the mapping $\sigma^F$ in Section 7 could have been chosen so that $\eta$ is an embedding. But for the formal side of the proof this is superfluous.) For $k \to \infty$ the sets $S_k$ converge to $\Lambda$ in the sense that

$$\lim_{k\to\infty} \sup_{x\in\Lambda} d^s(x, \sigma_k^s(x)) = 0$$

$$\lim_{k\to\infty} \sup_{x\in\Lambda} \quad (T_x(W_x^u), d_x\sigma_k^s(T_x(W_x^u))) = 0$$

Since $\mathcal{W}^s$ and $\mathcal{F}$ are transverse to $\Lambda$ this implies that there are an integer $k_0 \geq 0$ and a real $\alpha > 0$ such that for $k \geq k_0$, $x \in \Lambda$, $\tau = \pi^s f^{-k}(x)$, $y = \sigma_k^s(x)$, $v \in T_\tau S_k$, $w \in T_y W_y^s \cup T_y F_y$, $v, w \neq 0$ we have $(v, w) > \alpha$ .

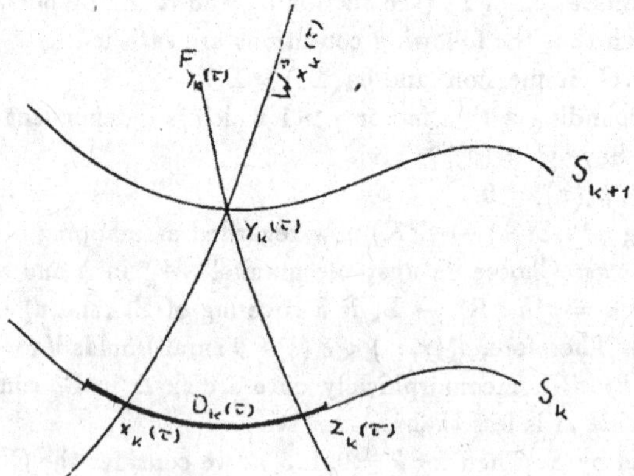


Fig. 4

If $k_0$ is chosen sufficiently large, then, since $f$ is expanding on the unstable manifolds of $\Lambda$ and $S_k$ is close to $\Lambda$, $f$ is expanding on $S_k$ in the sense that for some $\beta > 1$ and all

$\tau \in \Sigma$, $x = \eta_k(\tau)$, $v \in T_\tau S_k$ we have $|d_x f(v)| \geq \beta|v|$. Therefore $\Sigma_{k+1}$ is a dilatation of $\Sigma_k$ $(k = k_0, k_0 + 1, ...)$.
Since $\varphi^s$ is locally flattening we can choose for each $\tau \in \Sigma$ a neighbourhood $\Gamma_\tau$ which is the union of finitely many smooth disks each of which contains $\tau$ and is mapped by $\varphi^s$ homeomorphically onto the same smooth disk $\Delta(\tau)$. Since $\Sigma$ is compact we may assume that for some $\rho > 0$ and each $\tau \in \Sigma$ the distance from $\varphi^s(\tau)$ to $\partial\Delta(\tau) = Cl\,\Delta(\tau) \setminus \Delta(\tau)$ is at least $\rho$. Moreover, we may assume that $\rho$ is so small that each disk $\Delta(\tau)$ is mapped by $\eta$ diffeomorphically onto a smooth disk $D(\tau)$ in $S$. If $\Delta(\tau)$ is regarded a as subdisk of $\Sigma_k$, then it will be denoted by $\Delta_k(\tau)$, and $D_k(\tau) = \eta_k(\Delta_k(\tau)) = f^k(D(\tau))$. Now for $k$ sufficiently large we are going to define a $C^1$ mapping $\varphi_k : \Sigma_k \to \Sigma_k$ (see Fig. 4). First for $k \geq k_0$ and $\tau \in \Sigma_k$ we consider the points

$$x_k(\tau) = \eta_k\varphi^s(\tau) \in S_k,\; y_k(\tau) = f\eta_k(\tau) \in S_{k+1}.$$

and the disks $\Delta_k(\tau)$ in $\Sigma_k$ and $D_k(\tau)$ in $S_k$. If $k$ is large, then $y_k(\tau)$ is close to $x_k(\tau)$ uniformly in $\tau$. So we get a unique point $z_k(\tau)$ which is close to $y_k(\tau)$ in the leaf $F_{y_k(\tau)}$ of $\mathcal{F}$ and at which $F_{y_k(\tau)}$ intersects $D_k(\tau)$ transversely. Now we use that the restriction of $\eta_k$ to $\Delta_k(\tau)$ is a diffeomorphism onto $D_k(\tau)$ and define the mapping $\varphi_k : \Sigma_k \to \Sigma_k$ by

$$\varphi_k(\tau) = (\eta_k|_{\Delta_k(\tau)})^{-1}(z_k(\tau)).$$

Locally for $\tau'$ near $\tau$ in the sequence

$$\tau' \mapsto y_k(\tau') = f\eta_k(\tau') \mapsto z_k(\tau') \mapsto \varphi_k(\tau')$$

the first step is a $C^1$ immersion and the second a $C^1$ diffeomorphism. To show that the third one is also a $C^1$ diffeomorphism we remark that there is a subdisk $\Delta'(\tau)$ of $\Delta(\tau)$ and a neighbourhood $\Gamma$ of $\tau$ such that for each $\tau' \in \Gamma$ we have $\varphi^s(\tau') \in \Delta'(\tau) \subset \Delta(\tau')$. Then for $z \in D'_k(\tau) = \eta_k(\Delta'(\tau))$ we have

$$(\eta_k|_{\Delta_k(\tau)})^{-1}(z) = (\eta_k|_{\Delta_k(\tau')})^{-1}(z).$$

If $k$ is large and $\tau'$ is close to $\tau$, then $z_k(\tau')$ will lie in $D'_k(\tau)$, and we get

$$\varphi_k(\tau') = (\eta_k|_{\Delta_k(\tau)})^{-1}(z_k(\tau')).$$

This shows that for $\tau'$ in a neighbourhood of $\tau$ the mapping $z_k(\tau') \mapsto \varphi_k(\tau')$ is a diffeomorphism.
What we have obtained is a sequence $\varphi_k : \Sigma_k \to \Sigma_k$ of $C^1$ immersions ($k$ large). If we look at the disks $\Delta'(\tau)$ in the construction above we see that each $\varphi_k$ is locally flattening. The distance in $D_k(\tau)$ between the points $x_k(\tau)$ and $z_k(\tau)$ tends to 0 for $k \to \infty$ uniformly in $\tau$. This and the fact that $\eta_k$ when restricted to $\Delta_k(\tau)$ is an isometry implies

$$\lim_{k\to\infty} \sup_{\tau\in\Sigma_k} d_k^*(\varphi^s(\tau), \varphi_k(\tau)) = 0.$$

It remains to prove that for $k$ large the mappings $\varphi_k$ are expanding with a common expansion factor $\gamma > 1$. Since our metric in $\mathbf{R}^m$ is adapted we can find a real $\gamma' > 1$ such that for $x \in \Lambda$, $v \in T_x(W^u_x)$ we have $|d_x f(v)| \geq \gamma'|v|$. Now we remind that the sets $S_k$ together with their tangent spaces for $k \to \infty$ converge to $\Lambda$ in the sense described above and that the Riemannian metric in $\Sigma_k$ is defined via $\eta_k : \Sigma_k \to S_k$ by the metric in $S_k$. This implies that for $1 < \gamma < \gamma'$ and $k$ sufficiently large all mappings $\varphi_k$ are expanding with the factor $\gamma$.

# 10 The Final Step

Let $\Lambda$ be an $n$-dimensional expanding attractor of a $C^1$ diffeomorphism $f: \mathbf{R}^m \to \mathbf{R}^m$. In the sections 6, 7, 8 we have constructed a branched $n$-manifold $\Sigma$, a projection $\pi^s: \Lambda \to \Sigma$ and a locally flattening $C^0$ immersion $\varphi^s: \Sigma \to \Sigma$ such that
(1) the restriction of $\pi^s$ to an unstable manifold $W_x^u$ in $\Lambda$ is a $C^0$ immersion $\pi^s: W_x^u \to \Sigma$, and $\pi^s(W_x^u) = \Sigma$;
(2) $\varphi^s(\Sigma) = \Sigma$;
(3) the following diagram is commutative

$$\begin{array}{ccc} \Lambda & \xrightarrow{f} & \Lambda \\ \pi^s \downarrow & & \downarrow \pi^s \\ \Sigma & \xrightarrow{\varphi^s} & \Sigma \end{array}$$

Since each unstable manifold $W_x^u$ is homeomorphic to $\mathbf{R}^n$ we may choose a homeomorphism $h: \mathbf{R}^n \to W_x^u$ of $\mathbf{R}^n$ to some $W_x^u$. So we get a covering $\kappa = \pi^s h: \mathbf{R}^n \to \Sigma$.
Then in Section 9 we have defined a sequence $\Sigma_0, \Sigma_1, \ldots$ of branched $n$-manifolds each of which up to the metric coincides with $\Sigma$. Moreover, there is a $k_0 \geq 0$ such that for each $k \geq k_0$ a locally flattening $C^1$ immersion $\varphi_k : \Sigma_k \to \Sigma_k$ is defined. The projection $\pi^s$, the immersion $\varphi^s$ and the covering $\kappa$ can be regarded as mappings $\pi^s: \Lambda \to \Sigma_k$, $\varphi^s: \Sigma_k \to \Sigma_k$, $\kappa: \mathbf{R}^n \to \Sigma_k$, and $\kappa$ defines a distance $d_k^*$ in $\Sigma_k$ (see Section 2). If $k_0$ is sufficiently large, then for $k \geq k_0$ the following conditions are satisfied:
(4) $\Sigma_{k+1}$ is a dilatation of $\Sigma_k$.
(5) $\varphi_k : \Sigma_k \to \Sigma_k$ is expanding with a factor $\gamma > 1$ which does not depend on $k$.
(6) $\varphi^s: \Sigma_k \to \Sigma_k$ has a positive extension number with respect to the metric $d_k^*$.
(7) $\lim_{k\to\infty} \sup_{\tau\in\Sigma_k} d_k^*(\varphi^s(\tau), \varphi_k(\tau)) = 0$.
(Condition (6) and $\pi^s(W_x^u) = \Sigma$ were not mentioned explicitly in the preceding sections, but they can be directly deduced from the corresponding constructions.)
Now the lemma from Section 3 can be applied (for $\Sigma = \Sigma_{k_0}$, $\Sigma' = \Sigma_k$, $k$ large, $\varphi = \varphi^s$, $\varphi^* = \varphi_k$). So we get a $C^1$ immersion $\chi: \Sigma_k \to \Sigma_k$ such that

$$\begin{array}{ccc} \Sigma_k & \xrightarrow{\varphi^s} & \Sigma_k \\ \chi \downarrow & & \downarrow \chi \\ \Sigma_k & \xrightarrow{\varphi_k} & \Sigma_k \end{array}$$

is commutative, and with $\pi = \chi\pi^s : \Lambda \to \Sigma_k$, $\Sigma = \Sigma_k$, $\varphi = \varphi_k$ the commutative diagram

$$\begin{array}{ccc} \Lambda & \xrightarrow{f} & \Lambda \\ \pi \downarrow & & \downarrow \pi \\ \Sigma & \xrightarrow{\varphi} & \Sigma \end{array}$$

is a $W$-respresentation for $\Lambda$.

# References

[1] Bothe, H.G.: The ambient structure of expanding attractors I. Math. Nachr. 107 (1982), 327-348

[2] —: Geometrische Theorie differenzierbarer dynamischer Systeme, Mitteilungen d. Math. Ges. d. DDR 1988), 3-22

[3] Robinson, C., and R. Williams: Classification of expanding attractors: An example. Topology 15 (1976), 321-323

[4] Shub, M.: Stabilité globale des systèmes dynamiques. Astérisque 56 (1978)

[5] Smale, S.: Differentiable dynamical systems, Bull. Amer. Math. Soc. 73 (1967), 747-817

[6] Steenrod, N.: The topology of fibre bundles. Princeton 1951

[7] Williams, R.F.: One dimensional non wandering sets. Topology 6 (1967), 473-487

[8] —: Classification of one dimensional attractors. Proc. Symp. in Pure Math. Amer. Math. Soc. 14 (1970), 341-361

[9] —: Expanding attractors. Publ.Math. IHES 43 (1974), 169-203

# ON ABSOLUTELY FOCUSING MIRRORS

L. A. Bunimovich[1]

Fakultät für Physik, BiBoS
Universität Bielefeld
D-4800 Bielefeld 1

**Abstract**
We consider focusing curves $\Gamma^f$ (of class $C^\alpha$, $\alpha \geq 3$) such that each incoming infinitesimal beam of parallel rays focuses after hitting $\Gamma^f$ for the last time in the series of consecutive reflections from it. We call such curves absolutely focusing. We prove some characteristic properties of absolutely focusing curves and show that these remain absolutely focusing under small $C^3$-($C^4$)-perturbations if this component has constant (nonconstant) curvature. We also present examples of absolutely focusing curves and consider the applications of these curves to some classes of continuous fractions.

## 0. Introduction

A Hamiltonian system is called a plane billiard if its potential equals zero inside some closed domain Q and infinity on its boundary $\partial Q$. (From the general point of view billiards could be considered also as geodesic flows on manifolds with boundary with elastic reflections from its boundary). In order not to deal with some pathological situations (e.g. infinite number of reflections from $\partial Q$ in a finite time) we propose as usual, that $\partial Q$ consists of a finite number of curves (mirrors) $\Gamma_1, \ldots \Gamma_p$, $p \geq 1$ of class $C^3$. These curves are called regular components of the boundary, and their internal points are called regular points. Points of intersection of regular components are called singular. Billiard trajectories move along straight lines inside Q. When a trajectory hits the boundary $\partial Q$ it reflects from it elastically, i.e. "angle of incidence equals angle of reflection". The only difficulties here appear because of trajectories that hit singular points. It is easy to show [CSF] that the (invariant) phase volume of the set of such trajectories equals zero.

We assume that the boundary $\partial Q$ is equiped with the field of internal unit normal vectors $n(q)$, $q \in \partial Q$. Hence a curvature of $\partial Q$ is defined in every of its regular point q. We shall assume that each $\Gamma_i$, $i = 1, \ldots, p$ component of the

[1]On leave from: Institute of Oceanology of Ac. Sci. USSR, ul. Krasikova 23, 117218 Moscow, USSR

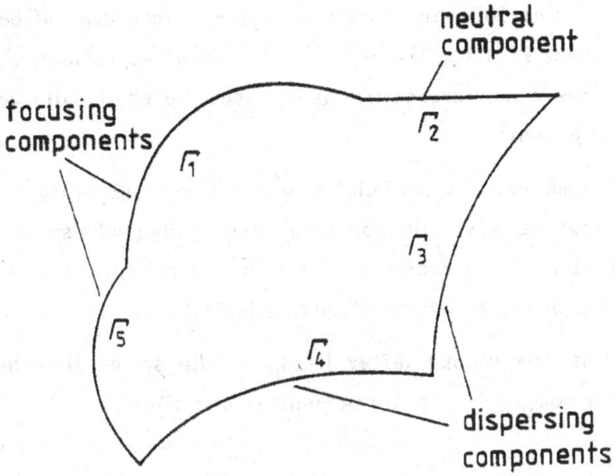


**Fig. 1**

boundary $\partial Q$ has curvature $k(q)$ of the same sign at all points $q \in \Gamma_i$ or its curvature is identically equal to zero. We shall call a regular component $\Gamma_i \subset \partial Q$ dispersing, focusing or neutral if $k(q) > 0$, $k(q) < 0$ or $k(q) = 0$ at all points for $q \in \Gamma$ respectively (Fig. 1).

A billiard is called a Sinai (or dispersing) billiard if all components of the boundary $\partial Q$ are dispersing. It is known [S1,BS] that such billiards are mixing and have positive Kolmogorov-Sinai entropy, if all regular components of $\partial Q$ intersect transversally. If $\partial Q$ has only neutral components (i.e. Q is a polygon) then the billiard in Q has zero entropy [BKM]. In case when $\partial Q$ also has focusing components the situation becomes essentially more complicated. First of all if $\partial Q$ is a convex $C^{6+\epsilon}$-smooth curve then the billiard in $\partial Q$ has caustics and therefore is non ergodic [L], [Do]. (In fact billiards in cicles and in ellipses are integrable Hamiltonian systems). On the other side there are numerous examples of ergodic billiards that have focusing components in $\partial Q$ [Bu1, Bu2, Bu3].

Therefore a natural question arises. Which focusing components can form a part of boundary of domains that generate ergodic billiards? This question was stated first in [Bu4]. In this paper a conjecture was formulated that such a component

$\Gamma^f$ has to satisfy to the following condition: for any infinitesimal beam of parallel rays which falls upon $\Gamma^f$ after the series of consecutive reflections from $\Gamma^f$ this beam will leave $\Gamma^f$ as a convergent (focusing) one. We shall call such $\Gamma^f$ an absolutely focusing component.

In [Bu4] a condition was formulated which ensure absolute focusing. (Focusing components that satisfy this condition were called admissible.) It was mentioned in [Bu4] that it is not necessary for focusing component to be admissible in order to be included in the boundary of mixing billiard.

This paper is the second (after [Bu3]) in the series that has the aim to prove the following conjecture, that was mentioned in [Bu4].

<u>Conjecture.</u> A focusing component $\Gamma^f$ can be a regular component of a boundary of mixing billiard iff $\Gamma^f$ is absolutely focusing.

In [Bu3] the general theorem was proven on mixing of some class of billiards, satisfying to some conditions. Two of these conditions ((G4) and (G5)) were mentioned in [Bu3] as crucial ones. The given paper deals with condition (G4) that is equivalent to the property of absolutely focusing (Theorem 1). Condition (G5) will be analized in details in the next paper [Bu6] where some results give the strong support for the above conjecture.

In a recent paper V. Donnay [D] introduced the class of focusing components that satisfy to the property that any infinitesimal family of parallel rays that falls on a focusing component $\Gamma^f$ focuses and goes through a conjugate point between each pair of consecutive reflections from $\Gamma^f$ and also focuses after the last time it hits $\Gamma^f$ in the series under consideration.

At the first sight this property looks much stronger then absolutely focusing but in fact these are equivalent (see Theorem 1 below). This was the reason to use in [Bu3] the property of absolutely focusing in this form (condition (G4)). We have to mention that V. Donnay has introduced the same condition as (G4) independently.

In the given paper we provide the complete characterization of absolutely focusing components (Theorems 1, 3 and Lemma 5). These allow us to show (Theorem 6) that a focusing component that is close to an absolutely focusing one in the $C^3$-($C^4$)-topology is also absolutely focusing if this component has constant (non-constant) curvature. It generalizes the analogous result of [D] on $C^6$-perturbations. We also found some unexpected application of billiards with absolutely focusing components to the theory of continued fractions and show some new results on their convergence (Theorems 10 and 11).

The structure of the paper is the following. In sect. 1 we give some general definitions and notions of the theory of two-dimensional billiards. Sect. 2 is devoted to the proof of Theorem 1 and of some other properties of absolutely focusing components. We also give some new examples of such mirrors. In sect. 3 we consider relations between billiards and continued fractions.

## 1. General Notions of the Theory of Billiards

Denote by M the restriction to Q of the unit tangent bundle of $\mathbb{R}^2$ and by $\pi:M \to Q$ the natural projection. The preimage $\pi^{-1}(q) = S^1(q)$, $q \in Q$, consists of unit vectors that are tangent to Q at the point $q \in Q$. Points $x = (q,v) \in M$ are unit tangent vectors and $q = \pi(x)$ is the natural projection. It is easy to see that M is a three-dimensional manifold with the boundary $\partial M = \bigcup_{i=1}^{p} \pi^{-1}(\Gamma_i) = \bigcup_{i=1}^{p} \partial M_i$. In every regular component $\partial M_i$ we introduce a coordinate system $(r,\varphi)$ where r is normalized arc length along $\partial Q$ and $\varphi$, $-\pi < \varphi \leq \pi$ is the angle between the vectors x and n(q), where $q = \pi(x) \in \partial Q$.

We introduce in M the measure $d\mu$ = const $dqd\omega$, where dq is the measure in Q induced by the Euclidean metric, $d\omega$ is the uniform measure on the sphere $S^1(q) = \pi^{-1}(q)$ and const is a normalizing factor. The one-parameter group $\{T^t\}$ is a flow in the sense of ergodic theory.

Let $V = \bigcup_{1 \leq i \leq j \leq p} (\Gamma_i \cap \Gamma_j) \subset \partial Q$, $W = \pi^{-1}(V)$ is the set of singular points.

Let $M_1 = \{x \in \partial M : (x,n(q)) \geq 0,\ q = \pi(x)\}$, $M_{1,i} = M_1 \cap \partial M_i$, $i = 1, 2, \ldots, p$. It is clear that each $\partial M_i$ is rectangular or a cylinder in the coordinates $(r,\varphi)$. For any point $x \in M$ we denote by $\tau^+(x)$ $(\tau^-(x))$ the nearest positive (negative) moment of reflection of the trajectory of x from $\partial Q$ and put $\pi_1^+ x = T^{\tau^+(x)+0}x$, $\pi_1^- x = T^{\tau^-(x)+0}x$.

For $z \in M_1$ consider

$$\tau(z) = \{\sup\{t > 0 : \text{for all } s,\ 0 \leq s \leq t,\ T^s z \in \partial M\} \text{ if it is positive; otherwise } \inf\{t > 0 : T^t z \in \partial M\}\}.$$

Since $\tau^+(x) < \infty$ the transformation $T_1 x = T^{\tau(x)+0}x$, $x \in M_1$, is defined and maps $M_1 \setminus \bigcup_{i=-\infty}^{+\infty} T^i W$ into itself. $T_1$ preserves the measure $\mu_1$, $d\mu_1$ = const $\cos\varphi dr d\varphi$ that is the projection of $\mu$ onto $M_1$ [CSF] (here const is a normalizing factor).

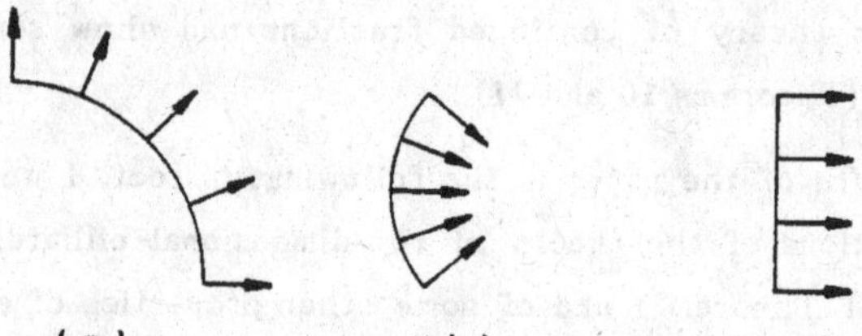


Fig. 2

Take a curve $\tilde{\gamma} \subset Q$ of class $C^2$. By a framing $\gamma$ of the curve $\tilde{\gamma}$ we mean a continuous section of the unit tangent bundle over $\tilde{\gamma}$ such that at each point $q \in \tilde{\gamma}$ a unit tangent vector $x \in \gamma$, $\pi(x) = q$, is orthogonal to $\tilde{\gamma}$ at the point q. According to this definition $\gamma$ corresponds to a beam of trajectories orthogonal to $\tilde{\gamma}$. It is clear that $\tilde{\gamma}$ has two framings. After choosing a framing we define a sign of the curvature of $\gamma$ at any of its points. If the curvature does not equal zero at all points of $\gamma$ then its value at a single point defines a framing uniquely.

Let $\tilde{\gamma} \in Q$ be an arbitrary smooth curve and let $\gamma$ be its framing. Denote by $\kappa_0(x)$ a curvature of $\gamma$ at a point $x \in \gamma$. We assume that there are no points of $\gamma$ which reflect from the boundary $\partial M$ during the time between 0 and t. It is easy to show (see [S1]) that

$$\kappa_t(x) = \frac{\kappa_0(x)}{1+\kappa_0(x)t} \tag{1}$$

where $\kappa_t(x)$ is the curvature of the curve $T^t\gamma$ at the point $T^tx$.

It follows from (1) that if $\kappa_0(x) > 0$ then $\kappa_t(x) > 0$ for any $t > 0$. In case $\kappa_0(x) < 0$ one has $\kappa_t(x) < 0$ for $0 \le t < 1/|\kappa_0(x)| = t_0$, and for $t > t_0$ the curvature $\kappa_t(x)$ is positive and decreases as $(t-t_0)^{-1}$.

We shall call a smooth nonintersecting curve $\gamma$ convex (concave) at a point $x \in \gamma$ if $0 < \kappa(x) < \infty$ $(-\infty < \kappa(x) < 0)$. A point x, where $\kappa(x) = \infty$ will be called a conjugate point. It is clear (see for instance [Bu1]) that convex, concave and flat curves generate divergent, convergent and parallel beams of trajectories respectively (Fig. 2).

The curvature of a curve jumps at the moment of reflection. Denote by $\kappa_-(x)$ and $\kappa_+(x)$ the curvature of the curve $\gamma$ at the point x in $\gamma$, $\pi(x) \in \partial Q$ at the moment just before and after the collision respectively. Then we obtain (see [BS]) that

$$\kappa_+(x) = \kappa_-(x) + \frac{2k^{(0)}(x)}{\cos \varphi(x)} \tag{2}$$

where $q = \pi(x)$, $k^{(0)}(x)$ is the curvature of $\partial Q$ at the point q, $\varphi(x)$ is the angle of incidence of the vector x.

## 2. Some Characteristic Properties of Absolutely Focusing Components

We shall start with the proof of equivalence of the property of a focusing component $\Gamma^f$ to be absolutely focusing and the property introduced in [Bu3, D].

Theorem 1. If $\Gamma^f$ is absolutely focusing component then each incoming infinitesimal beam of parallel trajectories focuses after each (consecutive) reflection from $\Gamma^f$ and goes through a conjugate point before its next collision from $\Gamma^f$ in this series.

Proof. Suppose that this assertion is not true. It means that there exists at least one series of consecutive reflections from $\Gamma^f$ such that corresponding infinitesimal beam of parallel trajectories does not go through a conjugate point between two of these reflections. Lets take all shortest series of reflection with this property. (Here the length of trajectory is the number k of its reflections from $\Gamma^f$.) Because $\Gamma^f$ is a focusing component these series hit $\Gamma^f$ at least twice, so $k \geq 2$. We now choose a series from this set that has the property that the first collision with $\Gamma^f$ is at a point q in $\Gamma^f$ where q is choosen first with minimal coordinate r and then among these is choosen with minimal angle $\varphi_i$ with the normal n(q). (Here we suppose in fact that some orientation of $\Gamma^f$ is chosen (Fig. 3).)

There are two possibilities depending if the series under consideration goes along $\Gamma^f$ monotonically with r (Fig. 3a) or not (Fig. 3b). We shall only consider the first one. For the second the proof is the same.

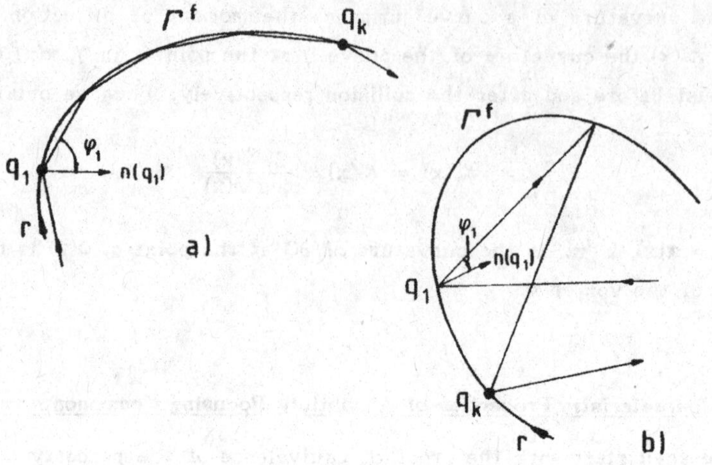


Fig. 3

According to our assumption all other series of reflections that start at $q_1$ with smaller angles with $n(q_1)$ go through conjugate points between any two consecutive reflections. Here there are two possibilities again depending on if the last point ($q_k$) of our trajectory coinside with the end of $\Gamma^f$ or not. We assume first that $q_k \notin \partial\Gamma^f$. Then in view of continuity of a transformation $T_1^m$ at the point $(q_1,\varphi_1)$, where $1 \leq m \leq k - 1$, $\varphi_1$ is the angle of reflection, our beam has a conjugate point exactly at a point of reflection $q_l$, where $l \leq k$. Consider now a continuous family of trajectories that go out of $q_1$ and have angles slightly less than $\varphi_1$. Because of our conditions these trajectories will hit $\Gamma^f$ at the point of $l$th reflection from $\Gamma^f$ it define dispersing infinitesimal beams of trajectories with arbitrary large curvature. But this means that after the following reflection from $\Gamma^f$ these beams do not go through conjugate points. So we have a contradiction.

Let now $q_k$ coincide with the end of $\Gamma^f$. Then we shall consider instead the series of reflection at the points $q_1, q_2, \ldots q_k$ the series $q_k, q_{k-1}, \ldots, q_1$. It is easy to see that for this series of reflections the infinitesimal beam of trajectories also cannot go through conjugate points between any two consecutive reflections. Therefore we can apply to this beam the same consideration as above.

We now have to consider the case when $q_1$ and $q_m$ are both endpoints of $\Gamma^f$, We take a continuous family of trajectories that have larger angles with $n(q_1)$ than $\varphi_1$. Then we get a series of reflections that after the last reflection from $\Gamma^f$ is not focusing. Hence we have a contradiction and Theorem 1 is proven.

If $q', q'' \in \Gamma \subset \partial Q$ we shall denote by $\Gamma_{(q',q'')}$ the part of $\Gamma$ with its ends at $q'$ and $q''$.

<u>Corollary 2.</u> If $\Gamma^f$ is an absolutly focusing curve then $\Gamma^f_{(q',q'')}$ is also absolutely focusing for any $q', q''$.

<u>Proof.</u> Suppose it is wrong. Then we get using Theorem 1 an immediate contradiction.

We shall now show how series of consecutive reflections from $\partial Q$ define a class of continued fractions [S1]. These continued fractions are closely related with the dynamics of billiards because they give solutions of Jacobi equations for these geodesic flows [S2].

Let $\Gamma^f$ be a focusing component and $q_1 \in \Gamma^f$. Each value $\varphi_1$, $-\frac{\pi}{2} < \varphi_1 < \frac{\pi}{2}$ defines a series of consecutive reflections of a parallel (local) beam of trajectories that starts the point $q_1$ and leaves it with the angle $\varphi_1$ (Fig. 3). Suppose that the length of this series (i.e. the number of reflections from $\Gamma^f$) equals k, $2 \le k < \infty$. Then for each integer n, $1 \le n \le k - 1$, we define the following (finite) continued fractions

$$\kappa^-_{k,n}(x_1) =$$

$$= \cfrac{1}{\tau_{k-1}(x_1) + \cfrac{1}{\cfrac{2k^{(0)}_{k-1}(x_1)}{\cos\varphi_{k-1}(x_1)} + \cfrac{1}{\tau_{k-2}(x_1) + \cfrac{1}{\cfrac{2k^{(0)}_{k-2}(x_1)}{\cos\varphi_{k-2}(x_1)} + \ddots + \cfrac{1}{\tau_{k-n}(x_i) + \cfrac{1}{\cfrac{2k^{0}_{k-n}(x_i)}{\cos\phi_{k-n}(x_i)}}}}}}}$$

where $k^{(0)}_j(x_1) = k^{(0)}(q_j)$, $\tau_j(x_1)$ is the length of the segment $[q_j, q_{j+1}]$, $j = 1, 2, \ldots, k - 1$. We also let $\kappa^-_{k,0}(x_1) = 0$. According to the laws of reflections (1), (2) for billiards $\kappa^-_{k,n}(x_1)$ is equal to the curvature of the local beam of trajectories that started from the point $q_{k-n} \in \Gamma^f$ with the angle $\varphi_{k-n}$ as a parallel beam (i.e. with zero curvature) just before its last reflection from $\Gamma^f$, where $x_{k-n} = (q_{k-n}, \varphi_{k-n}) = T^{k-n-1}x_1$.

Denote

$$\hat{\kappa}^-_{k,n}(x_1) = \cfrac{1}{\tau_{k-1}(x_1) + \cfrac{1}{\cfrac{2k^{(0)}_{k-1}(x_1)}{\cos\varphi_{k-1}(x_1)} + \ddots + \cfrac{1}{\cfrac{2k^{(0)}_{k-n+1}(x_1)}{\cos\varphi_{k-n+1}(x_1)} + \cfrac{1}{\tau_{k-n}(x_1)}}}}$$

From Theorem 1 follows easily

<u>Lemma</u> <u>3.</u> For each $1 \leq j \leq k - 1$ and $x \in M_1 \setminus \bigcup_{i=1}^{n} T^{-i} W$

$$\kappa^-_{k,j} < \hat{\kappa}^-_{k,j+1} < \kappa^-_{k,j+1} .$$

<u>Theorem</u> <u>4.</u> A focusing component $\Gamma_f$ is absolutely focusing one iff for any parallel (local) beam of trajectories that falls on $\Gamma^f$ at the point $q_1 \in \Gamma^f$ with the angle $\varphi_1$ and then goes through a series of k consecutive reflections from it one has

$$0 < \kappa^-_{k,n}(x_1) < \kappa^-_{k,n+1}(x_1) \tag{5}$$

where $x_1 = (q_1,\varphi_1)$, $1 \leq n \leq k - 2$, $x_j = T^{j-1}x_1$, $j = 1, 2, \ldots, k - 1$ and

$$0 < \kappa^-_{k,k-1}(x_1) < 2\,|k^{(0)}(x_k)|\,(\cos\varphi(x_k))^{-1} . \tag{6}$$

<u>Proof.</u> We assume first that $\Gamma^f$ is absolutely focusing. According to Lemma 2 $\Gamma^f_{(q_{k-m}, q_k)}$ is absolutely focusing for all $m = 1, 2, \ldots, k-1$, where $q_j = \pi(x_j)$, $j = 1, 2, \ldots, k$.

Now using the relation

$$\kappa^-_{k,1}(x_1) > 0 = \kappa^-_{k,0}(x_1) \tag{7}$$

we get

$$\kappa^-_{k,j+1}(x_1) > \kappa^-_{k,j}(x_1) \tag{8}$$

for $j = 1, 2, \ldots, k - 2$. For $j = 0$ (8) coinsides with (7). The relation (6) follows simply from the definitions of the property of absolutely focusing

If $x_1 \in M_1$ is such that $\pi(T^j x_1) \in \Gamma^f$ for $j = 0, 1, \ldots, k-1$, and $\pi(T^k x_1) \notin \Gamma^f$, then the sequence $\kappa^{-}_{k,n}(x_1) > 0$, $n = 1, 2, \ldots, k-1$ is an increasing one and relation (6) holds.

Suppose that $\Gamma^f$ is not absolutely focusing. Then there exists a series of consecutive reflections from $\Gamma^f$ such that a beam of trajectories being initially parallel does not go through a conjugate point between some two reflections. (One has to mention that in view of (6) all parallel beams of trajectories that have fallen on $\Gamma^f$ leave it being convergent ones). Thus one has $\kappa^{(-)}_{k_1,n_1-1}(x_1) \leq 0$ for some pair of integer $n_1$, $k_1 > n_1 \geq 1$. Take the maximal value of $n_1$, for which this is true. Then we have

$$\kappa^{(-)}_{k_1,k_1-1}(x_1) < \kappa^{(-)}_{k_1,k_1-n_1}(x_1) \,. \tag{9}$$

But this relation contradicts (5) and Theorem 4 follows.

Consider an arbitrary complete series of consecutive reflections from some focusing component $\Gamma^f$ with reflections at points $q_1, q_2, \ldots, q_k$ with corresponding angles $\varphi_j$, $1 \leq j \leq k$.

<u>Lemma 5.</u> $\Gamma^f$ is absolutely focusing iff there exist a set of numbers $\delta_2 > 0$ and $\delta_j > -1$, $3 \leq j \leq k-1$ such that

$$\tau_{j,j+1} = \tau(q_j, q_{j+1}) = \left|\frac{\cos\varphi_j}{2k_j^{(0)}}\right|\left(2 - \frac{\delta_j}{1+\delta_j}\right) + \left|\frac{\cos\varphi_{j+1}}{2k_{j+1}^{(0)}}\right|\left(2 + \delta_{j+1}\right) \tag{10}$$

and

$$\tau_{1,2} > \left|\frac{\cos\varphi_1}{2k_1^{(0)}}\right| + \left|\frac{\cos\varphi_2}{2k_2^{(0)}}\right|(2 + \delta_2)$$

$$\tau_{k-1,k} > \left|\frac{\cos\varphi_{k-1}}{2k_{k-1}^{(0)}}\right|\left(2 - \frac{\delta_{k-1}}{1 + \delta_{k-1}}\right) + \left|\frac{\cos\varphi_k}{2k_k^{(0)}}\right| . \tag{11}$$

Here $\tau(q_i, q_{i+1})$ is the length of the segment between $q_i$ and $q_{i+1}$.

<u>Proof.</u> We propose first that (10) and (11) hold. According to (2) at the moment just after the first reflection in the series under consideration our beam of trajectories has curvature $\frac{2k_1^{(0)}}{\cos\varphi_1}$. Hence according to the first relation in (11) it goes through a conjugate point before the second reflection. Further we obtain

from (1) and (2) that after the second reflection from $\Gamma^f$ (at $q_2$) this beam of trajectories will have the curvature

$$\kappa_2^{(+)} < \frac{2k_2^{(0)}}{\cos\varphi_2}\,\frac{1+\delta_2}{2+\delta_2} < 0 \ . \tag{12}$$

Analogously making use of (1), (2) and (10) we get consecutively that

$$\kappa_j^{(+)} < \frac{2k_j^{(0)}}{\cos\varphi_j}\,\frac{1+\delta_j}{2+\delta_j} < 0 \tag{13}$$

for $j = 3, 4, \dots k - 1$.

At last we get from (1), (2) and the second relation in (11) that

$$\kappa_k^{(+)} < 0 \ . \tag{14}$$

Hence any beam of parallel trajectories after a series of consecutive reflections from $\Gamma^f$ becomes convergent. This means that $\Gamma^f$ is absolutely focusing component.

We shall now prove the inverse statement. Suppose that $\Gamma^f$ is an absolutely focusing component. In view of Theorem 1 this means that any beam of parallel trajectories between each two consecutive reflections from $\Gamma^f$ goes through a conjugate point. Thus the first relation in (11) and the relation (10) hold with some $\delta_2 > 0$, $\delta_j > -1$, $3 \le j \le k - 1$. Further making use of the definition of absolutely focusing component and the relation (10) for $j = k - 2$ we get that the second relation in (11) holds as well. So Lemma 5 is proven.

We shall call a focusing curve $\Gamma^f$ such that the angle between the tangents at its ends is not more than $\pi$ a right one.

<u>Theorem</u> <u>6.</u> Let $\Gamma^f$ is a right and absolutely focusing component. Then any right focusing component of the same length $\Gamma^f$ that is sufficiently close to $\Gamma^f$ in the $C^3$ ($C^4$) topology is absolutely focusing too if $\Gamma^f$ has a constant (nonconstant) curvature.

<u>Proof.</u> Denote by $\theta_1$ the angle between the tangent to $\Gamma^f$ and the velocity vector at the point of the first reflection of some trajectory at $\Gamma^f$. It was shown in [L] that in coordinates $(r,\theta)$ for small (or close to $\pi$) $\theta$ the transformation T has the form

$$r_1 = r + 2R(r)\,\theta + \frac{4}{3} R(r)\,R'(r)\,\theta^2 + \bar{F}(r,\theta)\,\theta^3$$
$$\theta_1 = \theta - \frac{2}{3} R'(r)\,\theta^2 + \bar{G}(r,\theta)\,\theta^3 \tag{15}$$

where $|\bar{F}(r,\theta)|$ and $|\bar{G}(r,\theta)|$ are bounded from above and from below for all $(r,\theta) \in \Gamma^f$ if $T(r,\theta) = (\bar{r},\bar{\theta}) \in \Gamma^f$.

Suppose that the curvature of $\Gamma^f$ is nonconstant. We make now the following change of coordinates

$$(r,\theta) \to (r,\psi),\ \psi = R^{1/3}(r)\,\theta.$$

In these coordinates T becomes

$$r_1 = r + 2R^{2/3}(r)\,\psi + \frac{4}{3} R^{1/3}(r)\,R'(r)\,\psi^2 + \hat{F}(r,\psi)\,\psi^3$$
$$\psi_1 = \psi + \hat{G}(r,\psi)\,\psi^3, \tag{15'}$$

where $|\hat{F}(r,\psi)| < c_1$, $|\hat{G}(r,\psi)| < c_2$.

Take an arbitrary segment of a billiard trajectory between any two consecutive reflections at some points $q', q'' \in \Gamma^f$. It is easy to calculate [W] that

$$\tau(q', q'') = R(r') \sin\theta' + R(r'') \sin\theta'' - \int_{r'}^{r''} \sin\theta(r) \frac{dR(r)}{dr}\,dr =$$
$$= R(r') \sin\theta' + R(r'') \sin\theta'' - \int_{r'}^{r''} y(r) \frac{d^2R(r)}{dr^2}\,dr \tag{16}$$

where r is the length parameter in $\Gamma^f$ and $R(r)$ is the radius of curvature at a point $q = (r,\theta)$.

Denote $I_1(r', r'') = \int_{r'}^{r''} \sin\theta(r) \frac{dR(r)}{dr}\,dr$,

$$I_2(r',r'') = \int_{r'}^{r''} y(r) \frac{d^2R(r)}{dr^2}\,dr.$$

Consider now some "long" series (i.e., $\theta_i$ is small enough) of consecutive reflections from $\Gamma^f$ that occur at the points $q_i = (r_i,\theta_i)$, i = 1, 2, ..., N. Denote by $\tau_k^{(1)}$, $\tau_k^{(2)}$, $\tau_k^{(1)} + \tau_k^{(2)} = \tau_k$, times when an initially (at $q_1$) parallel beam of trajectories between kth and (k+1)th reflections in the series was convergent and divergent respectively. (Here $\tau_k$ is the whole time between these reflections).

Let us call $q_i$, $q_{i+1}$, $1 \le i \le N-1$, a null-segment if $dR(r)/dr = 0$ or $d^2R(r)/dr^2 = 0$ at some point $r \in (r_i, r_{i+1}) \subset \Gamma^f$.

The functions $dR(r)/dr$ and $d^2R(r)/dr^2$ do not change their signs in each connected component of the complement (in $\Gamma^f$) of the union of all null-segments. Making use of this fact it is easy to derive from (15), (15'), (16) that

$$\tau(q_{i_1}, q_{i_1+1}) \ge R(r_{i_1}) \sin \theta_{i_1} + R(r_{i_1+1}) \sin \theta_{i_1+1} \tag{17}$$

or

$$(R(r_{i_1+1})\sin \theta_{i_1+1})^{-1} \ge \tau^{-1}(q_{i_1}, q_{i_1+1}) + \tau^{-1}(q_{i_1+1}, q_{i_1+2})$$

if $r_{i_1}$, $r_{i_1+1}$, $r_{i_1+2}$ do not belong to different connected components.

Fix now some connected component $r_{i_1}$, $r_{i_1+k}$, $1 \le i_1 \le N$, $1 \le k \le \le N - i_1$, and suppose for instance that the first inequality from (17) holds in it. Let the parallel beam of trajectories that started reflections at the point $q_1$ with the angle $\theta_1$ comes to $q_{i_1}$ with the curvature $\kappa_{i_1} \le (\alpha_{i_1} R(r_{i_1}) \cos \varphi_{i_1} + R(r_{i_1+1}) \cos \varphi_{i_1+1})^{-1}$, where $\frac{1}{2} \le \alpha_{i_1} \le 1$.

Applying (1) and (2) consecutively to this series of reflections we get

$$\tau^{(2)}_{i_1+k-1} \ge R(r_{i_1+k}) \sin \theta_{i_1+k} + \frac{1}{\alpha_{i_1}^{-1} \dfrac{1}{R(r_{i_1})\sin\theta_{i_1}} + 2 \sum\limits_{\ell=i_1+1}^{i_1-k-1} \dfrac{1}{R(r_\ell) \sin \theta_\ell}}.$$

The case of connected component that correspond to the second inequality in (17) is quite analogous. In fact these two inequalities are symmetric with respect to the continuous fractions considered in sect. 2. Thus problems with the property of absolutely focusing can appear in null-segments only.

It can be shown analogously to [D] (Lemmas 5.1, 5.2) that $|\psi_k - \psi_1| < \psi_1/2$ for sufficiently small $\psi_1$ and $k = 1, 2, \dots N$; $N = 0(\psi_1^{-2})$ and that any series of consecutive reflections from $\Gamma^f$ has a length not more than $c_3\psi_1^{-1}$. Finally we have from (16) $|I_1(r',r'')| < c_4\, \theta_1^4$ and/or $|I_2(r',r'')| < c_5\, \theta_1^4$ if $(r',r'')$ is a null-segment and thus the theorem follows. If $\Gamma^f$ has a constant curvature then the change of coordinates is of course not needed (see also [Bu5]).

From the proof of theorem 6 follows immediately (compare with [D])

Corollary. If $\Gamma^f$ is focusing then there exist a number $l = l(\Gamma^f)$ such that small $C^4$-perturbations of any arc of $\Gamma^f$ with a length less than l are absolutely focusing. If $\Gamma^f$ has a constant curvature the same is true for $C^3$-perturbations.

Remark. In fact V. Donnay explored in [D] the same ideas. The reason that he considered $C^6$-perturbations is the use of Lazutkin coordinates that "are too much of a good thing ..." as was mentioned in [D]. Lazutkin really needed these coordinates in order to apply KAM-theory. For the study of ergodic properties however the adequate coordinates are given via (15). Stability of some specific subclass of absolutely focusing components under $C^4$-perturbations was shown also in [M2].

Consider all series of consecutive reflections from $\Gamma^f$ by an initially parallel beam of trajectories. Each such series is defined by an initial point $q_0 \in \Gamma$ and angle of incidence $\varphi_0$, $-\frac{\pi}{2} \leq \varphi_0 \leq \frac{\pi}{2}$. Denote by $\hat{\tau}(q_0,\varphi_0)$ the time upto conjugate point (beginning of defocusing) after the last reflection in the given series from $\Gamma^f$. Because $\Gamma^f$ is absolutely focusing $\hat{\tau}(q_0, \varphi_0)$ has some finite value.

It follows from (17) that $\hat{\tau}(q_0,\varphi_0) < \hat{\tau}(\Gamma^f) < \infty$ if the number of reflections $n(q_0,\varphi_0)$ from $\Gamma^f$ in the series under considerations is larger than $\hat{n} = \hat{n}(\Gamma^f) < \infty$. The proof of this fact was given in [D]. Another proof that gives a stronger property as well is given in [Bu6].

Let $N_n(\Gamma^f) = \{(q_0,\varphi_0) : (q_0,\varphi_0) \in \Gamma^f, n(q_0,\varphi_0) = n\}$, where $(q_0,\varphi_0)$ define the first reflection in a complete series of consecutive reflections from $\Gamma^f$. From the continuity of the function $\hat{\tau}(q_0,\varphi_0)$ we have for any fixed n

$$\sup_{(q_0,\varphi_0)\in\overline{N_n}(\Gamma^f)} \tau(q_0,\varphi_0) = \tau_n(\Gamma^f) < \infty.$$

Now applying this relation and Theorem 4 for all $n \leq \hat{n}(\Gamma^f)$ we get

Lemma 7. If $\Gamma^f$ is absolutely focusing component then

$$\hat{\tau}(\Gamma^f) = \sup_{q_0\in\Gamma^f,\ -\pi/2\leq\varphi_0\leq\pi/2} \hat{\tau}(q_0,\varphi_0) < \infty.$$

Because of Theorem 1 the statement of Lemma 7 is equivalent to Theorem 4.4 of [D].

The following theorem is proven absolutely analogously to Theorem 2 of [D] making use of Lemma 8 and of the same cone fields for absolutely focusing components from [D] and for dispersing components from [Bu3] (see [S1], [Bu1], [Bu2] as well).

<u>Theorem</u> <u>8.</u> Billiard in a bounded connected domain Q has positive Lyapunov exponents almost everywhere if:

(i) all focusing components $\Gamma_1^f, \Gamma_2^f, \dots \Gamma_k^f$ of the boundary $\partial Q$ are absolutely focusing

(ii) $\min\limits_{\Gamma_i^f, 1\le i\le k} \; \min\limits_{\substack{\pi(x)\in\Gamma_i^f \\ \pi(Tx)\notin\Gamma_i^f}} \tau(x,Tx) > 2\max\limits_{1\le i\le k} \hat{\tau}(\Gamma_i^f)$

where $\tau(x,Tx)$ is the time between the reflections at the points x and Tx.

(iii) the set of trajectories that hit focusing ($\partial Q^-$) or dispersing ($\partial Q^+$) part of the boundary $\partial Q$ has measure 1

(iv) dispersing components intersect each other and neutral components transversally

(v) $\partial Q^+ \cup \partial Q^- \neq \phi$.

We now give some examples of absolutely focusing curves. First we have to give the list of examples that are known to be absolutely focusing. All arcs of circles are absolutely focusing [Bu1] and also curves that are close to such arcs but that are shorter than semicircles [Bu5, M1, M2]. Another class of examples was introduced in [W]. This class is defined by the condition $\frac{d^2R(q)}{dr(q)^2} \le 0$ where R is a radius of curvature of $\partial Q$ at a point q and r(q) is the arc length parameter. It was shown in [W], [M1, M2] and [D] that some parts of ellipses could serve as focusing components as well. We shall generalize these results for some more rich class of arcs of ellipses. For the convenience of a reader we shall use the same notations as in [D].

Consider the ellipse $E(t) = \{x(t)=a\cos t,\ y(t)=\sin t\}$, $t \in [0,2\pi)$ and let $E_{[t_a,t_b]}$ denote the arc of the ellipse for which $t \in [t_a,t_b]$. The focal points of the ellipse are $x = \pm c = \sqrt{a^2-1}$, $y=0$. The only parameter of the problem is a because similar arcs of two different ellipses with the same excentricity are absolutely focusing or not simultaneously.

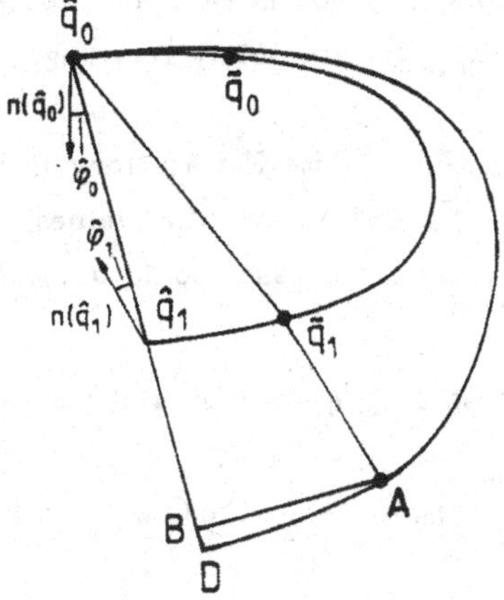


Fig. 4

It is known [CFS] that billiards in ellipses are absolutely integrable Hamiltonian systems. These billiards have two continuous families of caustics H-trajectories (E-trajectories) that cross the x axis with $-c < x < c$ ($|x| > c$) and are tangent to hyperbolas (ellipses) confocal with E(t). Therefore it is a well known fact that the elliptical arc $E_{[t_a,t_b]}$, $t_a \leq 0 < t_b \leq \pi$ is absolutely focusing if the line connecting $E(t_a)$ and $E(t_b)$ crosses the x axis at $|x| \geq c$ (for a proof see [D]). Previous results imply as well that $E_{[-\arcsin a^{-1}, \arcsin a^{-1}]}$ [M1] and $E_{[-\frac{\pi}{2},\frac{\pi}{2}]}$ for $a < 2^{1/4}$ [D] are focusing arcs.

Another well known fact [CFS, S2] is that there is a stable period-two trajectory of billiard corresponding to the smaller axis of an ellipse. Thus parts of ellipses that contain a neighborhood of the small axis can not serve as regular components at a boundary for billiards with chaotic behaviour. Therefore we shall consider only such arcs of ellipses that do not contain neighborhoods of the both ends of its small axis. Consider $E_{[t_a,t_b]}$ with $\pi \geq t_a > t_b \geq -\frac{\pi}{2}$. We denote the ends of this arc by $\hat{q}_0$, $\hat{q}_1$ and the angles that the segment $[\hat{q}_0, \hat{q}_1]$ makes with internal normals at these points by $\hat{\varphi}_0$ and $\hat{\varphi}_1$ (Fig. 4). Denote the radii of curvature at $\hat{q}_0$ and $\hat{q}_1$ by $\hat{R}_0$ and $\hat{R}_1$ correspondingly and consider the value

$$\Delta(\hat{q}_0, \hat{q}_1) = \tau(\hat{q}_0, \hat{q}_1) - \tfrac{1}{2}\hat{R}_0 \cos\hat{\varphi}_0 - \tfrac{1}{2}\hat{R}_1 \cos\hat{\varphi}_1 \ . \tag{18}$$

Lemma 9. If each H-trajectory has not more than two consecutive reflections in $E_{[t_a,t_b]}$ and $\Delta(\hat{q}_0,\hat{q}_1) > 0$ then $E_{[t_a,t_b]}$ is absolutely focusing.

Proof. Consider a segment $\tilde{q}_0$, $\tilde{q}_1$ of some H-trajectory in $E_{[t_a,t_b]}$ . Suppose that $|t_a| > |t_b|$ and $|\tilde{t}_a| > |\tilde{t}_b|$, where $\tilde{t}_a$ and $\tilde{t}_b$ are the values of the parameter t that correspond to $\tilde{q}_0$ and $\tilde{q}_1$. (All the other cases could be considered quite analogously to this one.)

We shall show first that $\Delta(\hat{q}_0, \tilde{q}_1) > 0$ as well. We have

$$\Delta(\hat{q}_0, \tilde{q}_1) = \tau(\hat{q}_0, \tilde{q}_1) - \frac{1}{2}\hat{R}_0 \cos \overline{\varphi_0} - \frac{1}{2}\tilde{R}_1 \cos \tilde{\varphi}_1$$

where $\overline{\varphi_0}$ ($\tilde{\varphi}_1$) is the angle between $[\hat{q}_0,\tilde{q}_1]$ and the normal $n(\hat{q}_0)$ ($n(\tilde{q}_1)$). Let us compare the ratios of the corresponding terms in the expressions for $\Delta(\hat{q}_0, \hat{q}_1)$ and $\Delta(\hat{q}_0, \tilde{q}_1)$.

Take the circle of curvature at $\hat{q}_0$ with radius $\hat{R}_0$ and consider its intersection $\mathfrak{D}$ and A with the lines $\hat{q}_0$, $\hat{q}_1$ and $\hat{q}_0$, $\tilde{q}_1$. Let B be the intersection of the line through $\hat{q}_0$, $\mathfrak{D}$ with the line that goes through A and is parallel to $\hat{q}_1$, $\tilde{q}_1$ (Fig. 4).

From the similarity of the triangles $\hat{q}_0$ $\hat{q}_1$ $\tilde{q}_1$ and $\hat{q}_0$ B A we get that $\hat{q}_0\hat{q}_1/\hat{R}_0 \cos\hat{\varphi}_0 < \hat{q}_0\tilde{q}_1/\hat{R}_1 \cos \overline{\varphi_0}$. Analogously it is easy to check that $\hat{q}_0\hat{q}_1/\hat{R}_1\cos\hat{\varphi}_1 < \hat{q}_0\tilde{q}_1/\tilde{R}_1\cos\tilde{\varphi}_1$.

Therefore $\Delta(\hat{q}_0,\tilde{q}_1) > 0$. Thus an infinitesimal beam of parallel trajectories that falls on $E_{[t_a,t_b]}$ at $\hat{q}_0$ and then reflected at $\tilde{q}_1$ also focuses after the last reflection.

Making use of the same construction (where $\hat{q}_0$, $\hat{q}_1$ and $\tilde{q}_1$ have to be replaced by $\tilde{q}_1$, $\hat{q}_0$ and $\tilde{q}_0$ correspondingly) we obtain that $\Delta(\tilde{q}_0,\tilde{q}_1) > 0$ as well. Hence Lemma 9 is proven.

Corollary The arc $E_{[-\frac{\pi}{2},\frac{\pi}{2}]}$ is absolutely focusing iff $a < \sqrt{2}$.

Proof. Let $\hat{q}_0$, $\hat{q}_1$ are the ends of semi-axis of some ellipse. Then $\Delta(\hat{q}_0,\hat{q}_1) > 0$ iff $a < \sqrt{2}$. Thus the Corollary follows from Lemma 9. This Corollary was formulated in [D] as a conjecture[2].

## 3. Some applications to the theory of continuous fractions.

In this section we discuss the relation of billiards in self-intersecting absolutely focusing curves on Euclidean plane to the problem of convergence of continuous fractions with elements of different signs.

Consider an infinite continuous fraction

$$\kappa = \cfrac{1}{a_1 + \cfrac{1}{a_2 + \cfrac{1}{a_3 + \ddots \cfrac{1}{a_n + \ddots}}}}$$

where $a_i$, $i = 1, 2, \ldots$ are some real numbers. The fraction $\kappa$ is called convergent if the series $\kappa_n = \frac{P_n}{Q_n}$ tends to a limit when $n \to \infty$, where $\frac{P_n}{Q_n}$ is the fraction that appears if we only consider the first n elements of $\kappa$ (i.e. we substitute $a_{n+1}$ by $\infty$).

In case when all elements $a_i$ have one and the same sign the problem of convergence of the sequence $\kappa_n$ is completely solved (see, for instance, [K]). However there are no sufficiently results for the general case.

Suppose that all elements $a_i$ of $\kappa$ have one sign if i is an odd number and the other one if i is even number. We attach to such continuous fraction an semiinfinite trajectory of some billiard.

Without any loss of generality we can assume that sgn $(a_i)$ is negative if i is odd. Consider now the following relations

$$a_i = 2\,\frac{k^{(0)}_{[i/2]+1}}{\cos\varphi_{[i/2]+1}} \tag{20}$$

if i is odd and

$$a_i = \tau_{i/2} \tag{21}$$

if i is even.

Take an arbitrary point $q_1$ on $\mathbb{R}^2$ and consider a small arc of a circle $S_1$ with radius $(k_1^{(0)})^{-1}$ that contains $q_1$. Consider now a straight segment $l_1$, with the length $\tau_1$ that makes the angle $\varphi_1$ with the tangent to $S_1$ at the point $q_1$. Denote the

---

[2]In the later version of [D] this was proven by using more sophisticated considerations than ours.

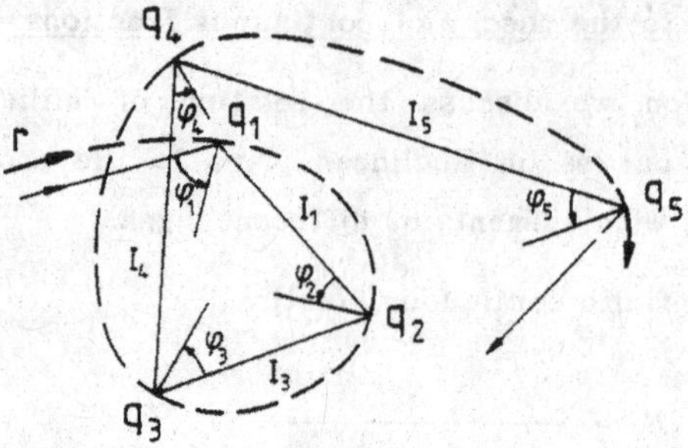


Fig. 5

second end of $I_1$ by $q_2$. Then we take at $q_2$ a small arc of a circle $S_2$ that has the curvature equal to $k_2^{(0)}$ and makes with the tangent at $q_2$ angle $\varphi_2$. Then we apply the same procedure as at the point $q_1$ and get the segment $I_2$ between $q_2$ and the point $q_3$ with an arc of the circle $S_3$ and so on (Fig. 5).

Theorem 10. If there exist a set of numbers $\delta_j \geq -1$, j = 1, 2, ..., such that

$$a_{2j} \geq |a_{2j-1}|^{-1} (2 + \delta_j) + |a_{2j+1}|^{-1} \left(2 - \frac{\delta_{j+1}}{1+\delta_{j+1}}\right) \tag{22}$$

then a continuous fraction $\kappa$ is the convergent one.

Proof. Consider the sequence $\theta_n = \frac{1}{\kappa_n} - a_1$, n = 1, 2, 3, ... . Applying Lemmas 3, 4 and Theorem 4 to a finite continuous fractions $\theta_k$ for any k = 1, 2, ... we get that the sequence $\{\theta_j\}$, j = 1, 2, ... k is increasing one. Further the relations (10) and (11) imply that this sequence is bounded from above by the value $\frac{4}{a_1} = \left|\frac{2\cos\varphi_1}{k_1^{(0)}}\right|$. Therefore the continuous fraction converges to the value

$$\kappa = \frac{1}{\frac{2\cos\varphi_1}{k_1^{(0)}} + \hat{\theta}} . \tag{23}$$

where $\hat{\theta} = \lim_{n\to\infty} \theta_n$.

Theorem 10 is proven.

Suppose now that all odd (even) elements of a continuous fraction $\kappa$ have one and the same sign and its even (odd) elements could have an arbitrary sign. Without any loss of generality we can suppose that all even elements are positive.

We can apply to such fraction the same change (20), (21) of variables.

Consider subsequence $a_{i'}$ of all even elements $a_j$, $j = 2, 4, \ldots$, such that at least one of neighboring elements of $a_j$ is negative.

<u>Theorem 11.</u> If there exists a set of numbers $\delta_{i'} \geq -1$ such that

$$a_{i'} \geq |a_{i'-1}|^{-1}(2 + \delta_{i'}) + |a_{(i+1)'+1}|^{-1}\left(2 - \frac{\delta_{(i+1)'}}{1+\delta_{(i+1)'}}\right)$$

then a continuous fraction $\kappa$ is the convergent one.

The proof of this theorem goes in the same way as of Theorem 10.

The same ideas can be applied to the proof of convergence for some more general classes of continuous fractions.

**Acknowledgement**

I want to thank S. Troubetzkoy for useful remarks and suggestions and improving my English. I want to express my sincere gratitude to Prof. Ph. Blanchard and to colleagues at BiBoS Research Center for their hospitality.

**References**

[BKM] Boldrighini C., Keane M., Marchetti F.: Billiards in polygons, Ann. Probab. 6, 532-540 (1978)

[Bu1] Bunimovich L.A.: On billiards closed to dispersing. Matem. Sbornik 95, 49-73 (1974)

[Bu2] Bunimovich L.A.: On the ergodic properties of nowhere dispersing billiards. Commun. Math. Phys. 65, 295-312 (1979)

[Bu3] Bunimovich L.A.: A Theorem on ergodicity of two-dimensional hyperbolic billiards. Commun. Math. Phys. 130, 599-621 (1990)

[Bu4] Bunimovich L.A.: Many-dimensional nowhere dispersing billiards with chaotic behaviour. Physica D 33, 58-64 (1988)

[Bu5] Bunimovich L.A.: On the stochastic dynamics of rays in resonators. Radiofizika 28, 1601-1602 (1985)

[Bu6] Bunimovich L.A.: On criterium of mixing for two-dimensional billiards (in preparation)

[BS] Bunimovich L.A., Sinai Ya.G.: On the main theorem of the ergodic theory of dispersing billiards. Matem. Sbornik 90, 415-431 (1983)

[CSF] Cornfeld I.P., Fomin S.V., Sinai Ya.G.: Ergodic theory. Springer Verlag, 1982

[D] Donnay V.J.: Using integrability to produce chaos: billiards with positive entropy. Preprint, Princeton University, 1990

[Do] Douady R.: Applications du théorème des tores invariants. Thesis, University of Parix VII, (1982)

[K] Khovanski A.N.: Application of continued fractions and their generalizations to problems of the theory of approximations. Groningen: Noordhoff, 1963

[L] Lazutkin V.F.: On the existence of caustics for the billiard ball problem in a convex domain, Math. USSR Izvestija 37, 185-215 (1973).

[M1] Markarian R.: Billiards with Pesin region of measure one. Commun. Math. Phys. 118, 87-97 (1988).

[M2] Markarian R.: Nonuniform hyperbolicity, quadratic forms and billiards. Preprint, 1990.

[S1] Sinai Ya. G.: Dynamical systems with elastic reflections. Ergodic properties of dispersing billiards. Russian Mathem. Surveys, 25, 137-189 (1970).

[S2] Sinai Ya.G.: Introduction to ergodic theory. Princeton University Press, 1976

[W] Wojtkowski M., Principles for the design of billiards with nonvanishing Lyapunov exponent. Commun. Mathem. Phys. 105, 391-414 (1986).

# Upper and lower class results for subsequençes of the Champernowne number

M. DENKER AND K.F. KRÄMER
UNIVERSITY OF GÖTTINGEN
INSTITUT FÜR MATHEMATISCHE STOCHASTIK, LOTZESTR. 13,
W–3400 GÖTTINGEN, GERMANY

Abstract: We determine upper and lower bounds for partial sums of subsequences of the dyadic Champernowne sequence, which are obtained from completely deterministic selection functions. This complements results by Shiokawa and Uchiyama.

## §1 NORMALITY OF GENERALIZED CHAMPERNOWNE NUMBERS

The classical Champernowne number .1 2 3 4 5 6 7 8 9 10 11... is normal to the base 10 ([2]). This sequence may be regarded as a generic point in the Bernoulli shift over ten symbols with identical probabilities, although it is not random at all. This phenomenon is true in any shift space over a finite symbol set. A sequence $(x_n)_{n\geq 1}$ in the shift space $\Sigma_p$ over $p \geq 2$ symbols is called normal if for any block $[b_1, ..., b_k]$

$$\lim_{n\to\infty} \frac{1}{n} \sum_{i=1}^{n} 1_{\{x_i=b_1,\dots,x_{i+k-1}=b_k\}} = p^{-k}.$$

(See [1], [2], [3], [4].) Equivalent definitions are given in [7], [8] and [9].

In this note we consider sequences of zeroes and ones which are normal to the base 2. Let

$$c = 0\ 1\ 00\ 01\ 10\ 11\ 000\ 001\ 010\ 011\ 100\ 101\ 110\ 111\ \dots$$

denote the dyadic Champernowne number. Let $B_n$ denote the collection of all $n$–blocks of length $n$. Then $c$ is obtained as a concatenation of all blocks in lexicographical order, first all 1-blocks, then all 2–blocks .... Formally

$$c = c_1^1 c_2^1 c_1^2 c_2^2 c_3^2 c_4^2 c_1^3 \dots c_1^n \dots c_k^n \dots$$

where $c_k^n = [c_{k,1}^n, ..., c_{k,n}^n]$ $(1 \leq k \leq 2^n)$ is a block of length $n$ of 0 and 1 so that

$$\sum_{j=1}^{n} c_{k,j}^n 2^{j-1} = k - 1.$$

A generalized Champernowne number $x$ is obtained rearranging the $n$-blocks within $c$ in any order. Formally, if for each $n \in \mathbb{N}$, $\pi_n$ is a permutation of $\{1, ..., 2^n\}$, then

$$x = c^1_{\pi_1(1)} c^1_{\pi_1(2)} c^2_{\pi_2(1)} \dots c^2_{\pi_2(4)} \dots c^n_{\pi_n(1)} \dots c^n_{\pi_n(2^n)} \dots$$

For $x \in \{0,1\}^{\mathbb{N}}$, denote $S_n(x) = \sum_{i \le j \le n} (1 - 2x_j)$. It is easy to see that a generalized Champernowne number is normal, and it is shown in [10] that

$$\limsup_{n \to \infty} \frac{\sqrt{\log n}}{n} |S_n(x)| \le \alpha \sqrt{\log 2}$$

$$\limsup_{n \to \infty} \frac{1}{\log n} |S_n(x)| \ge \frac{1}{2 \log 2},$$

where $\alpha > 0$ is determined by

$$3\alpha - \alpha \sqrt{\frac{2}{\pi}} \int_0^\alpha \exp[-u^2/2] du - \sqrt{\frac{2}{\pi}} \exp[-\alpha^2/2] = 0.$$

(There is only one such $\alpha > 0$.) For the dyadic Champernowne number one has

$$\limsup_{n \to \infty} \frac{\sqrt{\log n}}{n} |S_n(c)| = \frac{\log 2}{2}$$

$$\liminf_{n \to \infty} \frac{1}{\log n} |S_n(c)| = 0.$$

A function $f : \mathbb{N} \to \mathbb{R}_+$ is called slowly oscillating if

$$\forall s_n < t_n \; s_n \to \infty \; \lim_n \frac{s_n}{t_n} = 1 \Rightarrow \lim_n \frac{f(s_n)}{f(t_n)} = 1.$$

Theorem 1: ([10]) Let $f_i$ $(i = 1, 2)$ be slowly oscillating so that

$$\lim_{t \to \infty} \frac{f_i(t)}{\log t} = \infty \qquad \lim_{t \to \infty} \frac{\sqrt{\log t}\, f_i(t)}{t} = 0 \quad (i = 1, 2).$$

Then there exists a continuum of generalized Champernowne numbers $x$ such that

$$\limsup_{n \to \infty} \frac{S_n(x)}{f_1(n)} = 1 \quad \text{and} \quad \liminf_{n \to \infty} \frac{S_n(x)}{f_2(n)} = -1.$$

Remark: The function $f(t) = \sqrt{2t \log \log t}$ satisfies the assumption of the theorem. Hence there are generalized Champernowne numbers which also satisfy the law of the iterated logarithm.

We shall derive a similar result in Section 2.

## §2 Upper and lower bounds for the Champernowne number under completely deterministic selection functions

Let $t = (t_j)_{j \in \mathbb{N}}$, $x = (x_k)_{k \in \mathbb{N}} \in 0,1^{\mathbb{N}}$ be sequences of 0 and 1. Define

$$(x \circ t)_i = x_{k(i)}$$

where $i = \sum_{1 \leq j \leq k(i)} t_j$ for $i \in \mathbb{N}$, i.e. $k(i)$ is that index in $t$ for which the $i$–th 1 occurs. The function $i \to k(i)$ is called a selection function and $t$ will be called a selection sequence. Let $\Sigma_2 = \{0,1\}^{\mathbb{N}}$ and $\sigma$ be the shift on $\Sigma_2$. For $\theta \in \Sigma_2$, denote by $\mathcal{M}(\theta)$ the set of accumulation points (in the weak topology) of the measures $n^{-1} \sum_{0 \leq j < n} \delta_{\sigma^j(\theta)}$ ($\delta_y$ denotes the point mass in $y$). A selection sequence $t$ is called completely determinitic if for every measure in $\mathcal{M}(t)$ the metric entropy vanishes. It is shown in [6] and [11], that for any normal number $x$ and every completely deterministic selection sequence $t$, $x \circ t$ is normal as well, provided

$$\limsup_{n \to \infty} n (\sum_{j=1}^{n} t_j)^{-1} < \infty.$$

A function $f$ is called very slowly oscillating if it is slowly oscillating and

$$\lim_{n \to \infty} f(n) = \infty \quad \text{and} \quad \lim_{n \to \infty} \frac{f(n)}{n} = 0.$$

We prove

Theorem 2: Let $f_i$ ($i = 1,2$) be very slowly oscillating functions. Then there exists a continuum of completely deterministic selection sequences $\theta = (\theta_j)_{j \in \mathbb{N}}$ satisfying

$$\limsup_{n \to \infty} n (\sum_{j=1}^{n} \theta_j)^{-1} < \infty,$$

such that $c \circ \theta$ is normal to the base 2,

$$\limsup_{n \to \infty} \frac{S_n(c \circ \theta)}{f_1(n)} = 1 \quad \text{and} \quad \liminf_{n \to \infty} \frac{S_n(c \circ \theta)}{f_2(n)} = -1.$$

We use the following notation. Let $c_1^n, \ldots, c_{2^n}^n$ denote the $n$–blocks in lexicographical order as they appear in the dyadic Champernowne number $c$, i.e.

$$c_{m(n-1)+(k-1)n+1}, \ldots, c_{m(n-1)+kn} = c_k^n \quad (1 \leq k \leq 2^n),$$

where

$$m(l) = m(l-1) + l2^l, \; m(1) = 2 \quad (l \geq 2)$$

is the length of the word in $c$ where all blocks of length $\leq l$ appear. We also write $c_k^n = [c_{k,1}^n, \ldots, c_{k,n}^n]$. For the proof we need the following two elementary lemmas.

Lemma 1: There exists a constant $K$ such that: For every $n \in \mathbb{N}$ there exists a sequence $t \in \{0,1\}^{\mathbb{N}}$ with the following properties:
(1) $-2n \leq S_j(c \circ t) \leq 2n$ for all $j \in \mathbb{N}$.
(2) $\sum_{j=m(i-1)}^{m(i)} t_j \geq 2Ki2^i$ for all $i \geq 5n$.
(3) $\sum_{j=k}^{k+n-1} 1_{[01]\cup[10]}(t_j, t_{j+1}) \leq 1$ for every $k \in \mathbb{N}$.
Proof: Let

$$A_l = \{c_k^{5n} : 1 \leq k \leq 2^{5n};\ \sum_{j=1}^{2l}(1 - 2c_{k,j}^{5n}) = 0;\ \sum_{j=1}^{2l+s}(1 - 2c_{k,j}^{5n}) \neq 0\ \forall 1 \leq s \leq 4n - 2l\}$$

for $l = n, ..., 2n$. Using Theorem III.4.1 in [5] (p. 75), it is easy to see that $\mathrm{card} A_l = 2^n \binom{2l}{l}\binom{4n-2l}{2n-l}$.

Let $i \in \mathbb{N}$. Write $i = p5n + q$ with $p \geq 0$ and $1 \leq q < 5n$. Divide $c_k^i$ into $p$ $5n$–blocks and one $q$–block: $c_k^i = a_1 a_2 ... a_p b$. If $a_s \in A_l$ for some $2n \leq l \leq 4n$ and some $1 \leq s \leq p$, define $t_j = 1$ for $j - (m(i-1) + (k-1)i + (s-1)5n) \in \{1, ..., 2l\}$, and put $t_j = 0$ in all other cases.

By construction, if $t_{j_1} = t_{j_2} = 0$ and $u_r = \sum_{1 \leq l \leq t_{j_r}} t_l$ $(r = 1, 2)$ then $S_{u_2}(c \circ t) - S_{u_1}(c \circ t) = 0$. This implies (1). (3) is also obvious by construction. Using Stirling's formula, (2) follows from

$$\begin{aligned}\sum_{j=m(i-1)+1}^{m(i)} t_j &= p2^{i-5n} \sum_{l=n}^{2n} 2l \,\mathrm{card} A_l \\ &\geq p2^{i-5n} \sum_{l=n}^{2n} 2l \frac{2^{2l}2^{4n-2l}}{\pi\sqrt{l(2n-l)}\exp[(2l)^{-1} + (2n-l)^{-1}]} \\ &\geq 2Ki2^i\end{aligned}$$

for some constant $K$. (In fact $K^{-1} = 15\pi\exp(2)$.) ◇

Lemma 2: For every $n \in \mathbb{N}$ there exist sequences $\sigma^{\pm} \in \{0,1\}^{\mathbb{N}}$ such that
(1) $\liminf_{j\to\infty} j^{-1} S_j(c \circ \sigma^+) \geq (2n+1)^{-1}$ and
$\limsup_{j\to\infty} j^{-1} S_j(c \circ \sigma^-) \leq -(2n+1)^{-1}$.
(2) $\sum_{j=m(i-1)+1}^{m(i)} \sigma_j^{\pm} \geq 2K^{-1} i 2^i$ $(i \geq 4n)$.
(3) $\sum_{j=k}^{k+n-1} 1_{[01]\cup[10]}(\sigma_j^{\pm}, \sigma_{j+1}^{\pm}) \leq 1$ for every $k \in \mathbb{N}$.
Proof: The proof is similar to the previous one. For $i = p4n + q$, write $c_k^i$ as a concatenation of $p$ $4n$–blocks and 1 $q$–block. If such a $4n$–block $a$ (at the position $l$, say) satisfies $\sum_{j=1}^{2n+1}(1 - 2a_j) > 0$ put $\sigma_j^+ = 1$ for $j - m(i-1) - (k-1)i - (l-1)4n \in \{1, ..., 2n+1\}$. In all other cases set $\sigma_j^+ = 0$.

Since by construction $\sigma^+$ only chooses pieces of $c$ where the sum of the Rademacher function is $> 0$, we obtain for $j \geq 1$

$$j^{-1} S_j(c \circ \sigma^+) \geq j^{-1}\left[\frac{j}{2n+1}\right] - \frac{n}{j}$$

and hence (1). (3) is obvious by construction and (2) follows from

$$\sum_{j=m(i-1)+1}^{m(i)} \sigma_j^+ = 2^{i-2n-1} p(2n+1) \sum_{j=n+1}^{2n+1} \binom{2n+1}{j}$$
$$= 2^{i-2n-1} p(2n+1) 2^{2n} \geq \frac{1}{8} i 2^i.$$

The proof for $\sigma^-$ is analoguous. ◇

Proof of Theorem 2: We first define the family of selection sequences $\theta^r$, where $r \in \{0,1\}^{\mathbb{N}}$.

Choose $\rho(k)$ such that for $n \geq \rho(k)$, $f_j(n) \geq 4k^2$ ($j = 1,2$). Denote by $t^n$ and $\sigma^{\pm n}$ the sequences constructed in Lemmas 1 and 2. Let $T_s(t) = \sum_{j=1}^{s} t_j$ for $s \geq 1$ and $t \in \Sigma_2$.

To start with choose $l(2) \geq 10$ so large that $T_{m(l(2))}(t^1) \geq \max\{\rho(2), Km(l(2))\}$. Define $\theta_j^r$ for $j \leq m(l(2))$ by

$$\theta_j^r = \begin{cases} t_j^1 & 1 \leq j \leq m(l(2)) - 3l(2) \\ 1 & j = m(l(2)) - 2l(2) - s \quad \text{for } s = 1,2 \text{ if } r_1 = 0 \\ 1 & j = m(l(2)) - l(2) - s \quad \text{for } s = 1,2 \text{ if } r_1 = 1 \\ 0 & \text{otherwise} \end{cases}$$

Since $\lim_{j\to\infty} j/f_1(j) = \infty$ and by the choice of $\sigma^{+1}$, there is $\nu$ so that

$$\frac{S_{3\nu}(c \circ \sigma^{+1}) - S_{m(l(2))}(c \circ \sigma^{+1})}{f_1(T_{m(l(2))}(\theta^r))} \geq 1$$
$$\frac{S_{3(\nu-1)}(c \circ \sigma^{+1}) - S_{m(l(2))}(c \circ \sigma^{+1})}{f_1(T_{m(l(2))}(\theta^r))} < 1.$$

Choose $l^*(2)$ so that $T_{l^*(2)}(\sigma^{+1}) = 3\nu$ and define

$$\theta_j^r = \sigma_j^{+1} \quad (j = m(l(2)) + 1, ..., l^*(2)).$$

Similarly one finds $l^{**}(2) > l^*(2)$ replacing $\sigma^{+1}$ by $\sigma^{-1}$ and $f_1$ by $f_2$, and one can define $\theta_j^r$ for $l^*(2) < j \leq l^{**}(2)$.

Finally there is an $\bar{l}(2)$ such that $T_{\bar{l}(2)}(\sigma^{+1}) = 3\nu$ for some $\nu$ so that

$$0 \leq S_{3\nu}(c \circ \sigma^{+1}) - S_{l^{**}(2)}(c \circ \sigma^{+1}) + S_\mu(c \circ \theta^r) \leq 2,$$

where $\mu = \sum_{j=1}^{l^{**}(2)} \theta_j^r$. Then define $\theta_j^r = \sigma_j^{+1}$ for $l^{**}(2) < j \leq \bar{l}(2)$.

In order to complete the induction, assume that $l(2k-2) < l^*(2k-2) < l^{**}(2k-2) < \bar{l}(2k-2)$ and $(\theta_j^r)_{1 \leq j \leq \bar{l}(2)}$ are determined so that $T^{k-1} := T_{\bar{l}(2k-2)}(\theta^r)$ and $S^{k-1} := \sum_{j=1}^{\bar{l}(2k-2)} (c \circ \theta^r)_j$ are independent of $r \in \{0,1\}^{\mathbb{N}}$.

Choose $l(2k) > l(2k-2)$ such that the following conditions hold:

(1)

$$T := T^{k-1} + \sum_{j=\bar{l}(2k-2)+1}^{m(l(2k))} t_j^k \geq \max\{\rho(k+1); Km(l(2k))\}.$$

(2) $l(2k) \geq 5(k+1)$.
(3) $m(l(2k)) \geq \bar{l}(2k-2) + 3l(2k)$.
(4) $4(2k+1)f_s(T+2) < (1/k)(T+2)$.

Then we proceed as in the case $k = 1$ replacing the $3\nu$ appearing by $(2k+1)\nu$. This finishes the definition of the $\theta^r$, $r \in \{0,1\}^{\mathbb{N}}$.

Let $\theta^r$ be a sequence 'ust defined. Let $s \geq 1$ and choose $i$ so that $m(i) \leq s < m(i+1)$. Choose $k$ so that $l(2k) \leq i < l(2k+2)$. Then

$$\begin{aligned}\frac{1}{s}\sum_{j=1}^{s}\theta_j^r &\geq \frac{1}{m(i+1)}\sum_{j=1}^{m(i)}\theta_j^r \\ &= \frac{1}{m(i+1)}\left(\sum_{j=1}^{m(l(2k))}\theta_j^r + \sum_{j=m(l(2k))+1}^{m(i)}\theta_j^r\right) \\ &\geq \frac{1}{m(i+1)}\left(Km(l(2k)) + 2K\sum_{j=l(2k)+1}^{i-3} j2^j\right)\end{aligned}$$

by the choice of $l(2k)$. Therefore

$$\limsup_{s\to\infty} s\left(\sum_{j=1}^{s}\theta_j^r\right)^{-1} \leq 16K^{-1} < \infty.$$

We now prove that each $\theta^r$ is completely deterministic. Let $m \in \mathcal{M}(\theta^r)$. Then for a cylinder $a = [a_1, ..., a_l] \neq [0, ..., 0]$, $[1, ..., 1]$ (0 and 1 appear exactly $l$ times)

$$\begin{aligned}&\limsup_{s\to\infty}\frac{1}{s}\sum_{j=0}^{s-1} 1_a((\theta_\mu^r)_{\mu\geq j}) \\ &\leq \limsup_{s\to\infty}\frac{1}{s}\sum_{j=m(l(2k))+1}^{s-1} 1_{[01]\cup[10]}((\theta_\mu^r)_{\mu\geq j}) \leq \frac{1}{k+1}\end{aligned}$$

for every $k \geq 1$. Hence $m(a) = 0$ and for any $l \geq 1$

$$h_m = h_m(\{[0, ..., 0];\ [1, ..., 1]\}) \leq \frac{\log 2}{l}.$$

Hence $\theta^r$ is completely deterministic and, by the theorem of Kamae and Weiss ([6], [11]), every $c \circ \theta^r$ ($r \in \{0,1\}^{\mathbb{N}}$) is normal to the base 2.

It is left to show that

$$\limsup_{n\to\infty}\frac{S_n(c\circ\theta^r)}{f_1(n)} = 1.$$

(The proof for lim inf is analoguous and omitted.)

Let

$$\mathcal{E}_k = \{j \in \mathbb{N} : T_{m(l(2k))}(\theta^r) < j \leq T_{m(l(2k))}(\theta^r) + 4(2k+1)f_1\left(T_{m(l(2k))}(\theta^r)\right)\}.$$

By the choice (condition (4)) of $l(2k)$ we have

$$\lim_{j \in \mathcal{E}_k; j \to \infty} \frac{T_{m(l(2k))}(\theta^r)}{j} = 1,$$

hence

$$\lim_{j \in \mathcal{E}_k; j \to \infty} \frac{f_1\left(T_{m(l(2k))}(\theta^r)\right)}{f_1(j)} = 1.$$

It follows that

$$\begin{aligned} &\limsup_{j \to \infty} \frac{S_j(c \circ \theta^r)}{f_1(j)} \\ &= \limsup_{j \in \mathcal{E}_k; k \in \mathbb{N}; j \to \infty} \frac{S_j(c \circ \theta^r)}{f_1(T_{m(l(2k))}(\theta^r))} \\ &= 1 + \limsup_{k \to \infty} \alpha_k \frac{2k+1}{f_1(T_{m(l(2k))}(\theta^r))} = 1, \end{aligned}$$

for some $\alpha_k \in [0, 1]$, since $f_1(T_{m(l(2k))}(\theta^r)) \geq (2k+2)^2$. $\diamond$

## References


[1] Borel, E.: Les probabilités dénombrables et leurs applications arithmétiques. Rend. Circ. Math. Palermo **27**, (1909), 247–271.

[2] Champernowne, D.G.: The construction of decimals normal in the scale of ten. J. London Math. Soc. **52**, (1933), 254–260.

[3] Copeland, A.H.; Erdös, P.: Note on normal numbers. Bull. Amer. Math. Soc. **52**, (1946), 857–860.

[4] Davenport, H.; Erdös, P.: Note on normal decimals. Canad. J. Math. **4**, (1952), 58–63.

[5] Feller, W.: An Introduction to Probability Theory and its Applications. Vol. 1, 7th edition, Wiley, New York–London 1962.

[6] Kamae, T.: Subsequences of normal sequences. Israel J. Math. **16**, (1973), 121–149.

[7] Pillai, S.S.: On normal numbers. Proc. Indian Acad. Sci. Sect. A **10**, (1939), 13–15.

[8] Pillai, S.S.: On normal numbers. Proc. Indian Acad. Sci. Sect. A **12**, (1940), 179–184.

[9] Postnikov, A.G.: Arithmetic modelling of stochastic processes. Sel. Transl. in Statistics and Probability **13**, (1973), 41–122.

[10] Shiokawa, I.; Uchiyama, S.: On some properties of the dyadic Champernowne numbers. Acta Math. Acad. Sci. Hun. **26**, (1975), 9–27.

[11] Weiss, B.: Normal sequences as collectives. Proc. Symp. on Topological Dynamics and Ergodic Theory, Univ. of Kentucky, 1971.

# The Dichotomy of Hausdorff Measures and Equilibrium States for Parabolic Rational Maps

MANFRED DENKER
UNIVERSITY OF GÖTTINGEN
INSTITUT FÜR MATHEMATISCHE STOCHASTIK, LOTZESTR. 13,
W–3400 GÖTTINGEN, GERMANY
AND
MARIUSZ URBAŃSKI[1]
UNIVERSITY OF TORUŃ
INSTYTUT MATEMATYKI, UL. CHOPINA 12/18, 87–100 TORUŃ, POLAND

Abstract: Let $T : \overline{\mathbb{C}} \to \overline{\mathbb{C}}$ be a rational parabolic map of the Riemann sphere $\overline{\mathbb{C}}$ and let $f : \overline{\mathbb{C}} \to \mathbb{R}$ be a Lipschitz continuous function satisfying $P(T,f) > \sup_{x\in\overline{\mathbb{C}}} f(x)$, where $P(T,f)$ denotes the topological pressure. We show that the only equilibrium state for $T$ and $f$ is singular with respect to Hausdorff measures defined by functions of the upper class, and it is absolutely continuous with respect to Hausdorff measures defined by functions of the lower class.

## §1 INTRODUCTION

Let $T : \overline{\mathbb{C}} \to \overline{\mathbb{C}}$ be an analytic endomorphism of the Riemann sphere $\overline{\mathbb{C}} = \mathbb{C} \cup \{\infty\}$ of degree $d \geq 2$. Such a map is called rational since, restricted to $\mathbb{C}$, it can be represented by a quotient of two relatively prime polynomials of maximal degree $\geq 2$. Denote by $J(T)$ the Julia set of $T$, which is the set of points in $\overline{\mathbb{C}}$ where the family of positive iterates of $T$ is not normal in some neighbourhood. Basic facts about rational maps are contained in [2], [4], [13], [15], [16], [24], [25].

Rational maps $T$ can be classified according to the values of the derivative of $T$ on their Julia sets $J(T)$. A map $T$ is called hyperbolic or expanding if there exist $\lambda > 1$ and $n \geq 1$ such that for any point $x \in J(T)$, $|(T^n)'(x)| \geq \lambda$ (in the spherical metric). This notion is stronger than (positive) expansiveness which means that there exists $\beta > 0$ such that for any two distinct points $x, y \in J(T)$, $\sup_{n\geq 0} \text{dist}(T^n(x), T^n(y)) \geq \beta$ in the spherical metric (denoted by "dist"). Rational maps which are expansive but not expanding are called parabolic. It is shown in [8], that a rational map is parabolic if and only if the Julia set contains a rationally indifferent periodic point and no critical point. Thus the hyperbolic ones are those having neither a rationally indifferent periodic point

[1]Research supported by SFB 170, University of Göttingen

nor a critical point in $J(T)$. (A rationally indifferent point $x \in \overline{\mathbb{C}}$ is a periodic point (with some period $p$) such that $(T^p)'(x)$ is a root of 1.)

If $T : \overline{\mathbb{C}} \to \overline{\mathbb{C}}$ is hyperbolic (expanding) on its Julia set $J(T)$, then $T : J(T) \to J(T)$ is a mixing repeller, whence there are Markov partitions of arbitrarily small diameters, so that the projection of the associated topological Markov shift is Hölder continuous and almost everywhere 1–1 (see [40]). Thus the ergodic theory of subshifts of finite type applies (see [3], [5]). Especially, there is a dichotomy of equilibrium states and Hausdorff measures ([37]).

In all other cases the measurable dynamics of rational maps on their Julia set is not described by a well known theory. Recently, some progress has been made on the existence and uniqueness of equilibrium states and of conformal measures. Generally, rational maps are asymptotically entropy expansive, in particular for any continuous function there exists an equilibrium state ([26]). This measure is unique in many cases ([26], [28] and [7], also [36]).

In [7], one of the new ideas is a construction of conformal measures. The concept of conformal measures plays an important role in dynamics and ergodic theory. Patterson ([32], [33]) first emphasized it for limit sets of discrete groups acting on hyperbolic spaces and Sullivan ([41]) extended this notion to Julia sets of rational maps. It is introduced and studied for general dynamical systems in [6].

As in the case of a hyperbolic rational map, for a parabolic rational map there are Markov partitions of arbitrarily small diameters as well. However, the projection from the associated subshift to $J(T)$ is no longer Hölder but merely uniformly continuous ([9]). For this reason, the Renyi class of cylinder sets is not everything and it is necessary to apply the Schweiger formalism within Markov fibred systems. This theory is developed in full generality in [9] and [1]. Using this approach for the equilibrium states of Lipschitz continuous $f$ satisfying $P(T, f) > \sup_{x \in \overline{\mathbb{C}}} f(x)$, we show that probabilistic laws hold, in particular upper and lower class results, for Schweiger's jump transformation $T^*$ and for the unique $T^*$–invariant probability measure $q_f$ equivalent to the equilibrium measure $\mu_f$. It follows that the basic approach in [37] applies in the present situation to show that $\mu_f$ is either absolutely continuous or singular with respect to the Hausdorff measures defined respectively by functions of the lower and upper class. Precisely, first we prove that $\sigma^2$ the asymptotic variance of the function

$$\sum_{k=0}^{n-1}(f - P(T,f) + \kappa \log |T'|)(T^k(x)) \qquad (T^*(x) = T^n(x))$$

is positive, where $\kappa$ denotes the Hausdorff dimension of $\mu_f$. Let now $\phi$ be a positive, non–decreasing function. Set

$$\widetilde{\phi}(t) = t^\kappa \exp\left(\frac{\sigma}{\sqrt{\chi}}\phi(-\log t)\sqrt{-\log t}\right),$$

where $\chi$ denotes the Lyapunov exponent of $q_f$, and denote $H_{\widetilde{\phi}}$ the Hausdorff measure defined by $\widetilde{\phi}$ (see [14], [38]). Our main result is that $\mu_f$ is absolutely continuous with respect to $H_{\widetilde{\phi}}$ if $\phi$ belongs to the lower class and is singular with respect to $H_{\widetilde{\phi}}$ if $\phi$ belongs to the upper class.

This result has been motivated by recent investigations of harmonic measures on the boundary of an open topological disc in the complex plane (see [27], [37]). The paper [27] deals with the classical case (no dynamics involved) and [37] assumes the presence of holomorphic dynamics, treating the harmonic measure on the boundary of an RB-domain as a particular case of a measure which can be obtained as the image under a special coding of equilibrium states on subshifts of finite type. In the hyperbolic case, also the equilibrium states of Hölder continuous functions form a subclass of such images. Although it does not seem to be the case for parabolic maps, our result shows that still the full dichotomy appears. For expanding maps of the circle we studied the related problem in [6].

In Section 2 we give the necessary background to our result and show that equilibrium states can be represented as Markov fibred systems in the sense of [1]. In Section 3 we collect some probabilistic results for Schweiger's jump transformation. Finally, the dichotomy will be proved in Section 4.

## §2 Equilibrium states for parabolic rational maps

We begin with a brief survey on equilibrium states for general rational maps $T : \overline{\mathbb{C}} \to \overline{\mathbb{C}}$. Let $J(T)$ denote the Julia set of $T$. The first result has been derived in [26].

<u>Theorem 2.1:</u> (Lyubich) $T$ is asymptotically entropy expansive.

The notion of asymptotic entropy expansiveness was introduced in [31]. This property implies in particular that for every continuous function $f \in C(J(T))$ there exists an equilibrium state $\mu_f$ (see [31]), hence by the variational principle

$$P(T,f) = h_{\mu_f}(T) + \int_{\overline{\mathbb{C}}} f \, d\mu_f,$$

where $P(T,f)$ denotes the topological pressure of $f$ (see [45], [5]).

In particular, this applies to the topological entropy $h_{\text{top}}(T)$ (when $f = 0$), which, by a result of Gromov ([17]) and Lyubich ([26]), is equal to log $d$, where $d$ is the degree of the rational map. In this case, the uniqueness and the Bernoulli property has been obtained in [26], [28] and [29].

<u>Theorem 2.2:</u> (Lyubich, Mañé) The measure maximizing entropy is unique and isomorphic to a one–sided Bernoulli shift.

The uniqueness statement is due to Lyubich and Mañé, the Bernoulli property to Mañé alone. A general theorem on uniqueness of equilibrium states can be found in [7], another proof appeared later in [36].

Theorem 2.3: (Denker, Urbański) Let $f : \overline{\mathbb{C}} \to \mathbb{R}$ be a Hölder continuous function satisfying

$$P(T,f) > \sup_{x\in\overline{\mathbb{C}}} f(x).$$

Then there exists a unique equilibrium state for $f$.

The method of proof of the last theorem uses the techniques of conformal measures. In case of $f = 0$ (topological entropy), the maximal measure is already conformal; this makes the proof of the first part in Theorem 2.2 much simpler than that of Theorem 2.3. The definition of conformal measures is due to Patterson ([32]) in the case of discrete groups acting on hyperbolic spaces. It was extended to rational maps by Sullivan ([41]). The general definition for transformations can be found in [6]. Let $S$ be a measurable map acting on the measurable space $(X,\mathcal{B})$, and let $f : X \to \mathbb{R}$ be a measurable function. A measure $m$ (normalized) is called $f$–conformal, if for every special set $A \in \mathcal{B}$ (that is $S : A \to S(A)$ is a measurable isomorphism and $S(A) \in \mathcal{B}$)

$$m(S(A)) = \int_A f\, dm.$$

This definition is equivalent to the property that the Jacobian of the measure $m\circ S$ with respect to $m$ is given by $f$. In most cases, a conformal measure is determined by a fixed point of some Perron–Frobenius (transfer or dual) operator acting on a suitable space of measurable functions (for example, continuous, Hölder continuous or of bounded variation). In case of a rational map $T$, a $t$–conformal measure in the sense of Sullivan is a $|T'|^t$–conformal measure in the sense above, concentrated on the Julia set $J(T)$. Since $J(T)$ is completely invariant, a conformal measure for $T : J(T) \to J(T)$ is also conformal for $T : \overline{\mathbb{C}} \to \overline{\mathbb{C}}$. (This is not true for maps in general, if the subsystem is only forward invariant.) [6] contains a general construction principle to obtain conformal measures.

The existence of $t$–conformal measures is shown in [41].

Theorem 2.4: (Sullivan) For any rational map $T$ there exist $0 < t \le 2$ and a $t$–conformal measure $m_t$.

In case of a hyperbolic rational map it is also shown in [41] that there exists only one $t$–conformal measure and necessarily $t = h$, where $h = HD(J(T))$ denotes the Hausdorff dimension of the Julia set. The measure $m_h$ is, up to a multiplicative constant, equal to the $h$–dimensional Hausdorff measure and there exists an equivalent $T$–invariant measure $\mu_h$ which is the unique equilibrium state for $-h\log|T|$. Clearly $P(T,-h\log|T'|) = 0$ (see [40]). For a definition of Hausdorff measures and dimension we refer to [14] and [38].

Denote by $d(T)$ the dynamical dimension of $T$. This is defined as follows. Let $HD(\mu) = \inf\{HD(Y) : \mu(Y) = 1\}$ denote the Hausdorff dimension of the $T$–invariant, ergodic measure $\mu$ with positive metric entropy $h_\mu(T)$. Then $d(T)$ is the supremum of $HD(\mu)$ for such measures. We also define

$$s(T) = \inf\{t \ge 0 : P(T, -t\log|T'|) = 0\}$$

and

$$\delta(T) = \inf\{t \geq 0 : \exists t\text{–conformal measure } m_t\}.$$

Theorem 2.5: (Denker, Urbański, Ruelle, Sullivan) Let $T$ be hyperbolic or parabolic. Then

$$h = s(T) = \delta(T) = d(T).$$

In the hyperbolic case, this follows essentially from [41] and [40] (based on Bowen's work), and in the parabolic case it is in [8]. The first equation of the formula in Theorem 2.5 is known as the Bowen–Manning–McCluskey formula.

It also holds in some other cases of rational maps, for example, for subexpanding rational maps ([11]), which is a special case when the Julia set contains a critical point. Hence the definition of pressure is not canonical, but is well defined using the $K(V)$–method (cf. [43], [44]). For any rational map $T$ one knows that $d(T) = s(T) \leq \min\{h, \delta(T)\}$ ([10]).

Conjecture: Theorem 2.5 holds for any rational map.

Denote by $H_\phi$ the Hausdorff measure associated to the function $\phi : \mathbb{R}_+ \to \mathbb{R}_+$ ([14], [38]). If $\phi(t) = t^c$ for some constant $c$, we denote the Hausdorff measure $H_\phi$ by $H_c$. We also denote $\Pi_t$ the $t$–dimensional packing measure on $J(T)$ (see [42]). In case of a parabolic rational map, Theorem 2.4 can be formulated more precisely.

Theorem 2.6 (Aaronson, Denker, Urbański) If $T$ is parabolic, then there exists a non-atomic $h$–conformal measure $m_h$, where $h$ denotes the Hausdorff dimension of $J(T)$. The measure $m_h$ is the only non–atomic $t$–conformal measure for some $t \geq 0$. Moreover, if $h < 1$, then $H_h = 0$, if $h \geq 1$, then $m_h = cH_h$ for some constant $c \in \mathbb{R}_+$, if $h > 1$, then $\Pi_h(J(T)) = \infty$ and if $h \leq 1$, then $m_h = d\Pi_h$ for some constant $d \in \mathbb{R}_+$.

The first part of the theorem is proved in [8], [9] and [1]. The second part in [8] and the last part in [12]. For a large class of functions $f$ the existence of $f$–conformal measures for rational maps is derived in [7].

Theorem 2.7: (Denker, Urbański) Let $f : \overline{\mathbb{C}} \to \mathbb{R}$ be a Hölder continuous function satisfying

$$P(T, f) > \sup_{x \in \overline{\mathbb{C}}} f(x).$$

Then there exists a unique $\exp(P(T, f) - f)$–conformal measure $m_f$. The measure $m_f$ is non–atomic and the equilibrium state $\mu_f$ is the unique $T$–invariant equivalent measure and has a continuous Radon–Nikodym derivative.

In fact, Przytycki succeeded to strengthen the continuity part by determining the modulus of continuity for the density ([36]). His and our proof in [7] are based on the analysis of the associated Perron Frobenius operator

$$\widehat{T}\phi(x) = \sum_{T(y)=x} \phi(y)\exp(f(y) - P(T,f)).$$

This operator is almost periodic ([7], also [36]) for any rational map. Further properties in the case of a parabolic map will be given below.

In the remaining part of the paper, $T : \overline{\mathbb{C}} \to \overline{\mathbb{C}}$ denotes a parabolic rational map and let $f : \overline{\mathbb{C}} \to \mathbb{R}$ be Lipschitz continuous satisfying

$$P(T,f) > \sup_{x \in \overline{\mathbb{C}}} f(x).$$

Denote the set of rationally indifferent periodic points by $\Lambda$. It is well known that this set is finite and contained in the Julia set $J(T)$. For a partition $\mathcal{A}$ let

$$\mathcal{A}_0^n = \mathcal{A} \vee T^{-1}\mathcal{A} \vee \ldots \vee T^{-n}\mathcal{A}.$$

In [9] it has been shown that a parabolic rational map $T : J(T) \to J(T)$ admits Markov partitions of arbitrary small diameters (see [39] for a definition). Since $m_f$ is non-atomic, the result can also be formulated as

<u>Theorem 2.8:</u> Let $\beta > 0$ be an expansive constant such that for any pair of rationally indifferent periodic points $\omega \neq \omega' \in \Lambda$,

$$\mathrm{dist}(T(B(\omega, 2\beta)), \Lambda \setminus T(\{\omega\})) > 2\beta$$
$$\mathrm{dist}(B(\omega, 2\beta), B(\omega', 2\beta)) > 0.$$

Then there exists $0 < \delta < \beta$ and a Markov partition $\mathcal{A}$ such that

1) $\mathrm{diam} T(a) < \delta$ $(a \in \mathcal{A})$, in particular $T|_a$ is 1–1.
2) If $a \in \mathcal{A}$ and $T(a) \cap (J(T) \setminus B(\Lambda, \beta)) \neq \emptyset$, then all inverse branches $T_\nu^{-n}$ of $T^n$ : $\overline{\mathbb{C}} \to \overline{\mathbb{C}}$ are well defined and analytic on $B(T(a), 2\delta)$ for every $n \geq 1$.
3) If $\underline{a} \in \mathcal{A}_0^{n-1}$ satisfies $\mathrm{Int} T^n(\underline{a}) \cap (J(T) \setminus B(\Lambda, \beta)) \neq \emptyset$, then there exists a unique analytic inverse branch $T_\nu^{-n} : B(T^n(\underline{a}), 2\delta) \to \overline{\mathbb{C}}$ of $T^n$ such that $T_\nu^{-n}(T^n(\underline{a})) = \underline{a}$.
4) $m_f(\partial \mathcal{A}) = 0$.
5) $\max\{\mathrm{diam}\, \underline{a} : \underline{a} \in \mathcal{A}_0^n\} \to_{n \to \infty} 0$.

For completeness we state the following version of Köbe's Distortion Theorem which is used in the sequel (see [19]).

<u>Köbe Distortion Theorem:</u> (KDT) There exists a function $k : [0,1) \to [1,\infty)$ such that for $z \in \mathbb{C}$, $r > 0$, $t \in [0,1)$ and any univalent function $G : B(z,r) \to \overline{\mathbb{C}}$

$$\sup\{|G'(x)| : x \in B(z,tr)\} \leq k(t) \inf\{|G'(x)| : x \in B(z,tr)\}.$$

Using this theorem we immediately obtain the basic distortion property for parabolic rational maps, which permits to handle the system $(J(T), T, m_f)$ as a parabolic Markov fibred system (see the definition below). Denote by

$$S_n(\phi, T_0) = \sum_{i=0}^{n-1} \phi \circ T_0^i \qquad (n \geq 1)$$

the partial sums of $\phi$ with respect to a transformation $T_0$.

Lemma 2.9: There exists a constant $C$ with the following property: Let $x \in J(T)$ and suppose that for some $N \geq 1$, $T^N(x) \in J(T) \setminus B(\Lambda, \beta)$. Denote by

$$T_\nu^{-N} : B(T^N(x), 2\delta) \to \overline{\mathbb{C}}$$

the inverse branch of $T^N$ sending $T^N(x)$ to $x$. Then for all $y, z \in B(T^N(x), \delta)$

$$\frac{\exp[S_N(f - P(T,f), T)(T_\nu^{-N}(y))]}{\exp[S_N(f - P(T,f), T)(T_\nu^{-N}(z))]} \leq \exp(C\operatorname{dist}(y, z)).$$

Proof. Let $C_f$ denote a Lipschitz constant of $f$ and let $T_{\nu_i}^{-i}$ denote the unique holomorphic inverse branches defined on $B(T^N(x), 2\delta)$ sending $T^N(x)$ to $T^{N-i}(x)$ (see Theorem 2.8). Then for $y, z \in B(T^N(x), \delta)$

$$\begin{aligned} &|S_N(f - P(T,f), T)(T_\nu^{-N}(z)) - S_N(f - P(T,f), T)(T_\nu^{-N}(y))| \\ &\leq \sum_{i=0}^{N-1} |f(T^i(T_\nu^{-N} z))) - f(T^i(T_\nu^{-N}(y)))| \\ &= \sum_{i=1}^{N} |f(T_{\nu_i}^{-i}(z)) - f(T_{\nu_i}^{-i}(y))| \leq C_f \sum_{i=1}^{N} \operatorname{dist}(T_{\nu_i}^{-i}(z), T_{\nu_i}^{-i}(y)). \end{aligned}$$

If $N = \inf\{n \geq 0 : T^n(x) \in J(T) \setminus B(\Lambda, \beta)\}$, by Theorem 8.4 in [1] and Lemma 1 in [12], there exists a constant $K_0$ such that

$$\sum_{i=1}^{N} \operatorname{dist}(T_{\nu_i}^{-i}(z), T_{\nu_i}^{-i}(y)) \leq K_0 k(1/2) \operatorname{dist}(z, y) \sum_{i=1}^{N} i^{-(p+1)/p}$$

where $p \geq 1$ is some integer. Whence the lemma in this case with

$$C = K_0 k(1/2) C_f \sum_{n=1}^{\infty} n^{-(p+1)/p}.$$

In the general case, by Lemma 4.10 in [9], there exists $l_0$ such that for every

$$u \in J(T) \setminus \bigcup_{n \geq 0} T^{-n}\Lambda$$

$|(T^M)'(u)| \geq 2$ where $M$ denotes that integer for which $T^k(u)$ $(k \geq 1)$ visits $J(T) \setminus B(\Lambda, \beta)$ for the $l_0$-th time. Combinig this with the above estimate, it is easy to obtain the lemma. ◇

Let, for $n \in \mathbb{N}$ and $\underline{a} \in \mathcal{A}_0^{n-1}$,

$$\Delta(\underline{a}) = \Delta_{n-1}(\underline{a}) = \frac{dm_f \circ T^n}{dm_f}|_{\underline{a}} = \exp(nP(T,f) - S_n(f,T)).$$

Define $X = J(T) \setminus \bigcup_{n \geq 0} T^{-n}(\partial \mathcal{A})$ and

$$\mathcal{R} = \{a \cap X : \ a \in \mathcal{A}\},$$

and denote $\mathcal{R}_0^n = \mathcal{R} \vee T^{-1}\mathcal{R} \vee \ldots \vee T^{-n}\mathcal{R}$. Clearly $(\mathcal{R}, T, m_f)$ is a Markov fibred system, i.e.

$$\forall b \in \mathcal{R} \ \ T(b) = \bigcup_{B \in \mathcal{R}: B \cap T(b) \neq \emptyset} B,$$

and $T : b \to T(b)$ is invertible and nonsingular (with respect to $m_f$). Let

$$\mathcal{G}(C,T) = \{\underline{b} \in \mathcal{R}_0^n : \frac{\Delta_n(\underline{b})(x)}{\Delta_n(\underline{b})(y)} \leq C \ m \text{ a.e.}\}$$

and

$$\mathcal{R}_0 = \{\underline{b} \in \mathcal{R}_0^n : \ T^n(\underline{b}) \cap (J(T) \setminus B(\Lambda, \beta)) \neq \emptyset, \ n \geq 0\}.$$

Let

$$N_C(x) = \inf\{n \in \mathbb{N} : \ \underline{b}^n(x) \in \mathcal{R}_0\},$$

where $x \in \underline{b}^n(x) \in \mathcal{R}_0^{n-1}$ defines $\underline{b}^n(x)$ for $n \geq 1$. Then Schweiger's jump transformation $T^*$ is defined by

$$T^*(x) = T^{N_C(x)}(x).$$

It follows that $T^* = T^N$ on $\{N_C = N\}$ and that $\mathcal{R}^* = \bigcup_{N \geq 1} \mathcal{R}_0^{N-1} \cap \{N_C = N\}$ is a Markov partition for $T^*$. Hence $(\mathcal{R}^*, T^*, m_f)$ is also a Markov fibred system.

For a function $g : J(T) \to \mathbb{R}$ set $g^*(x) = g(x) + g(T(x)) + \ldots + g(T^{N_C(x)-1}(x))$. In particular, $P^*(x) = P(T,f)^*(x) = N_C(x)P(T,f)$.

Lemma 2.10: If $C$ is large enough, $(\mathcal{R}, T, m_f)$ has the Schweiger property with respect to $\mathcal{R}_0$. i.e.

1.) $\mathcal{R}_0 \subset \mathcal{G}(C,T)$.
2.) If $\underline{b} \in \mathcal{R}_0, \underline{\beta} \in \mathcal{R}_0^\infty$, then $\underline{\beta}\underline{b} \in \mathcal{R}_0$.
3.) $\bigcup_{\underline{b} \in \mathcal{R}_0} \underline{b} = J(T) \quad m_f$ a.e.

In particular, $\mathcal{R}_0$ generates the $\sigma$–algebra.

Proof. This follows immediately from Lemma 2.9.

Recall from [1] that a Markov fibred system is called aperiodic if the associated incidence matrix is aperiodic (equivalently that the associated Markov shift is topologically mixing), and is called parabolic if

$$N_C \circ T = N_C - 1 \quad \text{on } \{N_C \geq 2\},$$

$$|\mathcal{R}_0^1 \cap \{N_C = 2\}| < \infty,$$

$$T : \{N_C \geq 2\} \to T(\{N_C \geq 2\})$$

is invertible and

$$T(\{N_C = 1\} \setminus T(\{N_C = 2\})) = X.$$

As in [1] and [9] we can prove

Theorem 2.11: $(\mathcal{R}, T, m_f)$ is a finite, aperiodic, parabolic fibred system having the Schweiger property with respect to $\mathcal{R}_0$.
Proof. Note that by Lemmas 9.4 (1) and 9.5 in [1] $(\mathcal{R}, T, m_f)$ is a finite, aperiodic, parabolic fibred system. The Schweiger property is shown in Lemma 2.10. ◊

In [1] a general theory of Markov fibred systems is developed. Especially, the following results follow from the general theory.

Theorem 2.12: There exists a unique $T^*$–invariant probability measure $q_f$ equivalent to $m_f$. Moreover,

$$\mu'(B) = \sum_{k=0}^{\infty} q_f(T^{-k}B \cap \{N_C > k\})$$

is a $T$–invariant finite measure and $\mu_f = \mu'/\mu'(J(T))$.
Proof. The existence of $q_f$ follows from Lemma 2.1 in [1]. The uniqueness of the measure $\mu_f$ in Theorem 2.7 ([7]) implies that $\mu_f = \mu'/\mu'(J(T))$. Note that by Theorem 5.5 in [1] $\mu'$ is a finite measure since, by assumption on $f$, $\sum n \exp[S_n(f,T) - nP(T,f)] < \infty$. An alternative argument is based on the fact, that there exists only one $T^*$–invariant measure absolutely continuous with respect to $m_f$. ◊

Let $\mathcal{A}$ be the Markov partition described in Theorem 2.8. For $0 \leq s \leq 1$ and a function $g : J(T) \to \mathbb{R}$ define

$$C_g := \max_{A \in \mathcal{A}} \sup_{x,y \in \overline{A}} \frac{|g(x) - g(y)|}{(\mathrm{dist}(x,y))^s}$$

and

$$||g||_s = \begin{cases} C_g + ||g||_\infty & (s > 0) \\ ||g||_\infty & (s = 0). \end{cases}$$

Let $\mathcal{H}(s)$ denote the Banach space of all functions $g$ with $||g||_s < \infty$. Since the projection map $\mathcal{R}^{*\mathbb{N}} \to J(T)$ is Hölder continuous (by Lemma 4.10 in [9]), each function in $\mathcal{H}(s)$ defines a Hölder continuous function on $\mathcal{R}^{*\mathbb{N}}$ (the metric on $\mathcal{R}^{*\mathbb{N}}$ is e.g. given by $\sum_{n \geq 1} 2^{-n} 1_{b_n \neq b'_n}$).

Theorem 2.13: The dual (transfer) operator $\widehat{T}^*$ of Schweiger's jump transformation $T^*$ : $L_1(m_f) \to L_1(m_f)$ is given by

$$\widehat{T}^*\psi(x) = \sum_{T^*(y)=x} \psi(y)\exp\big(S_{N_C(y)}(f - P(T,f), T)(y)\big)$$

and acts, for any fixed $s \geq 0$, on $\mathcal{H}(s)$.

(1) $\lambda = 1$ is the only eigenvalue of modulus 1 for $\widehat{T}^* : \mathcal{H}(0) \to \mathcal{H}(0)$ and its eigenspace $E$ has dimension 1.

(2) $\widehat{T}^* = Q_0 + Q_1$, where $Q_1$ is the projection onto the eigenspace $E$, where

$$\sup_{n\geq 1} ||Q_0^n|| < \infty \text{ and where } Q_1 Q_0 = 0.$$

(3) $Q_0$ acts on each $\mathcal{H}(s)$ $(0 < s \leq 1)$ and satisfies

$$||Q_0^n||_s \leq M_s q_s^n$$

for some constants $M_s > 0$ and $q_s \in (0,1)$.

Remark: The analoguous statement is true for $\widehat{T}^*$ acting as an operator on the spaces of Hölder continuous functions on $\mathcal{R}^{*N}$.

Proof. Note that the representation is a well known fact.

We first show that $\sup_{n\geq 1} ||\widehat{T}^{*n}1||_\infty < \infty$.

By Lemma 2.9, and since on $\{N_C \geq 2\}$ all inverse branches of $T^*$ agree with inverse branches of $T$ (which map $\{N_C \geq 2\}$ into $\{N_C = 1\}$), there exists a constant $C_1$ such that for $x, y \in A \in \mathcal{R}$

$$\widehat{T}^{*n}1(x) \leq C_1 \widehat{T}^{*n}1(y) \quad (n \geq 1).$$

It follows that

$$1 = \sum_{A\in\mathcal{R}} \int_A \widehat{T}^{*n}1 \, dm_f \geq C_1^{-1} \sum_{A\in\mathcal{R}} \sup_{x\in A} \widehat{T}^{*n}1(x) m_f(A)$$

for every $n \geq 1$. Therefore

$$||\widehat{T}^{*n}1||_\infty \leq C_1 \max_{A\in\mathcal{R}} \frac{1}{m_f(A)} < \infty$$

for every $n \geq 1$. Hence $\sup_{n\geq 1} ||\widehat{T}^{*n}||_\infty < \infty$.

Fix $0 < s \leq 1$.

Let $d \geq 1$ be a constant so that for $u, v \in \underline{a} \in \mathcal{R}^* \vee T^{*-1}\mathcal{R}^* \vee \ldots \vee T^{*-n}\mathcal{R}^*$

$$\mathrm{dist}(u,v) \leq d|(T^{*n})'(u)|^{-1}\mathrm{dist}(T^{*n}(u), T^{*n}(v)).$$

Using [9], Lemma 4.10 (the expanding property of $T^*$), choose $m$ and $\gamma < 1$ so that

$$\sup_{n\geq 1} ||\widehat{T}^{*n}1||_\infty d^s = \gamma \inf_{x\in X} |(T^{*m})'(x)|^s.$$

Let $g \in \mathcal{H}(s)$ and let $\mathcal{Z} = \{z\}$ denote the collection of inverse branches of $T^{*m}$ defined on $a \in \mathcal{R}$. By distortion, for $x, y \in a \in \mathcal{R}$,

$$\sum_{z \in \mathcal{Z}} |g(z(x)) - g(z(y))| \exp\Big( S_m(f^* - N_C P(T,f), T^*)(z(x))\Big)$$
$$\leq C_g d^s |(T^{*m})'(z(x))|^{-s} (\mathrm{dist}(x,y))^s ||T^{*m}1||_\infty$$
$$\leq \gamma C_g (\mathrm{dist}(x,y))^s.$$

It follows that for $x, y \in a \in \mathcal{R}$

$$|\widehat{T}^{*m} g(x) - \widehat{T}^{*m} g(y)|$$
$$\leq \sum_{z \in \mathcal{Z}} |g(z(x)) - g(z(y))| \exp(S_m(f^* - N_C P(T,f), T^*)(z(x)))$$
$$- \sum_{z \in \mathcal{Z}} |g(z(y))| \exp(S_m(f^* - N_C P(T,f), T^*)(z(y)))$$
$$\Big(1 - \exp\Big( S_m(f^* - N_C P(T,f), T^*)(z(x)) - S_m(f^* - N_C P(T,f), T^*)(z(y))\Big)\Big)$$
$$\leq \gamma C_g (\mathrm{dist}(x,y))^s + C_2 ||g||_\infty ||\widehat{T}^{*m}1||_\infty C_f \mathrm{dist}(x,y),$$

where $C_2$ denotes some constant. It follows that $\widehat{T}^*$ acts on the space $\mathcal{H}(s)$ and satisfies

$$C_{\widehat{T}^{*m} g} \leq \gamma C_g + C_3 ||g||_\infty,$$

for some constants $C_3 > 0$. Since $||g||_\infty$ is the norm of $\mathcal{H}(0)$

$$||\widehat{T}^{*m}||_s \leq \gamma ||g||_s + C_4 ||g||_0$$

for some constant $C_4 > 0$.

By definition of $||\cdot||_s$, it follows from the Arzéla–Ascoli theorem, that a bounded set in $\mathcal{H}(s)$ ($s > 0$)is compact in $\mathcal{H}(0)$. Since $\mathcal{H}(s)$ is dense in $\mathcal{H}(0)$ we can apply the theorem of Ionescu–Tulcea and Marinescu ([22]) in the same way as in [20] to obtain the theorem. Note that there is only one eigenvalue of modulus 1 with multiplicity 1 by Theorem 2.12. ◇

Corollary 2.14: The density $dq_f/dm_f$ is Lipschitz continuous and bounded on $\mathcal{R}^{*N}$, consequently $d\mu_f/dm_f$ is also Lipschitz continuous (and bounded).
Proof. The first assertion follows immediately from the previous theorem. The second follows from the representation of the density $d\mu'/dm_f$ by the density $h$ of $dq_f/dm_f$ (see [9]):

$$\frac{d\mu'}{dm_f}(x) = h(x) + \sum_{n=1}^{\infty} h(x_n) \exp(S_n(f,T)(x_n) - nP(T,f)),$$

where $x_n \in \{N_C = n\}$ satisfies $T^n(x_n) = x$. ◇

It is not clear, whether Corollary 2.14 implies that the density $d\mu_f/dm_f$ is Lipschitz continuous on $J(T)$, since the Markov partition is only defined on $J(T)$.

## §3 Upper and lower class results for parabolic rational maps

Let $\Gamma$ be a finite or countable measurable partition of the probability space $(Y, \mathcal{F}, \mu)$ and let $S : Y \to Y$ be a measure preserving transformation. For $0 \le a \le b \le \infty$, set $\Gamma_a^b = \bigvee_{a \le l < b} S^{-l}\Gamma$. We say that $\mu$ is absolutely regular with respect to the filtration defined by $\Gamma$, if there exist $\beta(n) \downarrow 0$ such that

$$\int_X \sup_a \sup_{A \in \Gamma_{a+n}^\infty} |\mu(A|\Gamma_0^a) - \mu(A)| \le \beta(n).$$

$(\beta(n) : n \ge 1)$ are called the coefficients of absolute regularity.

In this section let $T : \overline{\mathbb{C}} \to \overline{\mathbb{C}}$ be a parabolic rational map and let $f : \overline{\mathbb{C}} \to \mathbb{R}$ be Lipschitz continuous with $P(T, f) > \sup_{x \in \overline{\mathbb{C}}} f(x)$. Let $(\mathcal{R}, T, m_f)$ denote the Markov fibred system of Theorem 2.11 and $q_f$ the $T^*$–invariant probability measure obtained in Theorem 2.12.

Theorem 3.1: The measure $q_f$ is absolutely regular with respect to the filtration defined by $\mathcal{R}^*$. The coefficients of absolute regularity decrease to 0 at an exponential rate.
Proof: This follows from Theorem 2.13 and [Ry, §3 Theorem 5]. A similar statement is proved in [20]. ◇

Recall some facts from probability theory on function spaces in the situation of Theorem 3.1. Let $g : J(T) \to \mathbb{R}$ be a measurable function so that $g \in L_{2+\gamma}(q_f)$ for some $\gamma > 0$ and such that

$$\xi_n = \|g - E_{q_f}(g|(\mathcal{R}^*)_0^{n-1})\|_{2+\gamma} \le M n^{-2-\gamma}$$

for some constant $M > 0$ and all $n \ge 1$. Denote the space of these functions by $\mathcal{L}^*$. A Hölder continuous function on $\mathcal{R}^{*\mathbb{N}}$ belongs to $\mathcal{L}^*$ if it has moments of order $> 2$ (since $\mathcal{R}^{*\mathbb{N}}$ is not compact). It follows from Theorem 3.1, [21] and [34] that the process $(g \circ T^{*n} : n \ge 1)$ satisfies the central limit theorem and an a.s. invariance principle if

$$\sigma^2 = \sigma^2(g) = \int_{J(T)} (g - q_f(g))^2 \, dq_f + 2 \sum_{n=1}^{\infty} \int_{J(T)} (g - q_f(g))(g \circ T^{*n} - q_f(g)) \, dq_f$$

is strictly positive, where $q_f(\psi) = \int_{J(T)} \psi \, dq_f$ for a function $\psi : J(T) \to \mathbb{R}$. (This series is always absolutely convergent.) The latter theorem means that one can redefine the process $(g \circ T^{*n} : n \ge 1)$ on some probability space on which there is defined a standard Brownian motion $(B(t) : t \ge 0)$ such that for some $\lambda > 0$

$$\sum_{0 \le j \le t} [g \circ T^{*j} - q_f(g)] - B(\sigma^2 t) = O(t^{\frac{1}{2} - \lambda}) \quad q_f \text{ a.e.}$$

Let $\phi : [1,\infty) \longrightarrow \mathbb{R}$ be a positive non–decreasing function. $\phi$ is said to belong to the lower class if

$$\int_1^\infty \frac{\phi(t)}{t}\exp(-\frac{1}{2}\phi(t)^2) < \infty$$

and to the upper class if

$$\int_1^\infty \frac{\phi(t)}{t}\exp(-\frac{1}{2}\phi(t)^2) = \infty.$$

Well known results for Brownian motion imply (see Theorem A in [34], or [23]).

Theorem 3.2: Let $g \in \mathcal{L}^*$. Then

$$q_f(\{x \in X : \sum_{j=0}^{n-1}[g(T^{*j}(x)) - q_f(g)] > \sigma\phi(n)\sqrt{n} \text{ for infinitely many } n \in \mathbb{N}\})$$
$$= \begin{cases} 0 & \text{if } \phi \text{ belongs to the lower class,} \\ 1 & \text{if } \phi \text{ belongs to the upper class,} \end{cases}$$

where $\sigma^2$ is as above and assumed to be strictly positive.

Let

$$L_n(x) = \min\{l : \sum_{i=0}^{l} N_C((T^*)^i(x)) \geq n\}.$$

By the ergodic theorem, $L_n/n \to A^{-1}$ a.e., where $A = \mu'(J(T))$ (cf. Theorem 2.12). Let $g : J(T) \to \mathbb{R}$ be measurable. Set, as before,

$$g^*(x) = g(x) + g(T(x)) + ... + g(T^{n-1}(x)) \qquad (x \in \{N_C = n\},\ n \geq 1).$$

We mention the following three results without proof. They are not used in the sequel and can be proved along the lines of the proof of the central limit theorem in [1] or by methods similar to those used below.

Theorem 3.3: If $g \in L_2(\mu_f)$, $g^* - N_C\mu_f(g) \in \mathcal{L}^*$ and if $\sigma^2 = \sigma^2(g^* - N_C\mu_f(g)) > 0$, then

$$\mu_f(\{x \in J(T) : \sum_{j=0}^{n-1}[g(T^j(x)) - \mu_f(g)] > \frac{\sigma}{\sqrt{A}}\phi(n)\sqrt{n} \text{ for infinitely many } n \in \mathbb{N}\})$$
$$= \begin{cases} 0 & \text{if } \phi \text{ belongs to the lower class,} \\ 1 & \text{if } \phi \text{ belongs to the upper class.} \end{cases}$$

Theorem 3.4: Let $g \in L_2(\mu_f)$ satisfy $g^* - N_C\mu_f(g) \in \mathcal{L}^*$ and $\sigma^2 = \sigma^2(g^* - N_C\mu_f(g)) > 0$. Then

$$\limsup_{n\to\infty} \frac{S_n(g - \mu_f(g), T)}{\sqrt{2\sigma^2 A^{-1} n \log\log n}} = 1 \qquad \mu_f \text{ a.e.}$$

Theorem 3.5: Let $g : J(T) \to \mathbb{R}$ be such that $g^* - N_C\mu_f(g) \in \mathcal{L}^*$ is bounded and $\sigma^2 = \sigma^2(g^* - N_C\mu_f(g)) > 0$. Then

$$\lim_{n\to\infty} \mu_f(\{\frac{\sqrt{A}}{\sigma\sqrt{n}} S_n(g - \mu_f(g), T) \le t\}) = \frac{1}{\sqrt{2\pi}} \int_{-\infty}^{t} \exp(-u^2/2)\, du$$

for all $t \in \mathbb{R}$.

All these results apply to Lipschitz continuous functions on $J(T)$ (but they are not needed in the sequel) according to

Theorem 3.6: If $g : J(T) \to \mathbb{R}$ is Hölder continuous (in the spherical metric) then $g^* \in \mathcal{L}^*$. In particular $N_C \in \mathcal{L}^*$.
Proof. By conformality and since $\rho := -P(T, f) + \sup_{x\in\overline{\mathcal{C}}} f(x) < 0$, we have

$$m_f(\{N_C = k\}) = \int_{T^k\{N_C=k\}} \exp(S_k(f - P(T, f), T))\, dm_f \le \exp(k\rho) \quad (k \ge 2).$$

Since $|g^*(x)| \le N_C(x)||g||_\infty$ it follows that for $r \ge 1$

$$\int_{J(T)} |g^*|^r\, dm_f \le ||g||_\infty^r \left(1 + \sum_{k=2}^{\infty} k^r \exp(k\rho)\right) < \infty.$$

Since by Corollary 2.14, $dq_f/dm_f$ is bounded, it follows that $g^* \in L_r(q_f)$ for any $r \ge 1$.

As in the proof of Lemma 2.9, for $x, y \in \underline{a} \in (\mathcal{R}^*)_0^{k-1} \cap \{N_C = n\}$, it follows from KDT that

$$|g^*(x) - g^*(y)| \le \sum_{i=0}^{n-1} |g(T^i(x)) - g(T^i(y))| \le K_0 C_g \operatorname{diam}(\underline{a})^s \sum_{i=1}^{n} i^{(p+1)s/p},$$

where $s$ denotes the exponent of Hölder continuity and where $C_g$, $p$ and $K_0$ denote constants. Since, by Lemma 4.10 in [9], $\operatorname{diam}(\underline{a}) \le \alpha^k$ for some $0 < \alpha < 1$, we obtain for $n \le k$ and some constant $K_1$ $|g^*(x) - g^*(y)| \le K_1 k^{3s}\alpha^{ks}$. Hence for $r \ge 1$ ($K_2$, $K_3$ denote constants),

$$\begin{aligned}
&\int_X |g^* - E_{q_f}(g^*|(\mathcal{R}^*)_0^{k-1})|^r\, dq_f \\
&\le \sum_{\underline{a}\in(\mathcal{R}^*)_0^{k-1}\cap\{N_C\le k\}} q_f(\underline{a})^{-1} \int_{\underline{a}} |\int_{\underline{a}} g^*\, dq_f - q_f(\underline{a}) g^*|^r\, dq_f \\
&\quad + K_2 \int_{\{N_C>k\}} N_C^r ||g||_\infty^r\, dq_f \\
&\le K_1 k^{3rs}\alpha^{ks} + K_2 \sum_{i=k+1}^{\infty} i^r \exp(i\rho) \; = \; O(k^{-r}).
\end{aligned}$$

Thus $g^* \in \mathcal{L}^*$. In particular $N_C \in \mathcal{L}^*$, since $N_C = 1^*$. $\diamond$

## §4 Equilibrium states and Hausdorff measures

Let $T : \overline{\mathbb{C}} \to \overline{\mathbb{C}}$ be a parabolic rational map, $f : \overline{\mathbb{C}} \to \mathbb{R}$ a Lipschitz continuous function satisfying $P(T, f) > \sup_{x \in \overline{\mathbb{C}}} f(x)$ and let $\mathcal{A}$ denote the Markov partition of Theorem 2.8 of diameter smaller than $\delta$. Let $X = J(T) \setminus \bigcup_{n \geq 0} T^{-n}(\partial \mathcal{A})$, $N_C$, $T^*$, $\mathcal{R}^*$ and $\mathcal{L}^*$ as introduced in the previous sections. We begin with a simple observation using KDT.

Lemma 4.1: Let $U$ be a ball containing $x \in X$ and let $n \geq 1$ satisfy

$$|(T^{*n+1})'(x)|^{-1} < \frac{2k(1/2)\mathrm{diam}(U)}{\delta} \leq |(T^{*n})'(x)|^{-1}$$

and $T^{*n}(x) \in \{N_C = 1\}$. Define $N$ by $T^N(x) = T^{*n}(x)$. Then:

1) There exists a unique holomorphic inverse branch

$$T_\nu^{-N} : B((T^{*n}(x), 2\delta) \to \overline{\mathbb{C}}$$

such that $T_\nu^{-N}(T^{*n}(x)) = x$ and

$$T_\nu^{-N}(B(T^{*n}(x), \delta)) \supset U.$$

2) There exists $\zeta > 0$ (not depending on $x$ and $U$) such that for some $y \in \overline{\mathbb{C}}$

$$T^N(U) \supset B(y, \zeta).$$

If $U$ is a ball with center $x$, then one can choose $y = T^N(x)$.

Proof. Since $x \in X$, $x \in \underline{b}$ for some $\underline{b} \in (\mathcal{R}^*)_0^{n-1}$. By definition of $T^*$ it follows that $T^{*n}(\underline{b}) = T(b) \subset B(T^{*n}(x), \delta)$ for some $b \in \mathcal{R}$ and

$$T(b) \cap (J(T) \setminus B(\Lambda, \beta)) \neq \emptyset.$$

Hence by Theorem 2.8 there exists a unique holomorphic inverse branch

$$T_\nu^{-N} : B(T^{*n}(x), 2\delta) \to \overline{\mathbb{C}}$$

mapping $T^{*n}(x)$ to $x$.

By KDT $T_\nu^{-N}(B(T^{*n}(x), \delta))$ contains the ball of radius $\geq k(1/2)^{-1}\delta|(T^{*n})'(x)|^{-1}$ and center $x$. Since $\mathrm{diam}(U) \leq (1/2)k(1/2)^{-1}\delta|(T^{*n})'(x)|^{-1}$,

$$U \subset B(x, k(1/2)^{-1}\delta|(T^{*n})'(x)|^{-1}).$$

This proves 1).

By KDT again, $T^N(U)$ contains a ball of radius

$$(1/2)\mathrm{diam}(U)k(1/2)^{-1}|(T^{*n})'(x)| \geq (1/4)\delta k(1/2)^{-2}|T'(T^{*n}(x))|^{-1} \geq \zeta > 0$$

for some constant $\zeta$, since $T^*(T^{*n}(x)) = T(T^{*n}(x))$ for $T^{*n}(x) \in \{N_C = 1\}$. This proves 2). $\diamond$

Lemma 4.2: There exist constants $C_1, C_2 \geq 1$ with the following property: For $x \in X$ there exists $r(x) > 0$ such that for all balls $U$ with center $x \in J(T)$ and $\mathrm{diam}(U) \leq r(x)$, there exists $n \geq 1$ such that

$$C_1^{-1}|(T^{*n+1})'(x)|^{-1} \leq \mathrm{diam}(U) \leq C_1|(T^{*n})'(x)|^{-1}$$

and

$$C_2^{-1}\exp\big(S_{n+1}(f^* - P^*, T^*)(x)\big) \leq m_f(U) \leq C_2 \exp\big(S_{n-1}(f^* - P^*, T^*)(x)\big).$$

Proof. Choose $r(x)$ so that $2k(1/2)r(x) \leq \delta|(T^*)'(x)|^{-1}$. Fix a ball with diameter $\leq r(x)$. Since $|(T^{*l})'(x)|^{-1} \downarrow 0$ there exists $n \geq 1$ such that

$$|(T^{*n+1})'(x)|^{-1} < \frac{2k(1/2)\mathrm{diam}(U)}{\delta} \leq |(T^{*n})'(x)|^{-1}.$$

1st case: Let $T^{*n}(x) \in \{N_C = 1\}$.
Let $N$ be defined by $T^N(x) = T^{*n}(x)$. By Lemma 4.1 $T^N : U \to T^N(U) \subset B(T^{*n}(x), \delta)$ is invertible. By Lemma 2.9 and conformality

$$\begin{aligned} m_f(U) &= \int_{T^N(U)} \exp\big(S_N(f - P(T,f), T)\big)\, dm_f \\ &\in [C^{-1}, C] m_f(T^N(U)) \exp\big(S_N(f - P(T,f), T)(x)\big). \end{aligned}$$

Since $T^N(U) \supset B(T^N(x), \zeta)$ (Lemma 4.1), there exists $\xi > 0$ (independent of $x$ and $U$) such that $m_f(T^N(U)) \geq \xi$. (Here we use the fact that a conformal measure is positive on non–empty open sets.) Whence the second claim holds in this case setting $C_2 \geq C\xi^{-1}$, since

$$S_N(f - P(T,f), T)(x) = S_n(f^* - P^*, T^*)(x),$$

and since $P(T,f) > \sup_{x \in \overline{C}} f(x)$.
2nd case: Let $T^{*n}(x) \in \{N_C \geq 2\}$.
Then $T^{*n-1}(x), T^{*n+1}(x) \in \{N_C = 1\}$ by the definition of $N_C$ and since the Markov fibred system $(\mathcal{R}, T, m_f)$ is parabolic.

Choose $N$ so that $T^N(x) = T^{*n+1}(x)$ and a ball $V \subset U$ with center $x$ and

$$\mathrm{diam}(V) = (1/2)\ k(1/2)^{-1}\delta|(T^{*n+1})'(x)|^{-1}.$$

By Lemma 4.1, $T^N : V \to T^N(V) \subset B(T^{*n+1}(x), \delta)$ is invertible and $T^N(V)$ contains a ball of radius $\zeta$ and center $T^N(x)$. Hence by Lemma 2.9 and conformality

$$m_f(U) \geq m_f(V) \geq C^{-1}\xi \exp\big(S_{n+1}(f^* - P^*, T^*)(x)\big).$$

Choose $N$ so that $T^N(x) = T^{*n-1}(x)$. Then $T^N : U \to T^N(U) \subset B(T^{*n-1}(x), \delta)$ is invertible by Lemma 4.1 and hence

$$m_f(U) \leq C \exp\big(S_{n-1}(f^* - P^*, T^*)(x)\big)$$

by Lemma 2.9 and conformality. $\diamond$

Some more notations are used in the sequel. Denote by

$$\chi = \chi_{q_f} = \int_{J(T)} \log |(T^*)'| \, dq_f = \mu'(J(T)) \int_{J(T)} \log |T'| \, d\mu_f = \mu'(J(T))\chi_{\mu_f} > 0$$

the Lyapunov characteristic exponents of $q_f$ and $\mu_f$ (cf. Theorem 2.12). Let

$$\kappa = \kappa_f = HD(\mu_f) = \inf\{HD(Y) : Y \subset X, \mu_f(Y) = 1\}$$

denote the Hausdorff dimension of $\mu_f$. Note that $HD(\mu_f) = HD(q_f)$ and that

$$\kappa = h_{q_f}(T^*)/\chi = h_{\mu_f}(T)/\chi_{\mu_f}$$

(see [30], [35]). Denote by $\mathcal{U}(x)$ the family of balls with center $x \in J(T)$.

For a function $f : J(T) \to \mathbb{R}$, let $g(f) : X \to \mathbb{R}$ be defined by

$$\begin{aligned} g(f)(x) &= \sum_{l=0}^{N_C(x)-1} (f - P(T, f) + \kappa_f \log |T'|)(T^l(x)) \\ &= f^*(x) - P^*(x) + \kappa_f \log |(T^*)'(x)|. \end{aligned}$$

Note that by Theorem 3.6, $f^*, P^*, \log |(T^*)'|$ and $g(f)$ belong to $\mathcal{L}^*$.

Lemma 4.3: Let $\eta > 0$ and let $\phi : [(\chi + \eta)^{-1}, \infty) \to \mathbb{R}_+$ belong to the upper (lower) class. Let $\psi : [(\chi + \eta)^{-1}, \infty) \to \mathbb{R}_+$ be a function such that

$$\phi(t)\psi(t) \to 0 \qquad \text{as } t \to \infty.$$

Then there exists an upper (lower) class function $\phi_+ : [1, \infty) \to \mathbb{R}_+$ ($\phi_- : [1, \infty) \to \mathbb{R}_+$) with the following properties:

(1) $\phi(t(\chi + \eta)) + \psi(t(\chi + \eta)) \le \phi_+(t) \quad (t \ge 1)$.

(2) $\phi(t(\chi - \eta)) - \psi(t(\chi - \eta)) \ge \phi_-(t) \quad (t \ge 1)$.

Proof. Since $\phi(t)\psi(t) \to 0$ as $t \to \infty$, there exists a constant $M$ such that $(\phi(t)+\psi(t))^2 \le \phi(t)^2 + M$. Let $\phi$ belong to the upper class. Then $t \to \phi(t/(\chi + \eta))$ also belongs to the upper class. Hence we may assume that $\chi + \eta = 1$. Define

$$\phi_+(t)^2 = \inf\{u(t)^2 : u \text{ is non-decreasing}, \, u(t) \ge \phi(t) + \psi(t)\}.$$

Then $\phi_+(t) \ge \phi(t) + \psi(t)$ for $t \ge 1$ and $\phi_+$ is non-decreasing. Since $\phi_+(t)^2 \le \phi(t)^2 + M$, we also get

$$\int_1^\infty \frac{\phi_+(t)}{t} \exp(-(1/2)\phi_+^2(t)) \, dt \ge \exp(-M/2) \int_1^\infty \frac{\phi(t)}{t} \exp(-(1/2)\phi^2(t)) \, dt = \infty.$$

The proof in case of a function of the lower class is the same. $\diamond$

Lemma 4.4: (Refined Volume Lemma) Suppose that $\sigma^2 = \sigma^2(g(f))$ is strictly positive. If $\phi$ belongs to the lower class, then for $\mu_f$ a.e. $x \in J(T)$

$$\limsup_{U\in\mathcal{U}(x)} \frac{m_f(U)}{\operatorname{diam}(U)^\kappa \exp\left[\sigma\chi^{-1/2}\phi(-\log\operatorname{diam}(U))\sqrt{-\log\operatorname{diam}(U)}\right]} = 0.$$

If $\phi$ belongs to the upper class, then for $\mu_f$ a.e. $x \in J(T)$

$$\limsup_{U\in\mathcal{U}(x)} \frac{m_f(U)}{\operatorname{diam}(U)^\kappa \exp\left[\sigma\chi^{-1/2}\phi(-\log\operatorname{diam}(U))\sqrt{-\log\operatorname{diam}(U)}\right]} = \infty.$$

Proof. Let $x \in X$ and $U \in \mathcal{U}(x)$ with $\operatorname{diam}(U) \le r(x)$. Then by Lemma 4.2 there exists $n \ge 1$ such that

$$\begin{aligned}
&(C_1^\kappa C_2)^{-1}\exp\Big(S_{n+1}(f^* - P^*, T^*)(x) + S_n(\kappa\log|(T^*)'|, T^*)(x)\Big)\\
&\le \frac{m_f(U)}{\operatorname{diam}(U)^\kappa}\\
&\le C_1^\kappa C_2\exp\Big(S_{n-1}(f^* - P^*, T^*)(x) + S_{n+1}(\kappa\log|(T^*)'|, T^*)(x)\Big).
\end{aligned}$$

By Theorem 3.6 $f^* - P^* \in L_{2+4\gamma}(q_f)$ for some $\gamma > 0$. Hence, if $\eta > 0$,

$$\begin{aligned}
&\sum_{n=1}^\infty q_f\big(|(f^* - P^*)\circ T^{*n}| \ge (1/2)\sigma[\phi(n(\chi-\eta))\sqrt{n}]^{1-\gamma}\big)\\
&\le \sum_{n=1}^\infty\big((1/2)\sigma[\phi(n(\chi-\eta))\sqrt{n}]^{1-\gamma}\big)^{-2-4\gamma}\int|f^* - P^*|^{2+4\gamma}\,dq_f < \infty.
\end{aligned}$$

Consequently, there exists $n_0(x)$ such that for $k \ge n_0(x)$

$$|(f^* - P^*)(T^{*k}(x))| \le (1/2)\sigma[(\phi(k(\chi-\eta))\sqrt{k}]^{1-\gamma}.$$

Therefore, if $n$ is large enough,

$$\begin{aligned}
&(C_1^\kappa C_2)^{-1}\exp\Big(-\sigma[\phi(n(\chi+\eta))\sqrt{n}]^{1-\gamma} + S_n(f^* - P^* + \kappa\log|(T^*)'|, T^*)(x)\Big)\\
&\le \frac{m_f(U)}{\operatorname{diam}(U)^\kappa}\\
&\le C_1^\kappa C_2\exp\Big(\sigma[\phi((n+1)(\chi-\eta))\sqrt{n+1}]^{1-\gamma} + S_{n+1}(f^* - P^* + \kappa\log|(T^*)'|, T^*)(x)\Big).
\end{aligned}$$

By the ergodic theorem and the choice of $n$ depending on $U$, $-\frac{1}{n}\log\operatorname{diam}(U) \to \chi$. Therefore, for all $n$ large enough,

$$-\log\operatorname{diam}(U) \le n(\chi+\eta), \quad\text{and}\quad -\log\operatorname{diam}(U) \ge (n+1)(\chi-\eta).$$

Since $\log|(T^*)'| \in \mathcal{L}^*$ by Theorem 3.6, the lower class result also applies to this function (Theorem 3.2). Let $\tau^2$ denote the asymptotic variance of $\log|(T^*)'|$. If $\tau^2 = 0$, then by [21], $\log|(T^*)'|$ is cohomologous to $\chi$ by a continuous coboundary (see [37], [46] or the proof of Theorem 4.5 below). It turns out that the following proof, where we assume $\tau^2 > 0$, becomes much simpler when $\tau^2 = 0$.

Since the functions

$$t \to t^{-1/2} 2\sqrt{(t\pm 1)\log\log(t\pm 1)} - \frac{\chi}{\tau} \pm \frac{\log C_1}{\tau}$$

belong to the lower class, we obtain for $n \geq n_1(x) \geq n_0(x)$

$$\begin{aligned}(n+1)\chi - 2\tau\sqrt{(n+1)\log\log(n+1)} &\leq S_n(\log|(T^*)'|, T^*)(x) - \log C_1 \\ = \log C_1^{-1}|(T^{*n})'(x)| \leq -\log \operatorname{diam}(U) &\leq \log C_1 |(T^{*n+1})'(x)| \\ = S_{n+1}(\log|(T^*)'|, T^*)(x) + \log C_1 &\leq n\chi + 2\tau\sqrt{n\log\log n}.\end{aligned}$$

It follows that for $n \geq n_1(x)$ and $a \in \mathbb{R}$

$$\begin{aligned}&(C_1^\kappa C_2)^{-1} \exp\Big(S_n(f^* - P^* + \kappa\log|(T^*)'|, T^*)(x)\Big) \\ &\quad \times \exp\Big(-\sigma\sqrt{n}(\phi(n(\chi+\eta))\sqrt{1 + 2\frac{\tau}{\chi^{-1}}\sqrt{(\log\log n)/n}} \\ &\quad + an^{-1/4} + \phi(n(\chi+\eta))^{1-\gamma}n^{-\gamma/2})\Big) \\ &\leq \frac{m_f(U)}{\operatorname{diam}(U)^\kappa} \exp(-\sigma\chi^{-1/2}\phi(-\log\operatorname{diam}(U))\sqrt{-\log\operatorname{diam}(U)})e^{\sigma a n^{1/4}} \\ &\leq C_1^\kappa C_2 \exp\Big(S_{n+1}(f^* - P^* + \kappa\log|(T^*)'|, T^*)(x)\Big) \\ &\quad \times \exp\Big(-\sigma\sqrt{n+1}(\phi((n+1)(\chi-\eta))\sqrt{1 - 2\frac{\tau}{\chi^{-1}}\sqrt{(\log\log n+1)/(n+1)}} \\ &\quad + an^{1/4}(n+1)^{-1/2} - \phi((n+1)(\chi-\eta))^{1-\gamma}(n+1)^{-\gamma/2}\Big).\end{aligned}$$

Let $\phi$ belong to the upper class and let

$$\psi(t) = \phi(t)(\sqrt{1 + \frac{2\tau}{\chi}\sqrt{\frac{\log\log t(\chi+\eta)^{-1}}{t(\chi+\eta)^{-1}}}} - 1) + \frac{a}{(t(\chi+\eta)^{-1})^{1/4}} + \frac{\phi(t)^{1-\gamma}}{(t(\chi+\eta)^{-1})^{\gamma/2}}.$$

We may assume that $\phi(t) = o(t^\lambda)$ for every $\lambda > 0$. Therefore $\phi(t)\psi(t) \to 0$ as $t \to \infty$. It follows from Lemma 4.3 that there exists $\phi_+$ belonging to the upper class such that

$$\phi_+(t) \geq \phi(t(\chi+\eta)) + \psi(t(\chi+\eta)).$$

Therefore, by Theorem 3.2,

$$\begin{aligned}0 &\le S_n(f^* - P^* + \kappa \log|(T^*)'|, T^*)(x) - \sigma\sqrt{n}\phi_+(n) \\ &\le S_n(f^* - P^* + \kappa \log|(T^*)'|, T^*)(x) - \sigma\sqrt{n}(\phi(n(\chi+\eta)) + \psi(n(\chi+\eta))) \\ &= S_n(f^* - P^* + \kappa \log|(T^*)'|, T^*)(x) - \sigma\sqrt{n}(\phi(n(\chi+\eta))\sqrt{1+\frac{2\tau}{\chi}\sqrt{\frac{\log\log n}{n}}} \\ &\quad + an^{-1/4} + \phi(n(\chi+\eta))^{1-\gamma}n^{-\gamma/2})\end{aligned}$$

for infinitely many $n$. It follows that

$$\frac{m_f(U)}{\operatorname{diam}(U)^\kappa}\exp\big(-\sigma\chi^{-1/2}\phi(-\log \operatorname{diam}(U))\sqrt{-\log \operatorname{diam}(U)}\big) \ge \exp(-\sigma a n^{1/4})$$

for infinitely many $n$. This proves the lemma for functions in the upper class, when $a$ is negative.

The proof for functions of the lower class is analoguous. $\Diamond$

The result about the relation between Hausdorff measures and equilibrium states is contained in the following theorem. Having the refined volume lemma, its proof is essentially standard, but we sketch the argument. For a function $\phi : [1,\infty) \longrightarrow \mathbb{R}_+$ define for sufficiently small $t > 0$

$$\widetilde{\phi}(t) = t^\kappa \exp\left(\frac{\sigma}{\sqrt{\chi}}\phi(-\log t)\sqrt{-\log t}\right).$$

<u>Theorem 4.5:</u>
(1) If $\phi$ belongs to the lower class, then

$$\mu_f \ll H_{\widetilde{\phi}}.$$

(2) If $\phi$ belongs to the upper class, then

$$\mu_f \perp H_{\widetilde{\phi}}.$$

<u>Remark:</u> Taking $\phi \equiv 0$ it follows from the theorem that the measure $\mu_f$ is orthogonal to the $\kappa$–dimensional Hausdorff measure $H_\kappa$ on $J(T)$.
<u>Proof.</u> Suppose first that $\sigma^2 = \sigma^2(g(f)) > 0$.

1) Let $\phi$ belong to the lower class. By Lemma 4.4, for a set $E$ with $\mu_f(E) > 0$, there exist a set $E' \subset E$ satisfying $\mu_f(E') > \mu_f(E)/2$ and a $\theta > 0$ such that for every $x \in E'$ and all balls $U$ of diameter less than $\theta$ and with center $x$,

$$m_f(U) < \widetilde{\phi}(\operatorname{diam}(U)).$$

Therefore

$$H_{\widetilde{\phi}}(E) \ge H_{\widetilde{\phi}}(E') \ge m_f(E').$$

Since $\mu_f \ll m_f$, $m_f(E') > 0$, whence $H_{\widetilde{\phi}}(E) > 0$ and $\mu_f$ is absolutely continuous with respect to $H_{\widetilde{\phi}}$.

2) Let $\phi$ belong to the upper class. For $n \geq 1$ and $\epsilon > 0$, by Lemma 4.4, there exists a set $E_n \subset J(T)$ such that $\mu_f(E_n) > 1 - \epsilon 2^{-n}$ and such that for $x \in E_n$ and some suitable ball $U_x$ of diameter $< 1/n$ with center $x$,

$$m_f(U_x) > n\tilde{\phi}(\mathrm{diam}(U_x)).$$

From the cover $(U_x : x \in E_n)$ of $E_n$ choose a subcover $(U_l : l \geq 1)$ of multiplicity $\leq K_0$, where $K_0$ does not depend on the cover (by Besicovic' Covering Theorem [18]). Since $\mathrm{diam}(U) < 1/n$,

$$H_{\tilde{\phi}}(E_n, \frac{1}{n}) \leq \frac{1}{n}\sum_{l=1}^{\infty} m_f(U_l) \leq \frac{K_0}{n} m_f(J(T)) = \frac{K_0}{n}.$$

Setting $F_\epsilon = \bigcap_{n\geq 1} E_n$ it follows that $H_{\tilde{\phi}}(F_\epsilon) = 0$ and $\mu_f(F_\epsilon) \geq 1 - \epsilon$. Finally the set $F = \bigcup_{l\geq 1} F_{1/l}$ satisfies $H_{\tilde{\phi}}(F) = 0$ and $\mu_f(F) = 1$.

3) It is left to show that $\sigma^2 > 0$.
Suppose that $\sigma^2(g(f)) = 0$. It is well known that $g(f)$ is cohomologous (in $L_2$) to a constant function with respect to $T^*$ (see [21]). Let $\psi \circ T^* - \psi$ denote this coboundary. Since $\kappa = \chi_{\mu_f}^{-1} h_{\mu_f}(T)$,

$$\begin{aligned}\int_{J(T)} g(f)\, dq_f &= \int f d\mu' - P(T,f)\mu'(J(T)) + \kappa \int \log|T'| d\mu' \\ &= \left(\int_{J(T)} f\, d\mu_f - P(T,f) + \kappa\chi_{\mu_f}\right)\mu'(J(T)) = 0.\end{aligned}$$

Hence $g(f)$ is cohomologous to zero with respect to $T^*$. By definition of $T^*$, if $x \in \{N_C = k\}$ for some $k \geq 2$, then a.e.

$$g(f)(x) = \sum_{i=0}^{k-1}[f(T^i(x)) - P(T,f) + \kappa \log|T'(T^i(x))|] = \psi(T^k(x)) - \psi(x)$$

$$g(f)(T(x)) = \sum_{i=1}^{k-1}[f(T^i(x)) - P(T,f) + \kappa \log|T'(T^i(x))|] = \psi(T^k(x)) - \psi(T(x)).$$

Therefore $f(x) - P(T,f) + \kappa \log|T'(x)| = \psi(T(x)) - \psi(x)$ for almost all $x \in X$, and $f - P(T,f)$ is cohomologous to $\kappa \log|T'|$ with respect to $T$ and with $L_2$–coboundary $\psi \circ T - \psi$. The function $\psi$ has a continuous version. This follows as in the proof of Lemma 1 in [37] (p. 14) using the argument in the proof of Lemma 2.9, since the preimages of every point in $J(T)$ are dense in $J(T)$ (see also [46]). It follows that $g(f)$ is cohomologous to 0 in the space of continuous functions. Hence $\psi$ is bounded, but for $x$ a rationally indifferent periodic point,

$$\psi(T^{k-1}(x)) - \psi(x) = S_k(f - P(T,f), T)(x) + O(1) \to -\infty.$$

This contradicts the assumption. $\diamond$

## REFERENCES


[1] Aaronson, J.; M.Denker; M. Urbański: Ergodic theory for Markov fibred systems and parabolic rational maps. Preprint.

[2] Blanchard, P.: Complex analytic dynamics on the Riemann sphere. Bull. Amer. Math. Soc. **11**, (1984), 85–141.

[3] Bowen, R.: Equlibrium states and the ergodic theory of Anosov diffeomorphisms. Lect. Notes in Math. **470**, (1975), Springer Verlag.

[4] Brolin, H.: Invariant sets under iteration of rational functions. Ark. f. Mat. **6**, (1965), 103–144.

[5] Denker M.; C. Grillenberger; K. Sigmund: Ergodic theory on compact spaces. Lect. Notes in Math. **527**, (1976), Springer Verlag.

[6] Denker, M.; M. Urbański: On the existence of conformal measures. to appear: Trans. Amer. Math. Soc.

[7] Denker, M.; M. Urbański: Ergodic theory of equilibrium states for rational maps. Nonlinearity **4**, (1991), 103–134.

[8] Denker, M.; M. Urbański: Hausdorff and conformal measures on Julia sets with a rationally indifferent periodic point. to appear: J. London Math. Soc.

[9] Denker, M.; M. Urbański: Absolutely continuous invariant measures for expansive rational maps with rationally indifferent periodic points. to appear: Forum Math.

[10] Denker, M.; M. Urbański: On Sullivan's conformal measures for rational maps of the Riemann sphere. to appear: Nonlinearity.

[11] Denker, M.; M. Urbański: Hausdorff measures on Julia sets of subexpanding rational maps. to appear: Isr. J. Math.

[12] Denker, M.; M. Urbański: Geometric measures for parabolic rational maps. to appear: Erg. Th. and Dynam. Syst.

[13] Devaney, R.: An introduction to chaotic dynamical systems. (1985), Benjamin.

[14] Falconer, K.J.: The geometry of fractal sets. (1985), Cambridge Univ. Press.

[15] Fatou, P.: Sur les équations fonctionelle. Bull. Soc. Math. France, **47**, (1919), 161–271.

[16] Fatou, P.: Sur les équations fonctionelle. Bull. Soc. Math. France, **48**, (1920), 33–94 and 208–314.

[17] Gromov, M.: On the entropy of holomorphic maps. Preprint, IHES.

[18] Guzmán, M.: Differentiation of integrals in $\mathbb{R}^n$. Lect. Notes in Math. **481**, (1974), Springer Verlag.

[19] Hille, E: Analytic Function Theory. Ginn and Company, Boston 1962.

[20] Hofbauer, F.; Keller, G.: Ergodic properties of invariant measures for piecewise monotonic transformations. Math. Zeitschrift, **180**, (1982), 119–140.

[21] Ibragimov, I.A.; Y.V. Linnik: Independent and stationary sequences of random variables. (1971), Wolters–Noordhoff Publ., Groningen.

[22] Ionescu–Tulcea, C.; Marinescu, G.: Théorie ergodique pour des classes d'operations non completement continues. Ann. Math. **52**, (1950), 140–147.

[23] Jain, N.C.; Jogdeo, K.; Stout, W.: Upper and lower functions for martingales and mixing processes. Ann. Probab. **3**, (1975), 119–145.

[24] Julia, G.: Mémoire sur l'iteration des fonctions rationelles. J. Math. Pure et Appl. Sér. **8.1**, (1918), 47–245.

[25] Keen, L.: Julia sets. In:Chaos and fractals, eds.: R. Devaney, L. Keen. Proc. Symp. in Appl. Math. **39**, (1989), 57–74.

[26] Lyubich, V.: Entropy properties of rational endomorphisms of the Riemann sphere. Ergod. Theory and Dynam. Sys. **3**, (1983), 351–386.

[27] Makarov, N.G.: On the distortion of boundary sets under conformal mappings. Proc. London Math. Soc. **51**, (1985), 369–384.

[28] Mañé, R.: On the uniqueness of the maximizing measure for rational maps. Bol. Soc. Bras. Mat. **14**, (1983), 27–83.

[29] Mañé, R.: On the Bernoulli property of rational maps. Ergod. Theory and Dynam. Syst. **5**, (1985), 71–88.

[30] Mañé, R.: The Hausdorff dimension of invariant probabilities of rational maps. Lect. Notes in Math. **1331**, (1988), 86–117, Springer.

[31] Misiurewicz, M.: Topological conditional entropy. Studia Math. **55**, (1976), 175–200.

[32] Patterson, S.J.: The limit set of a Fuchsian group. Acta Math. **136**, (1976), 241–273.

[33] Patterson, S.J.: Lectures on measures on limit sets of Kleinian groups. In: Analytic and Geometric Aspects of Hyperbolic Space. ed. D.B.A. Epstein. LMS Lect. Notes Ser. **111**, (1987), Cambridge Univ. Press.

[34] Philipp, W.; Stout, W.: Almost sure invariance principles for partial sums of weakly dependent random variables. Memoirs Amer. Math. Soc. **161 (2)**, (1975).

[35] Przytycki, F.: Hausdorff dimension of harmonic measure on the boundary of an attractive basin for a holomorphic map. Invent. Math. **80**, (1985), 161–179.

[36] Przytycki, F.: On the Perron–Frobenius–Ruelle operator for rational maps on the Riemann sphere and for Hölder continuous functions. Bol. Soc. Bras. Mat. **20**, (1990), 95–125.

[37] Przytycki, F.; M. Urbański; A. Zdunik: Harmonic, Gibbs and Hausdorff measures on repellers for holomorphic maps, I+II. Part I: Ann. Math. **130**, (1989), 1–40; Part II: to appear Studia Math.

[38] Rogers, C.A.: Hausdorff measures. (1970), Cambridge Univ. Press.

[39] Ruelle, D.: Thermodynamic formalism. Encycl. Math. Appl. **5**, (1978), Addison–Wesley.

[40] Ruelle, D.: Repellers for real analytic maps. Ergod. Theory and Dynam. Syst. **2**, (1982), 99–107.

[41] Sullivan, D.: Conformal dynamical systems. In: Geometric Dynamics. Lect. Notes in Math. **1007**, (1983), 725–752, Springer Verlag.

[42] Sullivan, D.: Entropy, Hausdorff measures old and new, and limit sets of geometrically finite Kleinian groups. Acta Math. **153**, (1984), 259–277.

[43] Urbański, M.: Hausdorff dimension of invariant subsets for endomorphisms of the circle with an indifferent fixed point. J. London Math. Soc. **40**, (1989), 158–170.

[44] Urbański, M.: On Hausdorff dimension of the Julia set with an indifferent rational periodic point. to appear: Studia Math.

[45] Walters, P.: An introduction to ergodic theory. (1982), Springer Verlag.

[46] Zdunik, A.: Parabolic orbifolds and the dimension of the maximal measure for rational maps. Invent. Math. **99**, (1990), 627–649.

# ON THE CONSTRUCTION OF GENERALIZED MEASURE PRESERVING TRANSFORMATIONS WITH GIVEN MARGINALS

Theodore P. Hill*
School of Mathematics
Georgia Institute of Technology
Atlanta, GA 30332 USA

and

Ulrich Krengel**
Institut für Mathematische Stochastik
Lotzestr. 13, Univ. Göttingen
W-3400 Göttingen, Germany

Abstract

Measure preserving transformations generate stationary processes and vice versa. Which processes $(X_i)$ correspond to the class of *generalized* measure preserving transformations? We give necessary conditions and show that they are sufficient for 2-valued processes as far as the marginals of $(X_0, X_1, X_2, X_3)$ are concerned. The general problem remains open. Our main tool is a construction of a class of generalized measure preserving transformations which may be of independent interest.

## §1. Introduction

The notion of generalized measure preserving transformation (gmp–transformation) was introduced in [K] as a mathematical model for the movement of sets of incompressible objects subject to interaction. Let $(\Omega, \mathcal{A}, \mu)$ be a probability space. A gmp–transformation is a map $\phi : \mathcal{A} \to \mathcal{A}$ which is order preserving and which preserves $\mu$. In other words, $A \subset B$ implies $\phi(A) \subset \phi(B)$, and $\mu(\phi(A)) = \mu(A)$ holds for all $A \in \mathcal{A}$.

A measure preserving transformation $\tau : \Omega \to \Omega$ induces a gmp–transformation $\phi_\tau$ by setting $\phi_\tau(A) = \tau^{-1}A$. In general, however, a gmp–transformation need not commute with the formation of unions or intersections. We only have

$$\phi(A \cap B) \subset \phi(A) \cap \phi(B) \quad (A, B \in \mathcal{A}) \tag{1.1}$$

and

$$\phi(A \cup B) \supset \phi(A) \cup \phi(B) \quad (A, B \in \mathcal{A}). \tag{1.2}$$

Let $(X_n)_{n \geq 0}$ be a real–valued stochastic process defined on a probability space $(\Omega', \mathcal{A}', P)$. It is well known that there exists a measure preserving $\tau$ on a suitable probability space $(\Omega, \mathcal{A}, \mu)$ and a measurable $f$ on $\Omega$ such that $(X_n)$ and $(f \circ \tau^n)$ have the same joint distributions if and only if the distribution of $(X_n)$ is stationary. We propose to study the corresponding problem for gmp–transformations.

As is shown in [K] and [LW], the map $f \to f \circ \tau$ can be extended to gmp–transformations by putting, for real–valued measurable $f$,

$$T_\varphi f(\omega) = \sup\{t \in \mathbb{R} : \omega \in \phi(\{f > t\})\}.$$

---

* Research partially supported by NSF Grant DMS-89-01267 and a Fulbright Research Grant.

** This research was done during a visit of the second author at the Georgia Institute of Technology, Atlanta. This visit was supported by the Deutsche Forschungsgemeinschaft.

$T_\varphi$ is nonlinear, in general, and satisfies $\{T_\varphi f \geq t\} = \phi(\{f \geq t\})$. If $\phi = \phi_\tau$, then $T_\varphi f = f \circ \tau$.

We say that a process $(X_n)_{n \geq 0}$ can be represented by a gmp–transformation if there exists a gmp–transformation $\phi$ on a suitable probability space $(\Omega, \mathcal{A}, \mu)$, and a measurable $f$ such that the sequence $(T_\varphi^n f)$ has the same joint distribution as $(X_n)$. Our question now is: Which processes can be represented by a gmp–transformation?

We shall consider the following conditions:

$M_\cap$) (Monotonicity condition for intersections): For all $n \in \mathbb{N}$ and all $t_0, \dots t_{n-1} \in \mathbb{R}$

$$P\left(\bigcap_{i=0}^{n-1} \{X_i \geq t_i\}\right) \leq P\left(\bigcap_{i=0}^{n-1} \{X_{i+1} \geq t_i\}\right).$$

$M_\cup$) (Monotonicity condition for unions): For all $n \in \mathbb{N}$ and all $t_0, \dots t_{n-1} \in \mathbb{R}$

$$P\left(\bigcup_{i=0}^{n-1} \{X_i \geq t_i\}\right) \geq P\left(\bigcup_{i=0}^{n-1} \{X_{i+1} \geq t_i\}\right).$$

We shall see below that these monotonicity conditions are necessary for a process to permit the representation by a gmp–transformation. This is a fairly simple consequence of (1.1) and (1.2). We do not know if all processes satisfying these two conditions can be represented by a gmp–transformation. In this direction we obtain only a very special result: If $(X_0, X_1, X_2, X_3)$ assumes only two values and the monotonicity conditions above hold, a representation for these 4-dimensional marginals is obtained. Even this special case requires a considerable argument. Our proof relies on a general method for constructing gmp–transformations which satisfy certain priority rules. This is presented in Section 2, and it seems of independent interest. Progress on the main problem seems to require new methods for the construction of gmp–transformations. It is clear that the class of gmp–transformations is very rich, but at present only few methods of construction are available.

Any gmp–transformation $\phi$ induces a transformation, also denoted by $\phi$, in the measure algebra $\bar{\mathcal{A}}$ obtained from $\mathcal{A}$ by identifying sets that differ only by null sets. It will be convenient to look mainly at the measure algebra, and we shall be satisfied with constructing $\phi$ on it.

## §2. Construction of gmp–Transformations by Priority Rules

In [K], examples of gmp–transformations on a *finite* set $\Omega$ with counting measure were obtained by prescribing certain priorities for the points of the space. We now introduce an extension of this idea to general measure spaces.

**Theorem 2.1.** *Let $(\Omega, \mathcal{A}, \mu)$ be a probability space supporting an ergodic invertible measure preserving transformation $\tau$. Let $\{\Omega_1, \Omega_2, \dots\}$ be a partition of $\Omega$ into finitely many or countably many (disjoint) measurable sets. For each $i = 1, 2, \dots$, let $\{\Omega_{i1}, \Omega_{i2}, \dots\}$ be a measurable partition of $\Omega$. Put $\Omega^0 = \emptyset$, $\Omega^i := \bigcup_{k=1}^{i} \Omega_k$,*

$\Omega_i^j := \bigcup_{k=1}^j \Omega_{ik}$, *and for any* $A \in \mathcal{A}$, $A^i = A \cap \Omega^i$. *There exists a gmp–transformation* $\phi$ *with the following property: For any* $i \geq 1$ *and* $j \geq 2$ *and any* $A \in \mathcal{A}$ *if*

$$\mu((\Omega_{ij} \cap \phi(A^i)) \backslash \phi(A^{i-1})) > 0$$

*then*

$$\mu(\Omega_i^{j-1} \backslash \phi(A^i)) = 0.$$

**Remark.** The heuristic meaning of the property above is the following: We imagine that the image $\phi(A)$ is constructed by first mapping $A \cap \Omega_1$, then mapping $A \cap \Omega_2$, then $A \cap \Omega_3$, etc. So, after $i$ steps one has constructed $\phi(A^i)$. If the part of the image constructed in the $i^{th}$ step contains a non-null subset of $\Omega_{ij}$, then $\Omega_i^{j-1}$ must already be filled up, i.e., $\Omega_i^{j-1}$ must be (mod null sets) a subset of $\phi(A^i)$. In other words, when a subset of $\Omega_i$ is mapped, one maps as much as possible into $\Omega_{i1}$, then as much as possible into $\Omega_{i2}$ etc. The sets $\Omega_{ij}$ are called priority sets for $\Omega_i$.

**Proof: Step 1.** (Construction of $\phi(A \cap \Omega_1)$). This step consists of a countable sequence of substeps. In the first substep, one tries to map as much as possible from $A \cap \Omega_1$ into $\Omega_{11}$, in the second substep one maps as much as possible from the remaining part of $A \cap \Omega_1$ into $\Omega_{12}$, etc. Formally, the first substep of Step 1 is defined by an inductive procedure. Set

$$B_0^1 = A^1 \cap \Omega_{11} = R_0^1.$$

We shall have $\phi(R_0^1) = B_0^1$. (On $\Omega_1 \cap \Omega_{11}$, $\phi$ is the identity map.) We imagine that we have two copies of $\Omega$, the original space and a second copy for the images. We paint that part of $A$ which has already been mapped red and the image blue in the second copy. So at this point, $R_0^1$ is painted red and $B_0^1$ blue. Now, we take the part of $A^1$ which is not yet painted and try to map it into the unpainted part of $\Omega_{11}$ by $\tau$. Formally

$$B_1^1 = \tau(A^1 \backslash R_0^1) \cap (\Omega_{11} \backslash B_0^1)$$
$$R_1^1 = \tau^{-1} B_1^1.$$

Then we continue with $\tau^2$ in the same way:

$$B_2^1 = \tau^2(A^1 \backslash (R_0^1 \cup R_1^1)) \cap (\Omega_{11} \backslash (B_0^1 \cup B_1^1))$$
$$R_2^1 = \tau^{-2} B_2^1.$$

It is clear how to continue.

Suppose that a subset $F$ of positive measure of $\Omega_{11}$ is not painted blue in this procedure. As $\tau$ is ergodic,

$$F^* = \bigcup_{k=0}^{\infty} \tau^{-k} F$$

is almost all of $\Omega$. Any point in $F^* \cap A^1$ must be painted red. In this case almost all of $A^1$ has already been mapped into $\Omega_{11}$ and therefore will not be mapped into $\Omega_{12} \cup \Omega_{13} \cup \ldots$ in the future steps.

If no such $F$ exists, $\Omega_{11}$ is painted blue and we perform the second substep of Step 1, replacing $\Omega_{11}$ by $\Omega_{12}$ and $A^1$ by $A^1 \backslash \bigcup_{m=0}^{\infty} R_m^1$, the unpainted part of $A^1$. If we write

$B^1_{m1}$ and $R^1_{m1}$ for the sets $B^1_m$ and $R^1_m$ constructed in the first substep above, the next sequence of steps could start with

$$A^1_2 = A^1 \setminus \bigcup_{m=0}^{\infty} R^1_{m1}$$

and

$$B^1_{02} = A^1_2 \cap \Omega_{12} = R^1_{02}.$$

Next put

$$\begin{aligned} B^1_{12} &= \tau(A^1_2 \setminus R^1_{02}) \cap (\Omega_{12} \setminus B^1_{02}) \\ R^1_{12} &= \tau^{-1} B^1_{12}, \end{aligned}$$

etc. One continues as above. In this substep either almost all of $\Omega_{12}$ is painted blue or almost all of $A^1_2$ is painted red. In the latter case, $\phi(A^1)$ shall be the union of all blue sets constructed so far; in the former case one must now repeat the construction with $\Omega_{13}$ and the unpainted part of $A^1$.

As $(\Omega_{11}, \Omega_{12}, \ldots)$ is a partition of $\Omega$, almost all of $A^1$ will be painted red after finitely many or denumerably many substeps. $\phi(A^1)$ shall be the union of all sets painted blue in this procedure.

**Step 2.** Step 2 is just the same as Step 1 except that $A \cap \Omega_1$ is now replaced by $A \cap \Omega_2$ and the sets $\Omega_{11}, \Omega_{12}, \ldots$ are replaced by

$$\Omega_{21} \setminus \phi(A^1), \Omega_{22} \setminus \phi(A^1), \ldots$$

(The set $\phi(A^1)$ has already been painted in Step 1 and is no longer available as an image.) If $B^2_{01}, B^2_{11}, B^2_{21}, \ldots$ is the family of all blue sets constructed in the first substep of Step 2, $B^2_{02}, B^2_{12}, \ldots$ the family of all blue sets constructed in the second substep of Step 2, etc. let $\phi(A \cap \Omega^2)$ be the union of $\phi(A \cap \Omega_1)$ and all these blue sets found in Step 2.

In Step 3 repeat the construction with $A \cap \Omega_3$ and with the sets $\Omega_{31} \setminus \phi(A \cap \Omega^2), \Omega_{32} \setminus \phi(A \cap \Omega^2)$, etc. Finally $\phi(A)$ is the union of all blue sets constructed in Steps 1,2,3,.... It is clear that $\mu(\phi(A)) = \mu(A)$ since $A$ is mod $\mu$ the disjoint union of red sets $R^j_{mi}$ and $\phi(A)$ the corresponding disjoint union of blue sets $B^j_{mi} = \tau^m R^j_{mi}$. If $\widetilde{A} \supset A$, then, at each stage of the construction, the total blue set for $\widetilde{A}$ is at least as large as the corresponding blue set for $A$. The unpainted part of $\widetilde{A}$ is, at each stage, at least as large as the unpainted part of $A$ in the corresponding step of the construction of $\phi(A)$. Hence $\phi(\widetilde{A}) \supset \phi(A)$. By the construction, if $\mu((\Omega_{ij} \cap \phi(A^i)) \setminus \phi(A^{i-1}) > 0$, then this means that a subset of $\Omega_{ij}$ of positive measure was painted blue in Step i. This happens only if almost all of $\Omega^{j-1}_i$ was painted blue before. Hence the gmp–transformation $\phi$ has the desired property. □

**Remark.** If $\mu(A \cap \Omega_i) \leq \mu(\Omega^j_i \setminus \phi(A^{i-1}))$, then $\phi(A^i) \setminus \phi(A^{i-1})$ is, mod $\mu$, a subset of $\Omega^j_i$. (If $\mu(\Omega_{ik} \cap (\phi(A^i) \setminus \phi(A^{i-1})))$ is positive for some $k > j$, then $\Omega^j_i \setminus \phi(A^{i-1})$ must have been painted blue.)

## §3. Representation of Processes by gmp–Transformations

**Theorem 3.1.** *The conditions* $(M_\cap)$ *and* $(M_\cup)$ *are necessary for a process* $(X_m)$ *to admit a representation by a gmp–transformation.*

**Proof.** Assume $(X_m)$ admits a representation on a probability space $(\Omega, \mathcal{A}, \mu)$. Then there exists a measurable $f$ and a gmp–transformation $\phi$ such that $(T_\varphi^m f)$ has the same joint distribution as $(X_m)$. Using $\{T_\varphi^k f \geq t\} = \phi^k(\{f \geq t\})$ $(k \geq 0)$ and (1.1) we obtain

$$\begin{aligned} P(X_0 \geq t_0, \ldots, X_{n-1} \geq t_{n-1}) &= \mu\left(\bigcap_{i=0}^{n-1} \phi^i(\{f \geq t_i\})\right) \\ &= \mu\left(\phi \bigcap_{i=0}^{n-1} \phi^i(\{f \geq t_i\})\right) \\ &\leq \mu\left(\bigcap_{i=0}^{n-1} \phi^{i+1}(\{f \geq t_i\})\right) \\ &= P(X_1 \geq t_0, \ldots, X_n \geq t_{n-1}). \end{aligned}$$

The symmetric argument with (1.2) shows that also $(M_\cup)$ is necessary. □

We do not know if the combined conditions $(M_\cap)$ and $(M_\cup)$ are sufficient. It even seems hard to answer this problem when the process takes only two values, say 0 and 1, and we ask only that for any fixed $n$ there exists a $\phi$ and $f$ such that $(X_i)_{i=0}^n$ and $(T_\varphi^i f)_{i=0}^n$ have the same joint distribution. In this case, $f$ is an indicator function $f = 1_{A_0}$, and we have $T_\varphi^k f = 1_{\phi^k(A_0)}$.

Most of the remainder of this paper will be devoted to showing that the combination $(M)$ of $(M_\cup)$ and $(M_\cap)$ is sufficient for this subproblem when $n \leq 3$.

Let $(\Omega, \mathcal{A}, \mu)$ be a nonatomic probability space supporting an ergodic invertible measure preserving transformation $\tau$ in $\Omega$. E.g., $(\Omega, \mathcal{A}, \mu)$ is the unit interval with Lebesgue measure. Clearly, we can assume $(\Omega', \mathcal{A}', P) = (\Omega, \mathcal{A}, \mu)$ replacing the original process $(X_i)_{i=0}^n$ by a process with the same distribution defined on $(\Omega, \mathcal{A}, \mu)$.

Let us reformulate the conditions $(M_\cap)$ and $(M_\cup)$ for 0-1–valued processes $X_i = 1_{A_i} (i = 0, \ldots, n)$ on $(\Omega, \mathcal{A}, \mu)$. For any nonempty subset $I = \{i_1, i_2, \ldots, i_k\}$ of $\{0, 1, \ldots, n-1\}$ put

$$A_{\cap I} := \bigcap_{\nu=1}^{k} A_{i_\nu}, \quad A_{\cup I} := \bigcup_{\nu=1}^{k} A_{i_\nu}$$

and

$$A_{\cap I+1} := \bigcap_{\nu=1}^{k} A_{i_\nu+1}, \quad A_{\cup I+1} := \bigcup_{\nu=1}^{k} A_{i_\nu+1}.$$

It is an exercise to show that $(M)$ is equivalent to the condition $(M')$ that for all $n \in \mathbb{N}$, both

$$\mu(A_{\cap I}) \leq \mu(A_{\cap I+1}) \quad \text{for all } I \subset \{0,\ldots,n-1\},$$

and

$$\mu(A_{\cup I}) \geq \mu(A_{\cup I+1}) \quad \text{for all } I \subset \{0,\ldots,n-1\}.$$

Note that $(M')$ implies $\mu(A_i) = \mu(A_{i+1})$ $(0 \leq i \leq n-1)$ by taking $I = \{i\}$.

By Theorem 3.1, $(M')$ is necessary for the existence of a gmp–transformation $\phi$ with $\phi^i(A_0) = A_i$ $(1 \leq i \leq n)$. It seems natural to conjecture that $(M')$ is also sufficient, but we can prove this only for $n \leq 3$. For larger $n$, the present approach gets extremely involved, and a new, possibly more canonical construction seems desirable. For $n \leq 3$, we actually prove a stronger result.

Let $B_0, B_1, \ldots$ be measurable sets. Define $B_{\cap I}$ and $B_{\cup I}$ just like $A_{\cap I}$ and $A_{\cup I}$.

**Theorem 3.2.** *Let $(\Omega, \mathcal{A}, \mu)$ be a nonatomic probability space supporting an invertible ergodic measure preserving transformation $\tau$ in $\Omega$. Let $A_0, \ldots, A_{n-1}$ and $B_0, \ldots, B_{n-1}$ be measurable sets, $n \leq 3$. The condition $(M'')$ that*

$$\mu(A_{\cap I}) \leq \mu(B_{\cap I})$$

*and*

$$\mu(A_{\cup I}) \geq \mu(B_{\cup I})$$

*hold for all $I \subset \{0,\ldots,n-1\}$ is necessary and sufficient for the existence of a gmp–transformation $\phi$ with $B_i = \phi(A_i)$ $(i = 0,\ldots,n-1)$.*

**Proof.** The necessity holds for all $n$ and is easily established using (1.1) and (1.2).

With the help of Theorem 2.1, the proof of the sufficiency is easy for $n = 2$: Map $\Omega_1 := A_0 \cap A_1$ into $\Omega_{11} := B_0 \cap B_1$. Then map $\Omega_2 := A_0 \backslash A_1$ into $\Omega_{21} := B_0$, then $\Omega_3 = A_1 \backslash A_0$ into $\Omega_{31} := B_1$, and finally $\Omega_4 := (A_0 \cup A_1)^c$ into $\Omega_{41} = \Omega$. We leave the details as an exercise. Note that $(M'')$ implies $\mu(A_0) = \mu(B_0), \mu(A_1) = \mu(B_1)$, and $\mu(A_0 \cap A_1) \leq \mu(B_0 \cap B_1)$.

$n = 3$: We consider the partition of $\Omega$ induced by the sets $A_0, A_1, A_2$. It consists of the sets

$$\begin{array}{ll} \Sigma_1 := A_0 \cap A_1 \cap A_2 & \Sigma_2 := A_0 \cap A_1 \cap A_2^c \\ \Sigma_3 := A_0 \cap A_1^c \cap A_2 & \Sigma_4 := A_0 \cap A_1^c \cap A_2^c \\ \Sigma_5 := A_0^c \cap A_1 \cap A_2 & \Sigma_6 := A_0^c \cap A_1 \cap A_2^c \\ \Sigma_7 := A_0^c \cap A_1^c \cap A_2 & \Sigma_8 := A_0^c \cap A_1^c \cap A_2^c. \end{array}$$

The partition $\{\Sigma_1^*, \Sigma_2^*, \ldots, \Sigma_8^*\}$ induced by the sets $B_0, B_1, B_2$ is defined in exactly the same way: $\Sigma_1^* := B_0 \cap B_1 \cap B_2$, etc. Set

$$\delta_i = \mu(\Sigma_i^*) - \mu(\Sigma_i) \quad (i = 1,\ldots,8).$$

(The actual sizes of the sets are less important than the differences $\delta_i$.) Assume $(M'')$ holds. In terms of the $\delta_i$'s this means that the following identities and inequalities are satisfied:

$$\begin{aligned} \delta_1+\delta_2+\delta_3+\delta_4&=0 \\ \delta_1+\delta_2+\delta_5+\delta_6&=0 \\ \delta_1+\delta_3+\delta_5+\delta_7&=0, \end{aligned} \tag{3.1}$$

$$\begin{aligned} \delta_1&\geq 0 \\ \delta_1+\delta_2&\geq 0 \\ \delta_1+\delta_3&\geq 0 \\ \delta_1+\delta_5&\geq 0 \end{aligned} \tag{3.2}$$

$$\delta_2+\delta_3+\delta_5+2\delta_1\geq 0 \tag{3.3}$$

(The last inequality follows from $\sum_{i=1}^{7}\delta_i\leq 0$ by subtracting the three identities in (3.1).)

The construction of $\phi$ shall be based on Theorem 2.1. The priorities will depend on the $\delta$'s. In all cases, we shall start by mapping $\Sigma_1$ into $\Sigma_1^*$ because $\Sigma_1$ is a subset of all $A_i$, so that $\phi(\Sigma_1)$ must be a subset of all $B_i$.

If we specify priority sets $\Omega_{k1},\Omega_{k2},\ldots,\Omega_{km}$ which do not cover all of $\Omega$, this shall mean that there is one additional set $\Omega_{k,m+1}$ which is the complement of the union of $\Omega_{k1},\ldots;\Omega_{km}$.

**Case 1.** $\delta_3\geq 0$. In this case, we start by mapping those $\Sigma_i$ which are contained in $A_1$. $\Sigma_3^*$ will be big enough to recieve all of $\Sigma_3$, and no part of $\Sigma_1^*$ will be needed as image of $\Sigma_3$. Therefore we can map $\Sigma_2$ into $\Sigma_1^*\cup\Sigma_2^*$. Possibly, $\Sigma_5$ is too big to be mapped into $\Sigma_5^*$, but we shall see that no harm is done mapping the surplus into $\Sigma_6^*$.

Formally, set

$$\Omega_1=\Sigma_1,\quad \Omega_{11}=\Sigma_1^*$$
$$\Omega_2=\Sigma_2,\quad \Omega_{21}=\Sigma_2^*,\quad \Omega_{22}=\Sigma_1^*$$

(so, tacitly, $\Omega_{23}=\Omega\backslash(\Omega_{21}\cup\Omega_{22})$.

$$\Omega_3=\Sigma_5,\quad \Omega_{31}=\Sigma_5^*,\quad \Omega_{32}=\Sigma_1^*,\quad \Omega_{33}=\Sigma_6^*$$
$$\Omega_4=\Sigma_6,\quad \Omega_{41}=B_1,\quad \Omega_{42}=B_1^c$$

(By now we know how to map subsets of $A_1$.) Next, set

$$\Omega_5=\Sigma_3,\quad \Omega_{51}=\Sigma_3^*$$
$$\Omega_6=\Sigma_4,\quad \Omega_{61}=B_0$$
$$\Omega_7=\Sigma_7,\quad \Omega_{71}=B_2$$
$$\Omega_8=\Sigma_8,\quad \Omega_{81}=\Omega.$$

This determines $\phi$. We have to check $\phi(A_i)=B_i$ for $i=0,1,2$:

($i=0$): $A_0=\Sigma_1\cup\Sigma_2\cup\Sigma_3\cup\Sigma_4$. $\Sigma_1\cup\Sigma_2$ is mapped into $\Sigma_1^*\cup\Sigma_2^*$ since $\delta_1+\delta_2\geq 0$. $\Sigma_3$ goes into $\Sigma_3^*$ since $\delta_3\geq 0$. $\Sigma_4$ goes into $B_0$ since $\mu(A_0)=\mu(B_0)$. Hence $\phi(A_0)=B_0$.

($i=1$): $\phi(A_1)=B_1$ is clear.

($i = 2$): Constructing $\phi(A_2)$, first $\Sigma_1$ is mapped into $\Sigma_1^*$. $A_2 \cap \Sigma_2$ is empty. Therefore, when $A_2$ is mapped, the priorities for $\Sigma_2$ do not matter. $\Sigma_5$ is mapped into $\Sigma_1^* \cup \Sigma_5^*$ since no part of this set was consumed by $\phi(A_2 \cap \Sigma_2)$. Thus, $\Sigma_5$ goes into $B_2$. $\Sigma_3$ goes into $\Sigma_3^* \subset B_2$; finally $\Sigma_7$ goes into $B_2$ since $\mu(B_2) = \mu(A_2)$. Thus, $\phi(A_2) \subset B_2$. As $\mu(\phi(A_2)) = \mu(A_2) = \mu(B_2)$, we have $\phi(A_2) = \phi(B_2)$.

The cases $\delta_2 \geq 0$ and $\delta_5 \geq 0$ are symmetric. Thus, we can assume $\delta_3 < 0$, $\delta_2 < 0, \delta_5 < 0$ in the sequel.

**Case 2.** $\delta_1 + \delta_2 + \delta_3 \geq 0$. This time, let the sets $\Omega_i$ and $\Omega_{ij}$ be defined by

$$\begin{array}{llll}
\Omega_1 = \Sigma_1, & \Omega_{11} = \Sigma_1^* & & \\
\Omega_2 = \Sigma_2, & \Omega_{21} = \Sigma_2^*, & \Omega_{22} = \Sigma_1^* & \\
\Omega_3 = \Sigma_5, & \Omega_{31} = \Sigma_5^*, & \Omega_{32} = \Sigma_1^*, & \Omega_{33} = \Sigma_6^* \cup \Sigma_2^* \\
\Omega_4 = \Sigma_3, & \Omega_{41} = \Sigma_3^*, & \Omega_{42} = \Sigma_1^*, & \Omega_{43} = \Sigma_5^* \cup \Sigma_7^* \\
\Omega_5 = \Sigma_4, & \Omega_{51} = B_0 & & \\
\Omega_6 = \Sigma_6, & \Omega_{61} = B_1 & & \\
\Omega_7 = \Sigma_7, & \Omega_{71} = B_2 & & \\
\Omega_8 = \Sigma_8, & \Omega_{81} = \Omega. & &
\end{array}$$

We must, again, check $\phi(A_i) = B_i (i = 0, 1, 2)$:

($i = 0$): $\Sigma_1 \cup \Sigma_2$ is mapped into $\Sigma_1^* \cup \Sigma_2^*$ again. Recall that we can assume $\delta_2 < 0$. Thus $\Sigma_2^*$ is covered by $\phi(\Sigma_2)$. Next, $\Sigma_3$ is mapped into $\Sigma_1^* \cup \Sigma_3^*$ since $\delta_1 + \delta_2 + \delta_3 \geq 0$ means that there is enough space left over in $\Sigma_1^* \cup \Sigma_1^* \cup \Sigma_3^*$; all this space must actually be in $\Sigma_1^* \cup \Sigma_3^*$ because $\Sigma_2^*$ is already covered. $B_0$ is large enough to receive the remaining part $\Sigma_4$ of $A_0$. Hence $\phi(A_0) \subset B_0$; and then $\mu(A_0) = \mu(\phi(A_0)) = \mu(B_0)$ yields $\phi(A_0) = B_0$.

($i = 1$): This is even simpler and therefore deleted.

($i = 2$): $\phi(\Sigma_1 \cup \Sigma_5)$ fits into $\Sigma_1^* \cup \Sigma_5^*$. Then any part of $\Sigma_3$ which is not mapped into $\Sigma_1^* \cup \Sigma_3^*$ is mapped into $\Sigma_5^* \cup \Sigma_7^* \subset B_0$. Finally, $\Sigma_7$ is mapped into $B_2$, too.

The cases $\delta_1 + \delta_2 + \delta_5 \geq 0$ and $\delta_1 + \delta_3 + \delta_5 \geq 0$ are symmetric. Thus it remains to study

**Case 3:** $\delta_1 + \delta_2 + \delta_3 < 0$, $\delta_1 + \delta_2 + \delta_5 < 0$, $\delta_1 + \delta_3 + \delta_5 < 0$. Recall that we can also assume $\delta_2 < 0, \delta_3 < 0, \delta_5 < 0$, and have $\delta_1 + \delta_2 \geq 0$, $\delta_1 + \delta_3 \geq 0$, $\delta_1 + \delta_5 \geq 0$, $\delta_1 > 0$ from (3.2). Let $E$ be a subset of $\Sigma_1^*$ having measure $\mu(E) = \delta_1$, and let $E_2, E_3, E_5$ be three subsets of $E$ with

$$\mu(E_i) = |\delta_i| \quad (i = 2, 3, 5)$$

such that no point of $E$ belongs to all three sets $E_2, E_3, E_5$. It is possible to find such subsets since (3.2) holds and

$$2\delta_1 + \delta_2 + \delta_3 + \delta_5 \geq 0.$$

Using $\delta_1 + \delta_2 + \delta_3 < 0$, we can assume that $E = E_2 \cup E_3$. (The sets $E_2, E_3$ need not be disjoint.)

As $\mu(\Sigma_5) - \mu(\Sigma_5^*) = -\delta_5 = \mu(E_5)$, and as $E_2 \cap E_5$ and $E_3 \cap E_5$ are disjoint, we can find three disjoint subsets $\Sigma_{51}, \Sigma_{52}, \Sigma_{53}$ of $\Sigma_5$ such that

$$\begin{aligned}\mu(\Sigma_{51}) &= \mu(\Sigma_5^*)\\ \mu(\Sigma_{52}) &= \mu(E_2 \cap E_5)\\ \mu(\Sigma_{53}) &= \mu(E_3 \cap E_5).\end{aligned}$$

Let $\phi$ be the gmp–transformation obtained by applying Theorem 2.1 with

$$\begin{aligned}
&\Omega_1 = \Sigma_1, \quad \Omega_{11} = \Sigma_1^* \backslash E\\
&\Omega_2 = \Sigma_2, \quad \Omega_{21} = \Sigma_2^*, \quad \Omega_{22} = E_2\\
&\Omega_3 = \Sigma_3, \quad \Omega_{31} = \Sigma_3^*, \quad \Omega_{32} = E_3, \quad \Omega_{33} = \Sigma_4^*\\
&\Omega_4 = \Sigma_{51}, \quad \Omega_{41} = \Sigma_5^*,\\
&\Omega_5 = \Sigma_{52}, \quad \Omega_{51} = E_2 \cap E_5, \quad \Omega_{52} = \Sigma_6^*, \quad \Omega_{53} = B_1 \backslash (E_3 \cap E_5)\\
&\Omega_6 = \Sigma_{53}, \quad \Omega_{61} = E_3 \cap E_5, \quad \Omega_{62} = \Sigma_7^*, \quad \Omega_{63} = B_2 \backslash \Sigma_7^*\\
&\Omega_7 = \Sigma_4, \quad \Omega_{71} = B_0\\
&\Omega_8 = \Sigma_6, \quad \Omega_{81} = B_1\\
&\Omega_9 = \Sigma_7, \quad \Omega_{91} = B_2\\
&\Omega_{10} = \Sigma_8, \quad \Omega_{10,1} = \Omega
\end{aligned}$$

Again, we check that $\phi(A_i) = B_i$ for $i = 0, 1, 2$.

($i = 0$): When $\phi(A_0)$ is constructed, first $\Omega_1 \cap A_0 = \Sigma_1$ goes into $\Sigma_1^* \backslash E$ and fills it up. Then $\Omega_2 \cap A_0$ is mapped. Part of it goes into $\Sigma_2^*$ and fills it up, and the surplus goes into $E_2 \subset \Sigma_1^*$. Now, $\Sigma_2^* \cup E_2$ is filled up.

Next, $\Omega_3 \cap A_0 = \Sigma_3$ is mapped. First $\Sigma_3^*$ is filled, then $E_3 \backslash E_2$ is filled (since $E_2$ was full already). The surplus goes into $\Sigma_4^*$. It fits into this set because $B_0 \backslash \Sigma_4^*$ is already filled up at this time.

$\Omega_4 \cap A_0, \Omega_5 \cap A_0$ and $\Omega_6 \cap A_0$ are empty. $\Omega_7 \cap A_0 = \Sigma_4$. None of this set is mapped to $B_0 \backslash \Sigma_4^*$ since that set is full by now. In view of $\mu(A_0) = \mu(B_0)$ there must be just enough space in $\Sigma_4^*$ left over to receive the image of $\Omega_7 \cap A_0$. Hence $\phi(A_0) = B_0$.

($i = 1$): $\Omega_1 \cap A_1 = \Sigma_1$ goes into $\Sigma_1^* \backslash E$. Next, $\Omega_2 \cap A_1 = \Sigma_2$ goes into $\Sigma_2^* \cup E_2$ and fills it up. $\Omega_3 \cap A_1$ is empty. $\Omega_4 \cap A_1 = \Sigma_{51}$ goes into $\Sigma_5^*$ and fills it. $\Omega_5 \cap A_1 = \Sigma_{52}$. The first priority for this set would be $E_2 \cap E_5$, but all of $E_2$ is already occupied. So, this set is mapped into $\Sigma_6^*$ and into $B_1 \backslash (E_3 \cap E_5)$. There is enough space in these sets since they form the remainder of $B_1$ except for $E_3 \cap E_5$ and the subset $\Sigma_{53}$ of $A_1$ remains to be mapped. (Recall that $\mu(\Sigma_{53}) = \mu(E_3 \cap E_5)$.) Thus, $\Sigma_{52}$ is mapped into $B_1 \backslash (E_3 \cap E_5)$. Next, $\Omega_6 \cap A_1 = \Sigma_{53}$ is mapped, and it gets its first priority since $E_3 \cap E_5$ was kept in reserve. The remaining part of $A_1$ is $\Sigma_6$. It fits into $B_1$. Hence, $\phi(A_1) = B_1$.

($i = 2$): $\Omega_1 \cap A_2 = \Sigma_1$ is mapped onto $\Sigma_1^* \backslash E$. $\Omega_2 \cap A_2$ is empty. $\Omega_3 \cap A_2 = \Sigma_3$ is mapped to $\Sigma_3^* \cup E_3$ and fills this set. $\Omega_4 \cap A_2 = \Sigma_{51}$ fits into $\Sigma_5^* \subset B_2$. $\Omega_5 \cap A_2 = \Sigma_{52}$ fits into $E_2 \cap E_5$ which is still unoccupied since $\Sigma_3$ was mapped to $\Sigma_3^* \cup E_3$ and this set is disjoint from $E_2 \cap E_5$. $\Omega_6 \cap A_2 = \Sigma_{53}$ has $E_3 \cap E_5$ as first priority. This set is occuppied, but the next priorities are in $B_2$. $\Omega_7 \cap A_2$ and $\Omega_8 \cap A_2$ are empty. $\Omega_9 \cap A_2 = \Sigma_7$ goes into $B_2$. Thus $\phi(A_2) = B_2$. $\square$

**Remarks:** (1) Clearly, only the measure algebra corresponding to $(\Omega, \mathcal{A}, \mu)$ matters. Thus, we can delete the assumption of existence of $\tau$ if $(\Omega, \mathcal{A}, \mu)$ is nonatomic and $\mathcal{A}$ countably generated.

(2) For $n = 2$,the present conditions are equivalent to the requirement that $\mu(B_i) = \mu(A_i)(i = 0, \ldots, n-1)$, and

$$\mu(A_{\cap I}) \leq \mu(B_{\cap I})$$

for all $I \subset \{0, \ldots, n-1\}$. For $n = 3$ however, the condition that $\mu(A_{\cup I}) \geq \mu(B_{\cup I})$ for all $I$ cannot be replaced by the condition that $\mu(B_i) = \mu(A_i)$ for all $i$, even when $B_i = A_{i+1}$. The following sets $A_i$ $(i = 0, \ldots, 3)$ in $\Omega = [0, 1]$ with Lebesgue measure $\mu$ can serve as an example: $A_0 := [0, .4]$, $A_1 := [.2, .6]$, $A_2 := [0, .2] \cup [.4, .6]$, $A_3 := [.1, .3] \cup [.4, .5] \cup [.6, .7]$. We have $\mu(A_i) = .4$ for all $i$, $\mu(A_i \cap A_j) = .2$ for all $i \neq j$ and $\mu(A_0 \cap A_1 \cap A_2) = 0 < \mu(A_1 \cap A_2 \cap A_3) = .1$. However, $\mu(A_0 \cup A_1 \cup A_2) = .6 < \mu(A_1 \cup A_2 \cup A_3) = .7$.

## References


[K] Krengel, U. *Generalized measure preserving transformations*, Proceed. Conf. on Almost Everywhere Convergence, Ohio State Univ. 1988, Academic Press (1989), 215-235.

[LW] Lin, M. and R. Wittmann. *Pointwise ergodic theorems for certain order preserving mappings in $L^1$*, to appear in Proceed. Conf. on Almost Everywhere Convergence II, Evanston, 1989.

# POSITIVE ENTROPY IMPLIES INFINITE $L^p$-MULTIPLICITY FOR $p>1$

Anzelm Iwanik
Institute of Mathematics, Technical University of Wrocław
Wybrzeże Wyspiańskiego 27, 50-370 Wrocław, Poland

Let T be an invertible measure preserving transformation (automorphism) of a Lebesgue probability space (X, B, m). The associated operator $U_T f = f \circ T$ acts as an invertible isometry on $L^p(m)$ $(1 \leq p < \infty)$. As in [1], we say that T has <u>$L^p$-simple spectrum</u> if $U_T$ admits an $L^p$-cyclic function f, which means that the functions of the form $U_T^k f$ $(k \in Z)$ span a dense subspace in $L^p(m)$. If there exists no finite collection $f_1, \ldots, f_r$ generating $L^p(m)$ in the sense that the set $\{U_T^k f_i : k \in Z,\ i = 1, \ldots, r\}$ is linearly dense in $L^p(m)$ then we say that T has <u>infinite $L^p$-multiplicity</u>.

The question whether Bernoulli automorphisms have $L^1$-simple spectrum was raised by J.-P. Thouvenot and seems to be still open. On the other hand it has been shown in [1] that if T is an ergodic automorphism of a (nontrivial) compact metrizable abelian group then T has infinite $L^p$-multiplicity for $p>1$. The proof in [1] relies on harmonic analysis and in particular exploits the notion of Sidon set and $\Lambda$-set in the dual group.

In the present note we give a purely measure-theoretic proof of the fact that the Bernoulli shift (1/2, 1/2) has infinite multiplicity for any $p>1$. Next, using a classical result of Sinai [3] we conclude that in fact the same is true for any measure preserving invertible transformation of positive entropy.

## 1. <u>Bernoulli shift</u>

Throughout this section T denotes the Bernoulli automorphism (1/2, 1/2), i.e., the 2-sided left shift $(Tx)_n = x_{n+1}$ on $X = \{0, 1\}^Z$ endowed with its product sigma-algebra B and product measure $m = \mu^Z$, where $\mu(0) = \mu(1) = 1/2$. We denote by $B_k$ the sub-sigma-algebra generated by the coordinate function $x_k$ and by $B_k^\infty$ the one generated by $x_k, x_{k+1}, \ldots$ Since $B_0$ and $B_1^\infty$ are stochastically independent and generate $B_0^\infty$, the restricted measure $m|B_0^\infty$ can be represented as a direct product

$$m|B_0^\infty = m|B_0 \otimes m|B_1^\infty .$$

Let $u_0 = 1$ and $u_1 = 2x_0 - 1$. Then $u_0$, $u_1$ form an orthogonal basis for

$L^2(m|B_0)$. Now choose an orthogonal basis $v_0, v_1, \ldots$ in $L^2(m|B_1^\infty)$ such that $v_0 = 1$ and $v_j(x) = 1$ or $-1$. Such a system clearly exists (e.g. the Walsh functions have the required properties). It is now obvious that the functions $u_i v_j$ form an orthonormal basis for $L^2(m|B_0^\infty)$. The functions $u_0 v_j$ form a basis for $L^2(B_1^\infty)$ while the remaining ones, $u_1 v_j$, span the orthogonal complement $L^2(B_0^\infty) \ominus L^2(B_1^\infty)$. For the sake of simplicity we shall write $e_j = u_1 v_j$. As $u_0 v_0 = 1$ we clearly have $\int e_j dm = 0$. Finally, denote

$$e_j^k = U_T^k e_j.$$

Since T is a K-automorphism with distinguished sigma-algebra $B_0^\infty$, it is clear that the functions $e_j^k$ ($j = 0, 1, \ldots;\ k \in Z$) form an orthonormal basis for $L_0^2(m) = 1^\perp$. Note also that $e_j^k(x) \in \{-1, 1\}$. Moreover, for a fixed j, the sequence $e_j^k$ ($k \in Z$) can be treated as a martingale difference sequence since

$$E(e_j^k | B_{k+1}^\infty) = 0$$

for any $k \in Z$. The following lemma plays the role of the corresponding property of $\Lambda$-sets in [1]. It is in fact a form of Khintchine inequality for the functions $e_j^k$ (j fixed) where the classical independence assumption for Rademacher functions has been replaced by the weaker martingale difference condition above. Since the proof differs only little (and in an obvious manner) from the classical proof of Khintchine inequality for Rademacher functions (see [4], V §8), it is omitted here.

<u>Lemma</u> 1. For any $2 < q < \infty$ there exists a finite positive constant $C_q$ such that if g belongs to the closed linear subspace spanned in $L^2(m)$ by the functions $e_j^k$ (j fixed) then $g \in L^q(m)$ and

$$\|g\|_q \leq C_q \|g\|_2.$$

Every function in $L^2(m)$ can be written as $f = \sum_j \sum_k a_j^k e_j^k + a_0$ ($a_0$, $a_j^k$ are complex numbers). For a finite set

$$J \subset \{0, 1, \ldots\}$$

we denote by $P_J$ the orthogonal projection

$$P_J f = \sum_{j \in J} \sum_k a_j^k e_j^k$$

of $L^2(m)$ onto the closed $L^2$-subspace $E_J$ spanned by $\{e_j^k : j \in J, k \in Z\}$. Clearly, $E_J$ is a $U_T$-invariant subspace and $U_T P_J = P_J U_T$.

<u>Lemma</u> 2. For every $1 < p < 2$ the projection $P_J$ extends uniquely to a continuous linear operator $\tilde{P}_J$ from $(L^p(m), \|\cdot\|_p)$ onto $(E_J, \|\cdot\|_2)$. Moreover, $U_T \tilde{P}_J = \tilde{P}_J U_T$.

*Proof*. Let f and g be in $L^2(m)$ and let $q = p/(p-1)$. By Lemma 1 we have $P_J g \in L^q(m)$ so

$$|(P_J f, g)| = |(f, P_J g)| \leq \|f\|_p \sum_{j\in J} \|P_j g\|_q$$

$$\leq \|f\|_p C_q \sum_{j\in J} \|P_j g\|_2$$

$$\leq C_q |J|^{1/2} \|g\|_2 \|f\|_p .$$

Consequently, $\|P_J\|_{p,2} \leq C_q |J|^{1/2}$ and the rest is clear.

*Proposition*. The Bernoulli shift (1/2, 1/2) has infinite $L^p$-multiplicity for every $p > 1$.

*Proof*. Suppose to the contrary that there exists a finite system $f_1, \ldots, f_r$ generating $L^p(m)$. Choose any J with $|J| > r$. Clearly the operator $U_T$ has Lebesgue spectrum of multiplicity $|J|$ on the Hilbert space $E_J$. On the other hand, assuming without loss of generality that $1 < p < 2$, we conclude from Lemma 2 that the functions

$$\tilde{P}_J f_1, \ldots, \tilde{P}_J f_r$$

generate $E_J$. Consequently, the multiplicity of $U_T$ restricted to $E_J$ does not exceed r, a contradiction.

## 2. Positive entropy

Using some classical ergodic theory we obtain the following stronger result.

*Theorem*. Any automorphism of positive entropy has infinite $L^p$-multiplicity for $p > 1$.

*Proof*. The proof is divided into 5 steps.

1. First assume that T is a Bernoulli automorphism with finite entropy $h(T) \geq \log 2$. By Ornstein's theory [2], the Bernoulli shift (1/2, 1/2) is a factor of T. The associated conditional expectation E commutes with the action, so if $f_1, \ldots, f_r$ generate $L^p$, the functions $Ef_1, \ldots, Ef_r$ must generate the $L^p$ space of the factor system, which is impossible by Proposition.

2. If T is Bernoulli with $h(T) < \log 2$ then let $n > 0$ be such that $h(T^n) = nh(T) \geq \log 2$. If the functions $f_1, \ldots, f_r$ generate $L^p$ for $U_T$ then the functions

$$f_1, \ldots, f_r, U_T f_1, \ldots, U_T f_r, \ldots, U_T^{n-1} f_1, \ldots, U_T^{n-1} f_r$$

generate $L^p$ for $U_{T^n}$, which is impossible by step 1 as $T^n$ is also Bernoulli.

3. Assume T is any ergodic automorphism with positive finite entropy. By Sinai [3], there exists a Bernoulli factor of T, so the result follows from steps 1 and 2.

4. If T is ergodic with infinite entropy then by the definition of h(T) it has a factor $T_0$ with $0 < h(T_0) < \infty$ . The result follows by step 3.

5. Finally, if T is not necessarily ergodic, but $h(T) > 0$, decompose T into its ergodic components. If the number of components is finite then there exists A with $0 < m(A) \leq 1$ such that $T|A$ is ergodic. The continuous projection $f - f|A$ maps $L^p(m)$ onto $L^p(m_A)$ (here $m_A = (m|A)/m(A)$ ) and commutes with the action of T, so the result easily follows from steps 3 and 4.

If the number of ergodic components is greater than r (possibly infinite) then no system $f_1, \ldots, f_r$ generates $L^p$ for $1 \leq p < \infty$. In fact let $A_1, \ldots, A_{r+1}$ be disjoint subsets of positive measure. Consider the r+1-dimensional vectors

$$w_i = \left( \int_{A_1} f_i dm, \ldots, \int_{A_{r+1}} f_i dm \right).$$

Since there are only r vectors, there exists $(a_1, \ldots, a_{r+1}) \in C^{r+1}$ which is nonzero and orthogonal to each $w_i$. Now let

$$h = \sum_{j=1}^{r+1} \bar{a}_j 1_{A_j} .$$

Clearly $0 \neq h \in L^\infty$, h is $U_T$-invariant, and $\int f_i h dm = 0$ for $i = 1, \ldots, r$; consequently,

$$\int U_T^k f_i h dm = 0 \qquad (k \in Z,\ i = 1, \ldots, r)$$

so $f_1, \ldots, f_r$ do not generate $L^p$ .

## References


[1] A. Iwanik, The problem of $L^p$-simple spectrum for ergodic group automorphisms, Bull. Soc. Math. France, to appear

[2] D. Ornstein, Ergodic Theory, Randomness, and Dynamical Systems, Yale Mathematical Monographs, 1974

[3] Ya. Sinai, On a weak isomorphism of transformations with invariant measure (in Russian), Mat. Sbornik 63 (1964), 23-42

[4] A. Zygmund, Trigonometric Series I, Cambridge University Press 1959

# ON MIXING GENERALIZED SKEW PRODUCTS

Zbigniew S. Kowalski
Institute of Mathematics, Technical University of Wrocław
Wybrzeże Wyspiańskiego 27, 50-370 Wrocław, Poland

Let us consider the one-parameter family $\{T_\mu\}_{\mu \in (a,b)}$ of transformations of the interval $[0,1]$ into itself, such that

$$T_\mu^{-1}(y) = (1-\mu)y+\mu g(y)$$

where $g \in C^2[0,1]$, $g(0) = 0$, $g(1) = 1$, and $a = (1-\sup g')^{-1}$, $b = (1-\inf g')^{-1}$. Moreover, let us assume that there exists exactly one point $y_\circ$, for which $g'(y_\circ) = 1$. Let $\sigma: X^N \to X^N$ be a one-sided $(p_1,\dots,p_s)$ - Bernoulli shift. Here $X = \{1,2,\dots,s\}$. We take $s$ functions $T_{\mu_1},\dots,T_{\mu_s}$, $\mu_i \neq \mu_j$, for $i \neq j$, such that $\sum_{i=1}^{s} p_i\mu_i = 0$ and we define the transformation

$$\bar{T}(x,z) = (\sigma(x), T_{\mu_{x(1)}}(z)).$$

$\bar{T}$ preserves the product measure $p \times m$, where $p$ is the product measure on $X^N$ and $m$ is the Lebesgue measure on $[0,1]$. Moreover, $\bar{T}$ is ergodic, for $s = 2$ and weakly mixing for $s \geqslant 3$ what has been observed in [1]. Our purpose is to prove that $\bar{T}$ is mixing. Let $P_{\bar{T}}$ denote the Frobenius-Perron operator for $\bar{T}$. Then

$$P_{\bar{T}}(g(x)f(y)) = \sum_{i=1}^{s} p_i g(ix)(T_{\mu_i}^{-1})'(y)f(T_{\mu_i}^{-1}y),$$

where $ix = (i,x_1,x_2,\dots)$, for $x = (x_1,x_2,\dots)$. Here $g \in L_1(p)$ and $f \in L_1(m)$. Particularly

$$P_{T}f(y) = \sum_{i=1}^{s} p_i(T_{\mu_i}^{-1})'(y)f(T_{\mu_i}^{-1}y) \quad \text{for } f \in L_1(m).$$

We obtain

<u>Lemma</u>. The iterations $P^n_T f$ converge weakly to $\int_0^1 f\,dm$ for every $f \in L_1(m)$.

Proof. We begin by introducing the auxiliary operator

$$H_T f(y) = \sum_{i=1}^{s} p_i f(T^{-1}_{\mu_i} y)\ .$$ Let $f \in C^1[0,1]$. Then

$$(H_T f)'(y) = \sum_{i=1}^{s} p_i (T^{-1}_{\mu_i})'(y) f'(T^{-1}_{\mu_i} y) = P_T f'(y)\ .$$ Hence

1) $(H^n_T f)'(y) = (P^n_T f')(y)$ for $n = 1,2,\ldots$ .

If a function $f$ belongs to $C^2[0,1]$ and $f'' \geqslant 0$, $f' \geqslant 0$ then

$$H_T f(y) = \sum_{i=1}^{s} p_i f(T^{-1}_{\mu_i} y) \geqslant f\Big(\sum_{i=1}^{s} p_i T^{-1}_{\mu_i} y\Big) = f(y)\ ,$$

by convexity of $f$. Therefore $H^{n+1}_T f \geqslant H^n_T f$, for $n = 1,2,\ldots$ . Let $f^*(y) = \lim_{n\to\infty} H^n_T f(y)$ for every $y \in [0,1]$. Due to $\sup H^n_T f \leqslant \sup f$ for $n = 1,2,\ldots$, $f^*$ is bounded. Since $\inf f' \leqslant P^n_T f' \leqslant \sup f'$ for $n = 1,2,\ldots$, the function $f^*$ is Lipschitzian one. By $f^* = H_T f^*$ and because of (1) we get $f^{*\prime} = P_T f^{*\prime}$ a.e. Therefore $f^{*\prime}$ is a density of $T$-invariant measure. By ergodicity of $T$, $f^{*\prime} = \text{const}$ a.e. Hence $\text{const} = f^*(1)-f^*(0) = f(1)-f(0)$ and $f^*(y) = (f(1)-f(0))y+f(0)$. By using the equality (1) for a function $f \in C^1[0,1]$ such that $f \geqslant 0$, $f' \geqslant 0$ we get

$$\lim_{n\to\infty} \int_0^y P^n_T f\,dm = \lim_{n\to\infty} H^n_T \Big(\int_0^y f\,dm\Big) = \Big(\int_0^1 f\,dm\Big) y\ ,$$

for every $y \in [0,1]$. This implies that $\lim_{n\to\infty} P^n_T f = \int_0^1 f\,dm$ in weak convergence. Due to the fact that the set $\{f : f \in C^1[0,1],\ f \geqslant 0,\ f' \geqslant 0\}$ is a linear dense subset of $L_1(m)$ we get

$$\lim_{n\to\infty} P^n_T f = \int_0^1 f\,dm$$ in weak convergence for every $f \in L_1(m)$.

<u>Theorem</u>. For every $h \in L_1(p \times m)$, $\lim_{n\to\infty} P^n_{\bar{T}} h = \int h \, dp \times m$ in weak convergence.

Proof. Let $h(x,y) = g(x)f(y)$, where $g(x) = 1_{A_{i_1 \ldots i_n}}(x)$ and $A_{i_1 \ldots i_n} = \{x : x_1 = i_1, \ldots, x_n = i_n\}$. Then

$$P_{\bar{T}}(g(x)f(y)) = p_{i_1} 1_{A_{i_2 \ldots i_n}}(x)(T^{-1}_{\mu_{i_1}})'(y) f(T^{-1}_{\mu_{i_1}} y)$$

Hence

$$P^n_{\bar{T}}(g(x)f(y)) = p_{i_1} \ldots p_{i_n} (T^{-1}_{\mu_{i_1}} \circ \ldots \circ T^{-1}_{\mu_{i_n}})'(y) f(T^{-1}_{\mu_{i_1}} \circ \ldots \circ T^{-1}_{\mu_n} y) =$$

$$f_1(y) \quad \text{and} \quad P^{n+k}_{\bar{T}}(gf) = P^k_{\bar{T}} f_1 \, , \quad \text{for } k = 1,2,\ldots \, .$$

By Lemma $\lim_{k\to\infty} P^{n+k}_{\bar{T}}(gf) = \lim_{k\to\infty} P^k_{\bar{T}} f_1 = \int f_1 \, dm = m(A_{i_1 \ldots i_n}) \int f \, dm = \int gf \, dp \times m$ in weak convergence. The linear density of the set $\{gf : g = 1_{A_{i_1 \ldots i_n}}$ for some sequence $i_1, \ldots, i_n$ and $f \in L_1(m)\}$ in $L_1(p \times m)$ implies the thesis .

<u>Conclusion</u>. The transformation $\bar{T}$ is mixing one .

<u>References</u>.

[1]. Z.S.Kowalski, Stationary perturbations based on Bernoulli processes, Studia Mathematica vol. 97 (1) (1990) pp. 53-57 .

# ERGODIC PROPERTIES OF THE STABLE FOLIATIONS

*François LEDRAPPIER*[*]

**Summary** : We describe some properties of the harmonic measures associated with the stable and the strong stable foliations of a geodesic flow on a negatively-curved manifold.

There is a close analogy between 1-dimensional foliations of compact manifolds and flows. In particular, qualitative behavior of the foliation could be described by ergodic notions such as invariant measures, ergodicity, ....

For higher dimensional foliations, invariant measures do not always exist. But always exist harmonic measures, i.e. measures which are invariant under the brownian motion on the leaves [G]. The question arises of an ergodic theory of such objects (see [H], [W]). Here we first want to understand a few examples.

With the help of recent work of Kaimanovich (see [K1], [K2]) we are able to describe the classical example of foliations with some complicated behavior of leaves, namely the stable and the strong stable foliations associated with the geodesic flow on a negatively-curved compact manifold.

Here we show that both foliations have a unique harmonic measure when endowed with the metric lifted from the manifold. We describe these measures and some of their properties. We also recall their role in the "rigidity" problem. Many of these properties were also observed by U. Hamenstädt, V. Kaimanovich, G. Knieper, C-B. Yue. See [K6], [Y2] for independent descriptions.

## I - General properties, notations.

*a) Brownian motion* :

We consider in all this paper a compact connected boundaryless negatively-curved manifold $M$, its universal cover $\tilde{M}$ and the absolute $\partial\tilde{M}$, i.e. $\partial\tilde{M}$ is the space of ends of geodesics in $\tilde{M}$. Let $T\tilde{M}$ be the unit tangent bundle of $\tilde{M}$.

---

* Laboratoire de Probabilités
Université Paris VI, Tour 56
4, Place Jussieu

75252 PARIS Cédex 05

For $\tilde{X}$ in $T\tilde{M}$, denote by $\gamma_X : \mathbb{R} \to \tilde{M}$ the unique geodesic in $\tilde{M}$ such that $(\gamma_{\tilde{X}}(0)\ ,\ \dot{\gamma}_{\tilde{X}}(0)) = \tilde{X}$. The geodesic flow $\tilde{\phi}_t$ is defined by :

$$\tilde{\phi}_t\, \tilde{X} = (\gamma_{\tilde{X}}(t)\ ,\ \dot{\gamma}_{\tilde{X}}(t))\ ,\ t \in \mathbb{R}.$$

We shall identify the unit tangent bundle $T\tilde{M}$ with $(\tilde{M}\times\partial\tilde{M})$ by associating to any unit vector $\tilde{X}$ the pair $(\gamma_{\tilde{X}}(0)\ ,\ \gamma_{\tilde{X}}(+\infty))$.

The set of $\tilde{Y}$ with $\gamma_{\tilde{Y}}(\infty) = \gamma_{\tilde{X}}(\infty)$ is the stable manifold $\tilde{W}^s(\tilde{X})$. Clearly, $\tilde{W}^s(\tilde{X})$ is identified with $\tilde{M}$. For $\tilde{X}$ in $T\tilde{M}$ we denote $\psi_{\tilde{X}} : \tilde{M} \to \mathbb{R}$ the Busemann function at $\gamma_{\tilde{X}}(+\infty)$ vanishing at $\gamma_{\tilde{X}}(0)$, $\psi_{\tilde{X}}$ is defined by $\psi_{\tilde{X}}(y) = \lim_{t\to+\infty} d(y,\ \gamma_{\tilde{X}}(t)) - t$. The set of $\tilde{Y}$ in $W^s(\tilde{X})$ with $\psi_{\tilde{X}}(\gamma_{\tilde{Y}}(0)) = 0$ is the strong stable manifold $\tilde{W}^{ss}(\tilde{X})$, identified with the level function of $\psi_{\tilde{X}}$.

We put on each (strong) stable leaf the metric lifted from (the restriction of) the original Riemannian metric on $\tilde{M}$. In particular the Laplace operator and the Brownian motion on each stable leaf is obtained by lifting Laplace operator and Brownian motion from $\tilde{M}$.

We recall some of their properties :

For $\tilde{x}$ in $\tilde{M}$, the Brownian motion starting from $\tilde{x}$ is the probability measure $\mathbb{P}_{\tilde{x}}$ on $C(\mathbb{R}_+\ ,\ \tilde{M})$ such that the coordinate process $\{\tilde{\omega}(t),\ t \geq 0\}$ is a Markov process with generator $\Delta$ and such that $\tilde{\omega}(0) = \tilde{x}$.

Almost every trajectory $\tilde{\omega}(t)$ converges in $\tilde{M} \cup \partial\tilde{M}$ towards a point $\tilde{\omega}(\infty)$ in $\partial\tilde{M}$ [P2]. The distribution of $\tilde{\omega}(\infty)$ is the harmonic measure $\tilde{\mu}_{\tilde{x}}$. For $x$ in $M$, we call the <u>spherical harmonic measure</u> $\mu_x$ the projection on $T_xM$ of the image of $\tilde{\mu}_{\tilde{x}}$ on $T_{\tilde{x}}\tilde{M}$, for any lift $\tilde{x}$ of $x$.

For $\tilde{x},\tilde{y}$ in $\tilde{M}$, the measures $\tilde{\mu}_{\tilde{x}}\ ,\ \tilde{\mu}_{\tilde{y}}$ have the same negligible sets and the Radon-Nykodym density $\dfrac{d\tilde{\mu}_{\tilde{y}}}{d\tilde{\mu}_{\tilde{x}}}(\xi)$ is given by a continuous function, the <u>Poisson kernel</u> $k(\tilde{x}\ ,\ \tilde{y}\ ,\ \xi)$ (see [AS1], [A] for further properties of $k$).

Finally we recall that there exists a positive number $\alpha$ such that for almost every trajectory in $C(\mathbb{R}_+\ ,\ \tilde{M})$

$$\lim_{t\to\infty} \frac{1}{t} \; d(\tilde{x}, \tilde{\omega}(t)) = \alpha$$

(see [V], [K1]).

*b) Geodesic flow :*

The projection $\pi$ from $\tilde{M}$ to $M$ extends to a projection from $T\tilde{M}$ to $TM$. The above defined $\tilde{\phi}$, $\tilde{W}^s$, $\tilde{W}^{ss}$ factorize on $TM$ into respectively the geodesic flow $\phi$, the stable manifold $W^s$, the strong stable manifold $W^{ss}$. The Liouville measure $\bar{m}$, i.e. the normalized volume element of the Sasaki metric on $TM$, is an invariant ergodic probability measure under the geodesic flow ([AS2]. We write it as

$$\int_{TM} f \, d\bar{m} = \int_M \left( \int_{T_xM} f(X) \, d\lambda_x \right) dm(x).$$

From the proof in [AS2], follows also that the corresponding measures $\lambda_{\tilde{x}}$, $\lambda_{\tilde{y}}$ on $\partial\tilde{M}$ have the same negligible sets.

In fact, the family $W^s$ form a continuous foliation of the compact manifold $TM$ and the absolute continuity of $W^s$ means that this foliation preserves the Lebesgue measure class on transversals, in particular on spheres.

*c) Margulis measures :*

Let $\tilde{x}$ be a point in $\tilde{M}$, $\tilde{S}_R(\tilde{x})$ the set of points $\tilde{y}$ such that $d(\tilde{x},\tilde{y}) = R$, $\tilde{V}_R(\tilde{x})$ the $(n-1)$-dimensional volume of $\tilde{S}_R(\tilde{x})$. Then there exists a positive number $h$ and a continuous function $c$ on $M$ such that, as $R$ goes to infinity :

$$\tilde{V}_R(\tilde{x}) \sim c(\pi\tilde{x}) \, e^{hR} \qquad [M2].$$

The number $h$ is the topological entropy of the geodesic flow $(TM, \phi_1)$. There exists a unique (up to multiplication by a constant) family of measures $\nu_X^{ss}$ on the strong stable manifolds $W^{ss}(X)$ such that

$$\phi_t \, \nu_X^{ss} = e^{-ht} \, \nu_{\phi_t X}^{ss} \qquad [M3].$$

The families $\nu^{ss}$ on $W^{ss}$ and $\nu^s$ on $W^s$ defined by $\nu^s = \nu^{ss} \otimes e^{-ht} \, dt$ are called the Margulis measures.

For $X$ in $TM$, the set of $-Y$, $Y$ in $W^s(-X)$, is the unstable manifold $W^u(X)$. The family of unstable manifolds $W^u$ form a continuous foliation which is transversal to $W^{ss}$. The corresponding Margulis measures $\nu^u$ can be characterized (up to a multiplicative constant) as the only family of measures on transversals to the foliation $W^{ss}$ which is invariant under the foliation [BM].

## II. Harmonic measure for the stable foliation.

Let $N$ be a compact space and $\mathcal{F}$ a $C^{0,2}$ foliation of $N$. We recall that a harmonic measure associated to a $C^{0,2}$ metric on $\mathcal{F}$ is a probability measure $m$ on $N$ which satisfies, for all $C^{0,2}$ function $f$

$$\int \Delta f \, dm = 0 ,$$

where $\Delta$ is the Laplace Beltrami operator along the leaf. (An object is $C^{0,2}$ if it is $C^2$ along the leaf and the first two leaf derivatives and the object itself depend continuously on the point). Equivalently a probability measure $m$ is harmonic if and only if one can find for each leaf $\ell$ a positive harmonic function $k_\ell$ on $\ell$ such that whenever one takes a measurable partition $\chi$ subordinate to $\mathcal{F}$, the conditional measures associated to $\chi$ are proportional to $k_{\ell(x)} dV_{\ell(x)}$, where $dV_\ell$ is the Riemannian volume on $\ell$ (see [G]).

We consider in this section the stable foliation $W^s$ of $TM$, endowed with the metric lifted $\tilde{M}$. Then, we have :

*Proposition 1. There exists a unique harmonic measure for the stable foliation. This measure $m^s$ can be described by :*

$$\int f \, dm^s = \int \left( \int_{T_x M} f(X) \, d\mu_x \right) dm(x)$$

Proof : Let $m^s$ a harmonic measure. First the projection of $m^s$ on $M$ is the normalized volume $m$.

In fact for any $f$ $C^2$ function on $M$, $f \circ \pi$ is a $C^{0,2}$ function on $TM$ and $m^s$ harmonic implies

$$\int \Delta f \, d(m^s \circ \pi^{-1}) = 0.$$

The only invariant measures for the Brownian motion on $M$ are proportional to the volume.

Then we construct the corresponding Markov process. For any $X$ in $TM$, we choose some lifted $\tilde{X}$ in $T\tilde{M}$, lift the Brownian motion on $\tilde{M}$ to the stable manifold $\tilde{W}^s(\tilde{X})$ and project to $W^s(X)$. This defines a probability measure $\mathbb{P}_X$ on $C(\mathbb{R}_+, TM)$ which is in fact carried by $(C(\mathbb{R}_+, W^s(X))$. Clearly a probability measure $m^s$ is harmonic if and only if it is invariant under this process. Write $m^s = \int m^s_x \, dm(x)$ where $m^s_x$ are conditionals of $m^s$ on spheres $T_xM$. We have for all $f$ in $C^{0,2}(TM)$, all positive $t$ ;

$$\int f \, dm^s = \int_{\substack{x \in M_0 \\ y \in \tilde{M} \\ \xi \in \partial\tilde{M}}} f(\tau(x,y,\xi)) \, p(t,x,y) \, m^s_x(d\xi) dy \, dm(x)$$

where $p(t,x,y)$ is the heat kernel on $\tilde{M}$, $\tau(x,y,\xi)$ is the vector $\tilde{Y}$ in $T\tilde{M}$ such that $\gamma_{\tilde{Y}}(0) = y$, $\gamma_{\tilde{Y}}(\infty) = \xi$ and $M_0$ a compact fundamental domain in $\hat{M}$. Let $\sigma(x,y)$ be the vector $\tilde{Z}$ in $T\tilde{M}$ such that $\gamma_{\tilde{Z}}(0) = y$, $\gamma_{\tilde{Z}}((d(x,y)) = x$. Given $\xi \in \partial\tilde{M}$, as $d(x,y)$ goes to infinity, $\sigma(x,y)$ and $\tau(x,y,\xi)$ get closer and closer, except for $y$ in some neighborhood of $\xi$. For large $d(x,y)$, this neighborhood can be choosen very small and thus the $p(t,x,y)dy$ measure of it can be made very small (see e.g. [KL]). So for large $t$, the above RHS is closer and closer to

$$\int_{\substack{x \in M_0 \\ y \in \tilde{M} \\ \xi \in \partial\tilde{M}}} f(\sigma(x,y)) \, p(t,x,y) \, m^s_x(d\xi) dy \, dm(x)$$

which is the same as

$$\int_{\substack{x \in \tilde{M} \\ y \in M_0}} f(\sigma(x,y)) \, p(t,x,y) \, dx \, dm(y).$$

This last expression converges, by definition of $\tilde{\mu}$ towards

$$\int_{y \in M_0} f(y,\xi) \, d\tilde{\mu}_y(\xi) \, dm(y)$$

which is the formula we wanted.

Remark 1. In the above proof we defined $\sigma(x,y)$ for $x,y$ in $\tilde{M}$ as the element $\tilde{Z}$ of $T\tilde{M}$ such that

$$\gamma_{\tilde{Z}}(0) = y \ , \ \gamma_{\tilde{Z}}((d(x,y)) = x.$$

For a continuous function $f$ on $TM$ and $x$ in $\tilde{M}$ define $f^s_x$ on $\tilde{M}$ by $f^s_x(y) = f(\pi\sigma(x,y))$. The above proof and the Markov property give

$$\lim_{t\to\infty} \int_{\tilde{M}} f^s_x(y)\ p(t,x,y)dy = \int f\ dm^s.$$

Remark 2. By the same argument, there exists a unique harmonic measure $m^u$ for the unstable foliation $W^u$. The measure $m^u$ can be written as

$$\int f\ dm^u = \int_M \left(\int_{T_xM} f(-X)\ d\mu_x\right) dm(x)$$

$$= \lim_{t\to\infty} \int_{\tilde{M}} f^u_{\tilde{x}}(y)\ p(t,\tilde{x},y)\ dy$$

where $\tilde{x}$ projects on $x$ and $f^u_{\tilde{x}}$ is defined by :

$$f^u_{\tilde{x}}(y) = f\left(\pi(-\sigma(\tilde{x},y))\right).$$

## III - Harmonic measure for the strong stable foliation.

We first describe properties of some measure $m^{ss}$ that we shall later identify as the harmonic measure for $W^{ss}$.

*Proposition 2. (G. Knieper [K6]). For any continuous function $f$ on $TM$ the averages*

$$\frac{1}{\mathrm{vol}\ \tilde{S}_R(x)} \int_{\tilde{S}_R(x)} f^s_x(y)\ d\mathrm{vol}(y)$$

*converge uniformly on $\tilde{M}$ towards $\int f\ dm^{ss}$, where $m^{ss}$ is the only probability measure which is locally the product of Lebesgue on $W^{ss}$ and $\nu^u$ on $W^u$.*

Proof : The key observation is that, since the strong stable foliation is uniquely ergodic [BM], the measure $m^{ss}$ is the only possible limit.

In fact any such limit should have conditional measures along strong stable manifolds which are proportional to Lebesgue. See [K6] for details.

*Corollary 1. Let* $f$ *belong to* $C^{0,1}(TM)$. *We have the following formula :*

$$\int_{TM}\left[Xf + (h-B)f\right] dm^{ss} = 0$$

*where* $X$ *is the geodesic spray on* $TM$ *and* $B$ *is the function on* $TM$ *given by factorization of* $\tilde{B}$ *on* $T\tilde{M}$ *, where*

$$\tilde{B}(\tilde{X}) = \Delta\psi_{\tilde{X}}\,(\gamma_{\tilde{X}}(0)).$$

Proof : We may assume that $f$ is in $C^{0,2}(TM)$. Fix $x$ in $\tilde{M}$ and write $\varphi(R)$ for

$$\varphi(R) = e^{-hR}\int_{\tilde{S}_R(x)} f^s_x(y)\, dvol(y).$$

We have, writing $B_R(x,y)$ for the mean curvature of the sphere $\tilde{S}_R(x)$ at $y$ :

$$\varphi'(R) = -h\varphi(R) + e^{-hR}\int_{\tilde{S}_R(x)} (-Xf)^s_x(y) + B_R(x,y)\, f^s_x(y)\, dvol(y)$$

and

$$\varphi''(R) = -2h\varphi'(R) + h^2\varphi(R) + e^{-hR}\int_{\tilde{S}_R(x)} F_R(y)\, dvol(y)$$

for some uniformly bounded $F_R(y)$.

By proposition 2, $\varphi, \varphi'$ converge as $R$ goes to infinity and $\varphi''$ is bounded. Thus $\lim_{R\to\infty} \varphi'(R) = 0$. By proposition 2 we find the formula if we take into account that $B_R(x,y) - B^s_x(y)$ goes to zero uniformly.

*Corollary 2. Let* $f,g$ *be* $C^{0,2}$ *functions on* $(TM, W^s)$ *and* $\Delta^s$ *denote the Laplace operator along stable leaves, we have*

$$\int(\Delta^s + hX)f\cdot g\, dm^{ss} = \int(\Delta^s + hX)g\cdot f\, dm^{ss}.$$

Proof : We first observe that since the measure $m^{ss}$ has conditional measures proportional to Lebesgue on $W^{ss}$ , the measure $m^{ss}$ is symmetric for the Laplace operator $\Delta^{ss}$ along strong stable leaves. Since we have :

$$\Delta^s = X^2 - BX + \Delta^{ss} ,$$

corollary 2 follows if the following integral is symmetric in $f,g$ :

$$\int(X^2 + (h-B)X)f\cdot g\, dm^{ss}.$$

We have :

$$(X^2 + (h-B)X)f \cdot g = (X + (h-B))(Xf \cdot g) - Xf \cdot Xg$$

and Corollary 2 follows from Corollary 1.

*Proposition 3. The measure* $m^{ss}$ *is the unique harmonic measure for the strong stable foliation.*

Proof : Recall that strong stable leaves have subexponential growth. By [K2] harmonic measures are "fully invariant", that is the conditional measures on leaves are proportional to Lebesgue and the transverse measure is invariant. As above by [BM], the only such probability measure is $m^{ss}$.

Remark : As in section 2, the properties of the harmonic measure $m^{su}$ for the strong unstable foliation are obtained by changing $X$ into $-X$.

## IV - APPLICATIONS :

### a) *Invariant measures, measures at infinity.*

We considered three "natural" probability measures on the unit tangent bundle of a negatively curved manifold : $\bar{m}$ , $m^s$ and $m^{ss}$. We considered them as harmonic for some foliation, but there are other equivalent points of view under which they could be more familiar.

In the decomposition $W^{ss} \times W^u$ , all three measures are equivalent to the product of the Lebesgue measure and some measure along unstable manifolds : Lebesgue for $\bar{m}$ , the Margulis measure $\nu^u$ for $m^{ss}$ and some measure $\mu^u$ for $m^s$. These measure classes on unstable manifolds are characterized as the classes of conditional measures along unstable leaves of some $\phi$-invariant measure : the Liouville measure, the measure of maximal entropy and the "harmonic" measure respectively, and one can use the ergodic theory of Anosov systems to study these conditional measures (see e.g. [H1] [L2] [K3] for examples of such constructions). Remark that in the same way $m^u$ ($m^{su}$) is, in the decomposition $W^s \times W^{su}$ equivalent to the product of some measure $\mu^s$ ($\nu^s$) on the stable manifolds and the Lebesgue measure on $W^{su}$. These measures are also characterized as conditional measures along stable leaves of the harmonic measure (of the measure of maximal entropy).

Alternatively, as explained in II and III, conditional measures on spheres of the measures $\bar{m}$ and $m^s$ correspond to natural families of measures "at infinity", i.e. measures on $\partial\tilde{M}$ : the visibility (or geodesic) measure $\tilde{\lambda}_x$ for $\bar{m}$ and the harmonic measure $\tilde{\mu}_x$ for $m^s$. In that sense, their properties reflect properties at infinity of $\tilde{M}$.

Remark that the conditional measures of $m^u$ along spheres are given by the images of $\mu_x$ by the symmetry $X \to -X$.

As will be detailed elsewhere, conditional measures on spheres of the measure $m^{ss}$ correspond to the Bowen-Margulis-Patterson-Sullivan measures at infinity (see [L3], [K3]).

*b) The rigidity problem.*

We shall use these measures to express some of the properties linked with "rigidity" properties of manifolds with negative curvature.

Theorem 1. *Let M be a surface with negative curvature. The following properties are equivalent.*

a) *the space M is locally symmetric.*

b) *the function B is constant*

c) *any two of the measures* $\bar{m}$ , $m^s$ , $m^{ss}$ , $m^u$ , $m^{su}$ *coïncide.*

d) *any two of the measures* $\bar{m}$ , $m^s$ , $m^{ss}$ , $m^u$ , $m^{su}$ *are equivalent.*

e) *for all x , the harmonic measure* $\mu_x$ *is quasi-invariant by* $X \to -X$.

Property a) means that M has constant curvature. Property e) is a much weaker form of symmetry of $\tilde{M}$. Theorem 1 is the combination of several facts.

That a <=> b follows from the Riccati equation satisfied by the function B

$$- XB + B^2 + K = 0$$

where K is the curvature of M at $\gamma_X(0)$.

That a) => c) => d) is clear and it follows from the discussion in section a) that $m^s$ is equivalent to $m^u$ if and only if the measures $\mu_x$ are symmetric in the sense of e).

Finally by the discussion in section a), that d) => a) amounts to the following facts : the curvature is constant as soon as the Liouville measure has maximal entropy [K4] , or the $\phi$-invariant harmonic measure is the

Liouville measure (i.e. harmonic measures are equivalent to geodesic measures at infinity [K5] or [L1]), or again the $\phi$-invariant harmonic measure has maximal entropy [L3].

Rigidity problem in this context is to generalize theorem 1 to higher dimensions. We now review known results in this direction.

Clearly a) => the other properties and most other "downwards" implications are also clear. We discuss only "upwards" implications.

That b) => a) in dimension 3 can be seen by considering Riccati equation (see also [H2]). A deep fact is that b) => a) in dimension 4 [H3]. The problem is open in higher dimensions.

That c) => b) is known in several cases : if $\bar{m} = m^{ss}$ or $m^{su}$, then by corollary 1, B is constant. Other known cases ($m^s = m^{ss}$ [L3] or $m^s = \bar{m}$ [Y]) use Kaímanovich theory of entropy (see proposition 5 below).

We still have that $m^s$ is equivalent to $m^u$ if and only if $\mu_x$ is symmetric.

But a very intriguing question is whether d) => b), see e.g. [H2].

*c) Entropy.*

There are different entropies associated to the geodesic flow, e.g. the metric entropy of the Liouville measure $h_{\bar{m}}$ and the topological entropy which coïncide with h [M1]. We have $h_{\bar{m}} \leq h$ with equality if and only if the measures $\bar{m}$ and $m^{ss}$, or $\bar{m}$ and $m^{su}$, are equivalent.

There are also the formulas

$$h_{\bar{m}} = \int B \, d\bar{m} \qquad \text{(see [AS2])}$$

$$h = \int B \, dm^{ss}$$

(apply corollary 1 to a constant function).

There are also different growth rates associated to the Brownian motion on $\tilde{M}$ : the diffusion length $\alpha$ (see I *a)*) and the Kaimanovich entropy $\beta$ defined as the limit for almost every trajectory in $C(\mathbb{R}_+, \tilde{M})$ of

$-\frac{1}{t}\,\ell n\, p(t,x,\tilde{\omega}(t))$ where $p(t,x,y)$ is the heat kernel on $\tilde{M}$, i.e. the density of the distribution of $\tilde{\omega}(t)$.

We have the formulas

$$\alpha = \int B\, dm^s$$

and

$$\beta = \int \|\nabla_y^s\, \ell n\, k(x,y,\xi)\|^2\, dm^s(y,\xi) \text{ (Kaimanovich) [K1].}$$

*Proposition 5. (Kaimanovich) [K1]. We have* $\alpha^2 \le \beta$ *with equality if and only if* $B$ *is constant.*

Proof : (G. Courtois, S. Gallot)

We have from the above formula :

$$\alpha = \int B\, dm^s = -\int\left(\int_{T_xM} (\mathrm{div}\ X)\, d\mu_x\right) dm(x).$$

Remark that $0 = \int \mathrm{div}\left(\int_{T_xM} X\, d\mu_x\right) dm(x)$ and that we can write as before

$$\mathrm{div}\left(\int_{T_xM} X\, d\mu_x\right) = \mathrm{div}\int_{T_x\tilde{M}} X_\xi \frac{d\tilde{\mu}_x}{d\tilde{\mu}_y}(\xi)\, d\tilde{\mu}_y(\xi)$$

$$= \mathrm{div}\int_{\partial\tilde{M}} X_\xi \cdot k(y,x,\xi)\, d\tilde{\mu}_y(\xi)$$

$$= \int_{T_xM} (\mathrm{div}\ X + \langle X\,,\, \nabla_x\, \ell n\, k(y,x,\xi)\rangle)\, d\mu_x.$$

Therefore

$$\alpha = \int \langle X\,,\, \nabla_x\, \ell n\, k(y,x,\xi)\rangle\, dm^s$$ and we have $\alpha^2 \le \beta$ with

equality only if for a.e. $\xi$, $k(\cdot,x,\xi)$ has the same level surfaces as $\psi_{\cdot,\xi}(x)$.

The only possible such function with $\Delta k = 0$ is $e^{-h\psi_\xi}$ and then $B$ has to be constant on a.e. stable leaf, and therefore constant.

*Corollary 3. We have* $\alpha \leq h$ *with equality if and only if* $B$ *is constant.*

Corollary 3 follows from proposition 5 and the observation that $\beta \leq \alpha h$ ([K1]).

Finally we would like to mention the existing definitions of the entropy of a foliation.

Ghys, Langevin and Walczak [GLW] define an entropy, with values 0, finite, or infinite, which measures how complicate the transverse behavior can be.

Walczak [W] studies the entropy of the geodesic flow of the foliation, which seems to combine the transverse structure and some growth rate of the individual leaves (see also [H4]).

Kaimanovich [K2] defines the entropy of a $C^{0,2}$ foliation with a harmonic measure $\mu$ as

$$h(\mathcal{F},\mu) = \int |\nabla^{\mathcal{F}} \log \gamma|^2 \quad du$$

where $\gamma$ is the density of local conditional measures of $\mu$ along leaves.

For the strong stable foliation $W^{ss}$, we have zero entropy with all definitions. For the stable foliation $W^s$, we have $h(W^s,m^s) = \beta$ and the relation with Walczak'entropy deserves further study. For a foliation into trajectories of a measure preserving flow, $h(\mathcal{F}^1,\mu) = 0$ while the geodesic flow of the foliation is the initial flow.

## REFERENCES


[A] **A. Ancona** : Negatively curved manifolds, elliptic operators and Martin boundary. *Ann. of Maths. 125 (1987), 495-536.*

[AS1] **M.T. Anderson and R. Schoen** : Positive harmonic functions on complete manifolds of negative curvature. *Ann. of Maths. 121 (1985) 429-461.*

[AS2] **D.V. Anosov and Ya. Sinaï** : Some smooth ergodic systems. *Russian math. survey 22:5 (1967) 103-167.*

[BM] **R. Bowen and B. Marcus** : Unique ergodicity for horocycle foliations. *Israël J. Maths. 26 (1977) 43-67.*

[G] **L. Garnett** : Foliations, the ergodic theorem and Brownian motion. *J. Func. Anal. 51 (1983) 285-311.*

[GLW] **E. Ghys, R. Langevin and P.G. Walczak** : Entropie géométrique des feuilletages. *Acta Math. 160 (1988) 105-142.*

[H1] **U. Hamenstädt** : An explicit description of the harmonic measure. *Math. Z. 205 (1990) 287-299.*

[H2] **U. Hamenstädt** : Metric and topological entropies of geodesic flow. *Preprint.*

[H3] **U. Hamenstädt** : *In preparation.*

[H4] **S. Hurder** : Ergodic theory of foliations and a theorem of Sacksteder. *Dynamical Systems Proceeding Spec. Year. College Park Md Springer L.N. maths 1342 (1988) 291-328.*

[K1] **V.A. Kaimanovich** : Brownian motion and harmonic functions on covering manifolds. An entropy approach. *Soviet math. Doklady* <u>*33*</u> *(1986) 812-816.*

[K2] **V.A. Kaimanovich** : Brownian motion on foliations : entropy, invariant measures, mixing. *Funct. Anal. Appl.* <u>*22*</u> *(1988).*

[K3] **V.A. Kaimanovich** : Invariant measures of the geodesic flow and measures at infinity on negetively curved manifolds. *Ann. I.H.P. (Physique Théorique),* <u>*53*</u> *(1990) 361-393.*

[K4] **A. Katok** : Entropy and closed geodesics. *Erg. Th. Dynam. Sys.* <u>*2*</u> *(1982) 339-365.*

[K5] **A. Katok** : Four applications of Conformal Equivalence to Geometry and Dynamics. *Erg. Th. & Dynam. Sys.* <u>*8*</u>* *(1988) 139-152.*

[K6] **G. Knieper** : Horospherical measure and rigidity of manifolds of negative curvature. *Preprint*.

[KL] **Y. Kifer, F. Ledrappier** : Hausdorff Dimension of Harmonic Measures on Negatively Curved Manifolds. *T.A.M.S. 318 (1990) 685-704.*

[L1] **F. Ledrappier** : Propriété de Poisson et courbure négative. *C.R.A.S. Paris 305 (1987) 191-194.*

[L2] **F. Ledrappier** : Ergodic properties of Brownian motion on covers of compact negatively-curved manifolds. *Bol. Soc. Bras. Mat. 19 (1988) 115-140.*

[L3] **F. Ledrappier** : Harmonic measures and Bowen-Margulis measures. *Isr. J. Maths. 71 (1990) 275-287.*

[M1] **A. Manning** : Topological entropy for geodesic flows. *Ann. Maths. 105 (1977) 81-105.*

[M2] **G.A. Margulis** : Applications of ergodic theory to the investigation of manifolds of negative curvature. *Funct. Anal. Appl. 3 (1969) 335-336.*

[M3] **G.A. Margulis** : Certain measures associated with U-flows on compact manifolds. *Funct. Anal. Appl. 4 (1970) 55-67.*

[P1] **J. Plante** : Foliations with measure preserving holomony. *Ann. Maths. 102 (1975) 327-361.*

[P2] **J-J. Prat** : Etude asymptotique et convergence angulaire du mouvement brownien sur une variété à courbure négative. *CRAS Paris 280 (1985) 1539-1542.*

[V] **N. Varopoulos** : Information theory and harmonic functions. *Bulletin Sci. Maths. 110 (1986) 347-389.*

[W] **P.G. Walczak** : Dynamics of the geodesic flow of a foliation. *Ergod. Th. & Dynam. Sys. 8, (1988) 637-650.*

[Y] **C-B. Yue** : Contribution to Sullivan's conjecture. *Preprint.*

[Y2] **C-B. Yue** : Brownian motion on Anosov foliation, integral formula and rigidity. *Preprint.*

# ERGODIC THEOREM ALONG A RETURN TIME SEQUENCE

Emmanuel LESIGNE
Département de Mathématiques
Université de Bretagne Occidentale
6,avenue Le Gorgeu 29287 BREST CEDEX - FRANCE

Abstract. We prove that return time sequences for dynamical systems which are abelian extensions of translations, are universaly good for the pointwise ergodic theorem. This can be used to prove the pointwise ergodic theorem along Morse sequence. This last result can also be proved by means of estimations of trigonometric sums.

Introduction. We study in this note the individual ergodic theorem for averages calculated along some increasing sequences of integers. It is possible to find in [BL] or in [K] (chap. 8) a presentation of the results known in 1985 on this subject. From that time, the most important improvements appear in Bourgain's works ([B1], [B2], [T]).

Definition. *An increasing sequence* $(n_k)_{k\geq 1}$ *of natural integers will be called a good sequence if it satisfies :*
*let* $(\Omega, \mathcal{T}, \mu)$ *be a probability space,* T *a measure preserving transformation on it, and* f *an element of* $L^1(\mu)$ *; then*

$$\lim_{K\to+\infty} \frac{1}{K}\sum_{k=1}^{K} f(T^{n_k}\omega) \quad \textit{exists for } \mu\textit{-almost all } \omega.$$

*The sequence* $(n_k)$ *will be called a very good sequence if moreover, when the dynamical system* $(\Omega, \mathcal{T}, \mu, T)$ *is ergodic,*

$$\lim_{K\to+\infty} \frac{1}{K}\sum_{k=1}^{K} f(T^{n_k}\omega) = \int f\, d\mu \qquad \textit{a.e.}$$

In [B2], J. Bourgain states the following theorem.

Theorem 1. *Let* $(X, \mathcal{C}, m)$ *be a probability space,* S *a measure preserving transformation on it, and* A *a non negligible element of* $\mathcal{C}$. *Then, for* m*-almost all* x *in* X, *the sequence* $\{n \in \mathbb{N} : S^n x \in A\}$ *is good.*

Because of the"m-almost all x", this theorem can never be used to see if a given non periodic sequence $(n_k)$ is good or not. Therefore it is natural to ask when is it possible to replace "m-almost all x" by "all x". According to Brunel and Keane ([BK]), this is possible when X is a compact abelian group with its uniform probability, S is a translation and A is a borelian with negligible boundary. The main aim of this note is to show that this result can be extended to abelian group extensions of translations (theorem 3).

Our principal tool will be the proof of theorem 1 given by Bourgain, Furstenberg, Katznelson and Ornstein in [BFKO]. In their proof appears a criterion to test the "good points x".

Outline of the note. In the first paragraph, we present the [BFKO] result. In the second one, we state and prove the theorem 3 ; we remark that this result applies to Morse sequence. In the last paragraph, we give a sketch of a different proof of the fact that the Morse sequence is very good ; we follow here the Reich's method ([R]).

This work originates in a question by C. Mauduit about Morse sequence.

## 1 - The [BFKO] argument

Remark. Using Birkhoff's ergodic theorem, it is not difficult to see that the statement of theorem 1 is equivalent to the following assertion.

Let $(X, \mathcal{A}, m, S)$ be an ergodic dynamical system and $u \in L^1(m)$. Then, for m-almost all x, we have :

let $(\Omega, \mathcal{T}, \mu, T)$ be a dynamical system and $f \in L^\infty(\mu)$ ; then, for $\mu$-almost all $\omega$,

$$\lim_{n\to+\infty} \frac{1}{n}\sum_{k=0}^{n-1} u(S^k x).\, f(T^k\omega) \text{ exists.}$$

The [BFKO] proof of theorem 1 is short but dense. The following statement is an attempt to summarize their method.

Theorem 2 ([BFKO])

1. *Let* $(u_n)_{n\geq 0}$ *be a bounded sequence of complex numbers such that*

$$(1)\quad \begin{cases} \forall \delta>0, \exists L_\delta>0, \forall L>L_\delta, \exists M_{\delta,L}>0, \forall M>M_{\delta,L}, \\ \frac{1}{M}\operatorname{card}\left\{ m\in[0,M[\ : \forall n\in[L_\delta,L[\cap\mathbb{N},\ \left|\frac{1}{n}\sum_{k=0}^{n-1} u_{m+k}\overline{u_k}\right|<\delta\right\} > 1-\delta. \end{cases}$$

*Then we have : if* $(\Omega, \mathcal{T}, \mu)$ *is a probability space,* T *a measure preserving transformation on it and* $f \in L^1(\mu)$,

$$\lim \frac{1}{n}\sum_{k=0}^{n-1} u_k.\, f(T^k\omega) = 0 \quad \textit{for } \mu\textit{-almost all } \omega.$$

2. *Let* $(X, \mathcal{A}, m)$ *be a probability space,* S *a measure preserving transformation on it and* $u \in L^\infty(m)$.

2.a. *Suppose that* u *is orthogonal, in* $L^2(m)$, *to all the S-eigenfunctions. Then we have, for m-almost all* x,

$$(2)\quad \begin{cases} x \textit{ is generic for } u \textit{ in } (X, \mathcal{A}, m, S) \textit{ and} \\ \textit{for } m\textit{-almost all } x',\ \lim \frac{1}{n}\sum_{k=1}^{n} u(S^k x).\,\overline{u(S^k x')} = 0. \end{cases}$$

2.b. *If* x *satisfies* (2), *then the sequence* $u_n = u(S^n x)$ *satisfies* (1).

2. Return time sequences for abelian extensions of translations

Theorem 3. *Let* G *and* H *be two compact metric abelian groups,* $\alpha$ *an element of* G, *and* $\varphi$ *a measurable map of* G *into* H. *We denote by* S *the transformation of* G x H *defined by*

$$S(g,h) = (g + \alpha, h + \varphi(g)).$$

*We suppose that :*

(a) *the map* $\varphi$ *is continuous outside a closed negligible part of* G.

(b) *the dynamical system* (G x H, S) *is uniquely ergodic.*

(c) *the cocycle* $\varphi$ *is weakly-mixing, that is to say every measurable function defined on* G x H *which is an eigenfunction for* S, *is in fact defined on* G.

*Then we have :*

*for every continuous function* u *on* G x H, *for all* $(g_0, h_0)$ *in* G x H, *the following is true :*

*let* $(\Omega, \mathcal{T}, \mu, T)$ *be a dynamical system and* $f \in L^1(\mu)$ ;

$$\text{for } \mu\text{-almost all } \omega, \ \lim \frac{1}{n} \sum_{k=0}^{n-1} u\left(S^k(g_0,h_0)\right).f(T^k\omega) \ \text{ exists.}$$

*If moreover the function* u *is orthogonal to the* S*-eigenfunctions, that is to say* $\int_H u(g,h)\, dh \equiv 0$, *then the limit is zero.*

Corollary. *Under the hypothesis of theorem* 3, *we have : if* A *is a borelian of* G x H,*with negligible boundary, then, for all* $(g_0, h_0)$ *in* G x H, *the sequence* $\{n \in \mathbb{N} : S^n(g_0, h_0) \in A\}$ *is good ; if moreover, for all* g, $\int_H \chi_A(g,h)\, dh$ *is equal to the measure of* A, *then this sequence is very good.*

This corollary is an easy consequence of theorem 3.

Application to the Morse sequence

Let $(a_n)_{n\geq 0}$ = (011010011001011010...) be the Morse sequence ; it is caracterised by $a_0 = 0$, $a_{2n} = a_n$, $a_{2n+1} = 1 - a_n$.

Denote by $(n_k)_{k\geq 1}$ = (1,2,4,7,8,11,...) the sequence of integers n such that $a_n = 1$; it is the sequence of natural integers whose writing in base two have an odd number of 1.

Consider the adding-machine $G = \{0, 1\}^{\mathbb{N}}$, $\alpha = (1,0,0,0,...)$, $H = \mathbb{Z}/2\mathbb{Z}$ and set, for $g = (\varepsilon_n)_{n\geq 0} \in G$,

$$\varphi(g) = \begin{cases} 0 \text{ if } \inf \{n : \varepsilon_n = 0\} \text{ is odd} \\ 1 \text{ if not} \end{cases}.$$

The preceding corollary, applied to $A = \{(g,h) : h = 1\}$ and $(g_0, h_0) = (0,0)$, allows us to assert that the sequence $(n_k)$ is very good.

---

The end of this paragraph is devoted to the proof of theorem 3. We shall use theorem 2 and the two following propositions.

Unique ergodicity of the skew-product

The study of unique ergodicity of a skew-product with uniquely ergodic base has been done by H. Furstenberg in [F1], under the hypothesis of cocycle continuity. In [C], J.P. Conze has noticed that this study extends under the hypothesis (a).

We denote by U the group of complex numbers of modulus one, and by $m_G$, $m_H$ the Haar probabilities of the compact abelian groups G, H.

Proposition 1 [C]. Let G, H be compact abelian groups, $\alpha$ an element of G and $\varphi$ a map of G into H. We suppose that $g \to g + \alpha$ is an ergodic translation of G and that $\varphi$ is continuous outside a closed negligible part of G.
We set $S(g,h) = (g + \alpha, h + \varphi(g))$.
Then the following assertions are equivalent :
(b) the dynamical system $(G \times H, S)$ is uniquely ergodic.
(b)' the dynamical system $(G \times H, m_G \times m_H, S)$ is ergodic.
(b)" for all continuous function u on $G \times H$ and for all $(g_0, h_0)$ in $G \times H$,

$$\lim \frac{1}{n} \sum_{k=0}^{n-1} u\left(S^k(g_0, h_0)\right) = m_G \times m_H(u).$$

(b)"' if $\sigma$ is a character of H such that there exists a measurable map $\psi$ from G into U with $\sigma(\varphi(g)) = \psi(g + \alpha) . \overline{\psi(g)}$ a.e., then $\sigma \equiv 1$.

Proposition 2. Under the hypothesis (a), (b) and (c) of theorem 3, we have : for almost all t in G, the dynamical system

(3) $$\left(G \times H, (g,h) \to (g + \alpha, h + \varphi(g+t) - \varphi(g))\right)$$

is uniquely ergodic.

Proof of proposition 2

Let t be an element of G such that the dynamical system (3) is not ergodic. According to proposition 1, we have : there exists $\sigma_t \in \hat{H}$, $\sigma_t \not\equiv 1$, and $\psi_t$ measurable map from G into U such that

$$\sigma_t (\varphi(g+t)). \overline{\sigma_t(\varphi(g))} = \psi_t (g+\alpha) . \overline{\psi_t(g)} .$$

Suppose that the conclusion of proposition 2 is not satisfied. Since $\hat{H}$ is countable, we have : there exists $\sigma \in \hat{H}$, $\sigma \not\equiv 1$, such that, for a non negligible set of t in G,

(4) there exists $\psi_t$ with $\sigma (\varphi(g+t)) . \overline{\sigma(\varphi(g))} = \psi_t(g + \alpha). \overline{\psi_t(g)}$ .

The set of t satisfying (4) is a subgroup of G, stable under the translation by $\alpha$. If it is not negligible it is equal to G. So we have : there exists $\sigma \in \hat{H}$, $\sigma \not\equiv 1$, such that, for all t in G, there exists $\psi_t$ with

$$\sigma (\varphi(g+t)) . \overline{\sigma(\varphi(g))} = \psi_t (g+\alpha) . \overline{\psi_t(g)} .$$

I claim that it is possible to choose $(\psi_t)_{t\in G}$ so that the map $(g,t) \to \psi_t(g)$ is measurable on G x G. I do not give here the detailed proof of this fact.

Consider now the unitary operator $U_1$ of $L^2(G)$ defined by

$$U_1 f(g) = \sigma(\varphi(g)).\ f(g+\alpha)$$

and the unitary operator $U_2$ of $L^2(G)$ defined by

$$U_2 f(g) = \overline{\sigma(\varphi(g))}\ .\ f(g+\alpha)$$

The unitary operator $U_1 \otimes U_2$ acts in $L^2$ (G x G). We define a function F on G x G by $F(g,g') = \psi_{g'-g}(g)$.

We have $\big((U_1 \otimes U_2)\,F\big)(g,g') = \psi_{g'-g}(g+\alpha).\ \overline{\sigma(\varphi(g'))}\ .\sigma(\varphi(g))$ and therefore $(U_1 \otimes U_2)\,F = F$.

The fact that the unitary operator $U_1 \otimes U_2$ admits some non zero invariant vector implies that the unitary operator $U_1$ admits some eigenvector (this is classical ; see for example, [F2] lem. 4-16). If $\sigma(\varphi(g)).\ f(g+\alpha) = \lambda.f(g)$ with f in $L^2(G)$, $f \not\equiv 0$, and $\lambda$ in U, the function $(g,h) \to f(g)\sigma(h)$ is an eigenfunction for the transformation S on G x H ; this contradicts the hypothesis (c) and proves the proposition 2.

<u>Proof of theorem 3.</u>

Let us remark first that, by the ergodic maximal inequality, it suffices to consider functions f which are bounded.

It is not difficult to verify that the set of continuous functions u for which the conclusion of the theorem is true, is a linear subspace, closed in the uniform convergence topology. If u is a character of G, the conclusion of the theorem is an immediate consequence of Birkhoff's theorem. So, to prove the theorem, it suffices to find a family of continuous functions, satisfying the condition (2) for all $x = (g_0, h_0)$, and generating, with $\hat{G}$, a dense linear subspace of the space of continuous functions on G x H.

We consider the functions u of the form $u(g,h) = \gamma(g)\,.\,\sigma(h)$ with $\gamma \in \hat{G}$, $\sigma \in \hat{H}$, $\sigma \not\equiv 1$. We are going to show that these functions satisfy (2), for all $x = (g_0, h_0)$.

Fix $\gamma \in \hat{G}$, $\sigma \in \hat{H}$, $(\sigma \not\equiv 1)$, $g_0 \in G$ and $h_0 \in H$.

According to proposition 1, the point $(g_0, h_0)$ is generic for $u = \gamma \otimes \sigma$.

We have

$$\frac{1}{n}\sum_{k=1}^{n} u\,(S^k(g_0, h_0))\,.\,\bar{u}\,(S^k(g,h)) =$$

$$\frac{1}{n}\sum_{k=1}^{n} u\left(g_0 + k\alpha,\ h_0 + \sum_{j=0}^{k-1} \varphi(g_0+j\alpha)\right).\ \bar{u}\left(g + k\alpha,\ h + \sum_{j=0}^{k-1} \varphi\,(g + j\alpha)\right) =$$

$$\left(\frac{1}{n}\sum_{k=1}^{n} \sigma\left(\sum_{j=0}^{k-1} \varphi(g_0 + j\alpha) - \varphi\,(g+j\alpha)\right)\right).\ \gamma\,(g_0 - g)\,.\,\sigma(h_0 - h)$$

We set $g = g_0 + t$ and $\varphi_t(g) = \varphi(g) - \varphi(g+t)$ and we obtain

$$\left|\frac{1}{n}\sum_{k=1}^{n} u\,(S^k\,(g_0,\,h_0))\;.\;\bar{u}\,(S^k(g,h))\right| = \left|\frac{1}{n}\sum_{k=1}^{n}\prod_{j=0}^{k-1}\sigma\left(\varphi_t\,(g_0+j\,\alpha)\right)\right|.$$

Now the proposition 2 associated with the assertion (b)" of proposition 1, insures that : for almost all t, for all $g_0$,

$$\lim \frac{1}{n}\sum_{k=1}^{n}\prod_{j=0}^{k-1}\sigma\left(\varphi_t\,(g_0+j\,\alpha)\right)=0$$

(Consider, for the dynamical system (3), the ergodic averages of the function $(g,h) \to \sigma(h)$).

This ends the proof of theorem 3.

## 3. Trigonometric sums estimations

We give now, without proof, two results that can be used to prove that Morse sequence is very good.

The following theorem is due to J.I. Reich [R].

Theorem 4. *Let* $(v_n)$ *be a bounded sequence of complex numbers. Suppose that, for all* $\alpha \in \mathbb{R}$, *there is* $C(\alpha)$ *and* $\varepsilon(\alpha)$ *in* $]0,+\infty[$, *such that* $\left|\sum_{k=0}^{n-1} v_k\, e^{ik\alpha}\right| \leq C(\alpha)\, n^{1-\varepsilon(\alpha)}$.

*Then, if* $(\Omega, \mathcal{T}, \mu)$ *is a probability space,* T *a measure preserving transformation on it and* $f \in L^1(\mu)$,

$$\lim \frac{1}{n}\sum_{k=0}^{n-1} v_k .\, f\circ T^k = 0 \quad \textit{almost everywhere.}$$

This result applies to Morse sequence, via the following proposition.

Proposition 3. Denote by $(a_n)$ the Morse sequence.

There exists $\varepsilon > 0$ such that, for all $\alpha \in \mathbb{R}$ and all integer $n > 0$,

$$\left|\sum_{k=0}^{n-1} (-1)^{a_k}\, e^{ik\alpha}\right| \leq 3\; n^{1-\varepsilon}.$$

---

## REFERENCES


[B1] J. Bourgain : "Pointwise ergodic theorems for arithmetic sets"
Publ. Math. IHES, 69, 1989, 5-45.

[B2] J. Bourgain : "Temps de retour pour les systèmes dynamiques"
C.R. Acad. Sci. Paris, t. 306, série I, 1988, 483-485.

[BFKO] J. Bourgain, H. Furstenberg, Y. Katznelson and D. Ornstein : "Return times of dynamical systems"
Appendix to [B1].

[BK] A. Brunel and M. Keane : "Ergodic theorems for operator sequences"
Z. Wahrsch. Verw. Gebiete, 12, 1969, 231-240.

[BL] A. Bellow and V. Losert : "The weighted pointwise ergodic theorem and the individual ergodic theorem along subsequences"
Trans. AMS, Vol. 288, n° 1, (1985) 307-345.

[C] J.P. Conze : "Equirépartition et ergodicité de transformations cylindriques"
Séminaire de Probabilités (I). Université de Rennes - 1976.

[F1] H. Furstenberg : "Strict ergodicity and transformations of the torus"
Amer. J. Math., 83 (1961), 573-601.

[F2] H. Furstenberg : *Recurrence in Ergodic Theory and Combinatorial Number Theory*
Princeton University Press (1981).

[K] U. Krengel : *Ergodic Theorems*
De Gruyter. Studies in Mathematics 6 (1985).

[R] J.I. Reich : "On the individual ergodic theorem for subsequences"
Ann. Proba. 5 (1977), 1039-1046.

[T] J.P. Thouvenot : "La convergence presque sûre des moyennes ergodiques suivant certaines sous-suites d'entiers"
Sém. Bourbaki, 1989-90 n° 719.

# Some limit theorems for Markov operators and their applications

Jan Malczak
Department of Computer Science, Jagiellonian University
30-501 Cracow, Kopernika str. 27

## 1. Introduction

In studying evolution of densities two problems may attract our attention: the asymptotical stability of the process admitting a stationary density and mixing (sweeping) properties of stochastic (Markov) operators. Using some ideas of Foguel [Fog69] we analyze these two problems. Another, stronger version of mixing has been discussed in literature ([KrS69, Lin71, OrS70] and others).

Our goal is to show that for some semigroups of Markov operators, which describe the evolution of the systems the following situation occurs: either there is an invariant density to which other initial densities are attracted as time evolves; or time varying processes do not have a stationary density, but a subinvariant function, and then have a sweeping property. A very important application of this alternative can be seen while studying asymptotic behavior of solutions of parabolic partial differential equations for which the solutions are given by a kernel operator (the fundamental solution). In this case the asymptotic behavior of the solutions depend on the existence of a stationary solution and its summability. Theorem 5.1 says that if there exists a stationary solution then either the (unique) solution is integrable and all solutions are asymptotically stable, or the solutions have a sweeping property on any set on which the stationary solution is integrable. This theorem is applied to study limit behavior of dynamical systems with the presence of noise (in Section 6). The presence of noise in combination with dynamics leads to a situation in which one may describe the global behavior of the system by the evolution of densities. That evolution is described by the Fokker-Planck (parabolic) partial differential equation. The steady-state solutions to the Fokker-Planck equation are known as stationary densities. Using our method we prove the global asymptotic stability of the solutions of the Fokker-Planck equation (see [MLL90]).

## 2. Existence of an invariant density and sweeping for Markov operators

Let (X, Σ, m) be a $\sigma$-finite measure space. A linear operator $P : L^1(m) \mapsto L^1(m)$ is called a Markov operator if $P(D) \subset D$, where $D = \{f \in L^1(m) : f \geq 0, \quad \|f\| = 1\}$ is the set of densities and $\| \, \|$ stands for the norm in $L^1(m)$. By a standard procedure using monotone sequences of integrable functions, any linear positive operator on $L^1(m)$

extends (uniquely) beyond $L^1(m)$ to act on arbitrary nonnegative (possible infinite) measurable functions.
Let a Markov operator be given. A density f is called *stationary* if $Pf = f$. Consequently, a nonnegative measurable function f is called subinvariant if $Pf \le f$. If a Markov operator P has a positive subinvariant function $f_*$ then we can define the Markov operator $\tilde{P} : L^1(X,\Sigma,\mu) \mapsto L^1(X,\Sigma,\mu)$ by letting

$$\tilde{P}h = \frac{P(f_* h)}{f_*} \tag{2.1}$$

where $d\mu = f_* dm$. Clearly $\tilde{P}1 \le 1$, so by the Riesz-Thorin convexity theorem P acts as a positive contraction on any $L^p(\mu)$, $1 \le p \le \infty$. We denote by $\tilde{U}$ the $L^2(\mu)$-adjoined of $\tilde{P}$ as well as its monotone extensions to all the spaces $L^p(\mu)$. Now applying the well-known complex Hilbert space technique to $\tilde{P}$ (see [Fog69, Chapter III]), we define

$$K = \{f \in L^2(\mu) : \| \tilde{P}^n f \|_2 = \| \tilde{U}^n f \|_2 = \| f \|_2, \quad n = 1,2,...\}. \tag{2.2}$$

Then K is a closed sublattice of $L^2(\mu)$ and the operator $\tilde{P}$ is unitary on K. For every $f \in K^\perp$, $\tilde{P}^n f \mapsto 0$ and $\tilde{U}^n f \mapsto 0$ weakly in $L^2(\mu)$. Now let

$$\Sigma_1(\tilde{P}) = \{A \in \Sigma : 1_A \in K\}. \tag{2.3}$$

Then $\Sigma_1(P)$ is a subring of $\Sigma$ on which $\tilde{P}$ and $\tilde{U}$ act as automorphisms. Moreover, K is the closed span in $L^2(\mu)$ of $\{1_A : A \in \Sigma_1\}$ and if $X_1 \in \Sigma$ is minimal in $\Sigma$ such that $A \subset X_1$ (mod $\mu$) for every $A \in \Sigma_1$ then $X_1$ is $\tilde{P}$-invariant. The set $X_1$ will be referred to as the deterministic part of $\tilde{P}$. Finally $X_2 = X \setminus X_1$.

Now, let a family $\mathcal{A} \subset \Sigma$ be given. A Markov operator $P : L^1(m) \mapsto L^1(m)$ is called a sweeping operator on $\mathcal{A}$ if

$$\lim_{n\to\infty} \int_A P^n f dm = 0, \quad A \in \mathcal{A}, \quad f \in D \tag{2.4}$$

The following theorem can be derived from [Fog69, Chapter VIII]

**Theorem 2.1.** Let $P : L^1(m) \mapsto L^1(m)$ be a Markov operator. Suppose that there is a family $\mathcal{A} \subset \Sigma$ such that $\bigcup_n A_n = X$ for some sequence $\{A_n\} \subset \mathcal{A}$. Assume also that P has a positive subinvariant function $f_*$ such that $\int_A f_* dm < \infty$ for all $A \in \mathcal{A}$. If in addition, $\Sigma_1(\tilde{P})$ defined by (2.3) is atomic, then either P has an invariant density or P is a sweeping operator on $\mathcal{A}$.

**Remark.** If a Markov operator is given by the integral formula

$$(Pf)(x) = \int_X K(x,y) f(y) m(dy), \quad f \in L^1(m), \tag{2.5}$$

where K is stochastic kernel, i.e. $K : X \times X \mapsto R_+$ is jointly measurable and

$$\int_X K(x,y) m(dx) = 1, \quad y \in X,$$

then $\tilde{P}$ has the form

$$(\tilde{P}h)(x) = \int_X \tilde{K}(x,y) h(y) \mu(dy), \quad h \in L^1(\mu), \tag{2.6}$$

$d\mu = f_* dm$, $\tilde{K}(x,y) = \frac{K(x,y)}{f_*(x)}$ and $f_*$ is positive and subinvariant for P. It is known [Fel65] that $\Sigma_1(\tilde{P})$ defined for the integral operator (2.6) is atomic.

**Proof of Theorem 2.1.** Let K, $\Sigma_1, X_1, X_2$ be defined as above. We will use the results contained in [Fog69, Chapter VIII]. For every $A \subset X_2$, $\mu(A) < \infty$ and $h \in L^1(\mu)$

(2.7)
$$\lim_{n\to\infty}\int_A \tilde{P}^n h d\mu = \lim_{n\to\infty}\int_A \tilde{U}^n h d\mu = 0.$$

Moreover $\Sigma_1$ consists of sums of atoms: $W_1, W_2, \cdots$. Each atom has a finite measure $\mu$. Let $W_k$ be an atom of $\Sigma_1$. Two possibilities arise: either all the atoms produced by $\tilde{P}^n 1_{W_k}$ are distinct, or there is a smallest index m with $\tilde{P}^m 1_{W_k} = 1_{W_k}$. In the former case $W_k$ is called wandering and we have

(2.8)
$$\lim_{n\to\infty}\int_{W_k} \tilde{P}^n h d\mu = \lim_{n\to\infty}\int_{W_k} \tilde{U}^n h d\mu = 0, \quad h \in L^1(\mu).$$

In the latter case all atoms produced by $\tilde{P}^j 1_{W_k}$ are distinct for $0 \le j \le m-1$, and $W_k$ is called *cyclic of order m*. Then $h_* = \frac{1}{m\mu(W_k)}\Sigma_0^{m-1}\tilde{P}^j 1_{W_k}$ is an invariant density for $\tilde{P}$ and consequently $h_* f_*$ is an invariant density for P. To end the proof we have to verify that P is a sweeping operator on $\mathcal{A}$ assuming that all atoms are wandering. Let $f \in D$ and $A \in \mathcal{A}$ be fixed. Thus $h = \frac{f}{f_*} \in L^1(\mu)$. Therefore

(2.9)
$$\int_A P^n f dm = \int_A \tilde{P}^n h d\mu = \int_{A\cap X_2} \tilde{P}^n h d\mu + \int_{A\cap X_1} \tilde{P}^n h d\mu$$

From (2.7) it follows that the first integral on the right hand side of (2.9) converges to zero. To evaluate the second one, pick $\varepsilon > 0$. There exist a number $M > 0$ and a finite sequence of atoms $W_1, W_2, ..., W_k$ from $\Sigma_1(\tilde{P})$ such that

$$\int_{X_1} (h - M \cdot 1_B)^+ d\mu \le \varepsilon,$$

where $B = \bigcup_1^k W_i$ and $f^+(x) = max(0, f(x))$. Thus

$$\int_{X_1\cap A} \tilde{P}^n h d\mu \le \int_{X_1} \tilde{P}^n[(h - M\cdot 1_B)^+]d\mu + M\int_{X_1\cap A} \tilde{P}^n 1_B d\mu \le \varepsilon + M\int_B \tilde{U}^n 1_{X_1\cap A} d\mu.$$

Since $1_{X_1\cap A} \in L^1(\mu)$, by (2.8) the last integral converges to zero. □

## 3. Alternative: asymptotical stability or sweeping for integral operators

A Markov operator $P : L^1(m) \mapsto L^1(m)$ is called *asymptotically stable* if there is a unique stationary $f_* \in D$ and if $\|P^n f - f_*\| \mapsto 0$ as $n \mapsto \infty$ for every $f \in D$.

Now, we examine a limit behavior of Markov operators admitting a stationary density. We have the following

**Theorem 3.1** Let the Markov operator $P : L^1(m) \mapsto L^1(m)$ defined by (2.5) have a positive invariant density $f_*$. Assume that for every $A, B \in \Sigma$ with positive finite measure the following condition is satisfied:

(3.1)
$$m(A - supp P^n 1_B) \mapsto 0, \quad n \mapsto \infty$$

Then P is asymptotically stable.

**Proof.** *Uniqueness of invariant density.* Assume there are two stationary densities for P, namely, $f_2$ and $f_2$. Set $f = f_1 - f_2$, so we have $Pf = f$. Let further $f = f^+ - f^-$, so that, if $f_1 \neq f_2$, then neither $f^+$ nor $f^-$ are zero. Note that $Pf^+ = f^+$ and $Pf^- = f^-$. Further, there exist sets A, B with positive measure such that $\alpha 1_A \le f^+$, and $\beta 1_B \leq f^-$. Thus, we have $supp\ P^n 1_B \subset supp\ f^-$, which contradicts the condition (3.1).

*Asymptotical stability.* Define, as in the proof of Theorem 2.1, an operator $\tilde{P} : L^1(\mu) \mapsto L^1(\mu)$ by letting $\tilde{P}h = \frac{P(f_* h)}{f_*}$, where $d\mu = f_* dm$. Note that $\tilde{P}1 = \frac{P(f_*)}{f_*} = 1$ and $\mu(X) = 1$. Using (3.1) it is easy to prove that $\Sigma_1(\tilde{P}) = \{\emptyset, X\}$. Thus the operator $\tilde{P}$ is a Harris operator with trivial $\Sigma_1(\tilde{P})$. Then, by [Fog69, Chapter VIII, Theorem E, p.98], we have

$$\| \tilde{P}^n h - 1_X \|_{L^1(\mu)} \mapsto 0, \quad n \mapsto \infty, h \geq 0, \quad \int_X h d\mu = 1 \tag{3.2}$$

Now, pick $f \in D)$, then the function $h = \frac{f}{f_*}$ belongs to $L^1(\mu)$ and from (3.2) we get $\| P^n f - f_* \|_{L^1(m)} \mapsto 0$ as $n \mapsto \infty$. □

A very useful consequence of the preceding results is

**Corollary 3.2.** Let $P : L^1(m) \mapsto L^1(m)$ be a Markov operator given by (2.5). Suppose that there is a family $\mathcal{A} \subset \Sigma$ such that $\bigcup_n A_n = X$ for some sequence $\{A_n\} \subset \mathcal{A}$. Assume that P admits a positive subinvariant function $f_*$ such that $\int_A f_* dm < \infty$ for all $A \in \mathcal{A}$. If (3.1) is satisfied, then either $P$ is asymptotically stable or $P$ is a sweeping operator on $\mathcal{A}$.

**Remark.** The main advantage of our formulation of the classical alternative: either there is an invariant density or P is sweeping with respect to some $\mathcal{A}$'s (see [Fog66]), is that in our case a subinvariant function $f_*$ is chosen in advance and a family $\mathcal{A} \subset \Sigma$ consists of all $A \in \Sigma$ on which $f_*$ is integrable. Also, the condition (3.1) is immediately satisfied if a kernel $K(x, y)$ in (2.5) is positive. These facts simplify applications very much.

**Example.** Let $X = R_+$ with the standard Borel measure. Consider the integral operator $P : L^1[0, \infty) \mapsto L^1[0, \infty)$ of the form

$$(Pf)(x) = \int_x^\infty \psi(\frac{x}{y}) f(y) \frac{dy}{y}, \quad x \geq 0 \tag{3.3}$$

where $\psi : [0, 1] \mapsto R$ is an integrable function such that

$$\psi(z) \geq 0, \quad and \int_0^1 \psi(z) dz = 1 \tag{3.4}$$

Operator (3.3) appears in the right hand side of the Chandrasekhar-Münch equation describing the fluctuations in the brightness of the Milky Way (see Examples 7.9.2 and 11.10.2 in [LaM85]). Here we will discuss the properties of (3.3) independently of this equation. An immediate calculation shows that $f_*(x) = \frac{1}{x}$ is an invariant function for (3.3). On the other hand, we show that there is no invariant density for (3.3). Let $V : R_+ \mapsto R$ be a nonnegative, measurable and bounded function. We have

$$\int_0^\infty V(x) Pf(x) dx = \int_0^\infty V(x) dx \int_x^\infty \psi(\frac{x}{y}) f(y) \frac{dy}{y} = \int_0^\infty f(y) dy \int_0^y \psi(\frac{x}{y}) V(x) \frac{dy}{y}$$

or substituting $\frac{x}{y} = z$

$$\int_0^\infty V(x) Pf(x) dx = \int_0^\infty f(y) dy \int_0^1 \psi(z) V(zy) dz$$

Assuming that for some density $f_* : Pf_* = f_*$ we get

$$\int_0^\infty V(x)f_*(x)dx = \int_0^\infty f(y)dy \int_0^1 \psi(z)V(zy)dz$$

or

$$\int_0^\infty f_*(y)dy \int_0^1 \psi(z)[V(y) - V(zy)]dz = 0 \tag{3.5}$$

Now choose $V : [0,\infty) \mapsto R$ to be positive, bounded and strictly increasing (e.g. $V(z) = \frac{z}{1+z}$) then $V(y) - V(zy) > 0$ for $y > 0, 0 \leq z < 1$ and the integral

$$I(y) = \int_0^1 \psi(z)[V(y) - V(zy)]dz$$

is strictly positive for every $y > 0$. In particular the product $f_*(y)I(y)$ is a non-negative and nonvenishing function. This shows that the equality (3.5) is impossible. Thus by Theorem 2.1 for every $\psi$ satisfying (3.4) the operator (3.3) is sweeping to zero, i.e. $\int_c^\infty P^n f dx \mapsto 0$, $c > 0$ for every $f \in L^1[0,\infty)$. This seems to be quite interesting since it was proved in [LaM85] that the semigroup of Markov operators generated by the Chandrasekhar-Münch operator is asymptotically stable.

## 4. Stochastic semigroups

Let (X,Σ,m) be a $\sigma$-finite measure space. A family of Markov operators $\{P^t\}_{t\geq 0}$ is called a stochastic semigroup if $P^{t_1+t_2} = P^{t_1} \circ P^{t_2}$ and $P^0 = 1$ for all $t_1, t_2 \geq 0$. A stochastic semigroup $\{P^t\}_{t\geq 0}$ is called asymptotically stable if there exists a unique $f_* \in D$ such that $P^t f_* = f_*$ for all $t \geq 0$ and $\lim_{t\mapsto\infty} \|P^t f - f_*\| = 0$ for all $f \in D$.

Now let a family $\mathcal{A} \subset \Sigma$ be given. A stochastic semigroup $\{P^t\}_{t\geq 0}$ is called a sweeping semigroup on $\mathcal{A}$ if $\lim_{t\mapsto\infty} \int_A P^t f dm = 0$ for any $f \in D$ and $A \in \mathcal{A}$.

The following lemmas shows the relationship between the asymptotical stability and sweeping of discrete semigroup $\{P^n\}_{n\in N}$ and the semigroup $\{P^t\}_{t\geq 0}$.

**Lemma 4.1.** Let $\{P^t\}_{t\geq 0}$ be a semigroup of Markov operators. Assume there exist $t_0 > 0$ and a unique $f_* \in D$ such that $P^n f_* := P^{nt_0} f_* = f_*$ and $\|P^n f - f_*\| \mapsto 0$ for $f \in D$ if $n \mapsto \infty$. Then $P^t f_* = f_*$ for all $t \geq 0$ and $\lim_{t\mapsto\infty} \|P^t f - f_*\| = 0$ if $f \in D$.

**Proof.** First we show that $P^t f_* = f_*$ for all $t \geq 0$. Fix $t' > 0$ and set $f_1 := P^{t'} f_*$. Therefore

$$\|P^{t'} f_* - f_*\| = \|P^{t'}(P^{nt_0}) f_* - f_*\| = \|P^{nt_0}(P^{t'} f_*) - f_*\| = \|P^n f_1 - f_*\| .$$

Since $\lim_{n\mapsto\infty} \|P^n f_1 - f_*\| = 0$, we must have $\|P^{t'} f_* - f_*\| = 0$, and hence $P^{t'} f_* = f_*$. At the end, to show asymptotical stability pick a function $f \in D$, so that

$$\|P^t f - f_*\| = \|P^t f - P^t f_*\|$$

is a nonincreasing function. Since for $t_n = nt_0$ we have $\lim_{n\mapsto\infty} \|P^{t_n} f - f_*\| = 0$, we have a nonincreasing function that converges to zero on a subsequence and, hence $\lim_{t\mapsto\infty} \|P^t f - f_*\| = 0$. □

A stochastic semigroup $\{P^t\}_{t\geq 0}$ is continuous if $\|P^t f - f_*\| \mapsto 0$ if $t \mapsto 0$ for all $f \in L^1(m)$.

**Lemma 4.2.** Let $A \subset \Sigma$ be given. Assume that a stochastic semigroup $\{P^t\}_{t\geq 0}$ is continuous. If there exists $s > 0$ such that $\{P^{ns}\}_{n\in N}$ is a sweeping discrete semigroup on $A$, then $\{P^t\}_{t\geq 0}$ is a sweeping semigroup on $A$.

**Proof.** Note that for any fixed integer k, the discrete stochastic semigroup $\{P^{\frac{s}{k}n}\}_{n\in N}$ is sweeping on $A$. From the assumptions for given $\varepsilon > 0$ there exists $\delta > 0$ such that $\|P^r f - f\| < \frac{\varepsilon}{2}$ if $0 \leq r < \delta$ and $f \in L^1(m)$. Choose an integer k so large that $\frac{s}{k} < \delta$. Put $\rho := \frac{s}{k}$. For any fixed $t > 0$ there exists the integer $n(t)$ such that $t = \rho n(t) + r(t)$ with $0 \leq r(t) < \delta$.
Now for $f \in D$ we have

$$\int_A P^t f dm = \int_A (P^t f - P^{\rho n(t)} f) dm + \int_A P^{\rho n(t)} f dm.$$

Further

$$\int_A (P^t f - P^{\rho n(t)} f) dm \leq \int_A P^{\rho n(t)} [(P^{r(t)} f - f)^+] dm \leq \|P^{r(t)} f - f\| \leq \frac{\varepsilon}{2}.$$

Therefore $\int_A P^t f dm \leq \frac{\varepsilon}{2} + \int_A P^{\rho n(t)} f dm$. Since $n(t) \mapsto \infty$ if $t \mapsto \infty$, then the term $\int_A P^{\rho n(t)} f dm < \frac{\varepsilon}{2}$ for t sufficiently large. □

## 5. Asymptotic behavior of solutions of parabolic equations

In order that our results may be applied to the solutions of the equation

(5.1) $$\frac{\partial u}{\partial t} = \sum_{i,j=1}^{d} \frac{\partial}{\partial x_i}(a_{ij}(x)\frac{\partial u}{\partial x_j}) - \sum_{i=1}^{d} \frac{\partial}{\partial x_i}(b_i(x)u) =: Lu, \quad t > 0, \quad x \in R^d,$$

with the initial condition

(5.2) $$u(0, x) = f(x), \quad x \in R^d$$

we assume that the coefficients $a_{ij}$ and $b_i$ are sufficiently regular that the problems (5.1), (5.2) has exactly one bounded solution on the half space $t \geq 0$, $x \in R^d$. For the broad historic background and classical conditions the reader is referred to the monographs [Fri64] and [Eid69] (see also [IKO62, Eid59, Cha70, ArB67a, Arb67b, Has80, Ris84]). Assuming sufficient regularity conditions, for every continuous bounded initial function f there exists unique classical solution of (5.1) and (5.2). The term classical means that for every $T > 0$, $u(t, x)$ is bounded for $t \in (0, T]$, $x \in R^d$. Moreover, $u(t, x)$ has continuous derivatives $u_t$, $u_{x_i}$, $u_{x_i x_j}$ and satisfies equation (5.1) for every $t > 0$, $x \in R^d$; and $\lim_{t\mapsto 0} u(t, x) = f(x)$. The solution $u(t, x)$ is given by the formula

(5.3) $$u(t, x) = \int_{R^d} \Gamma(t, x, y) f(y) dy, \quad x \in R^d, \quad t > 0$$

where $\Gamma$ is the fundamental solution of (5.1). The function $\Gamma(t, x, y)$, defined for $t > 0$, $x, y \in R^d$, is continuous, positive and differentiable with respect to t, is twice differentiable with respect to x, and satisfies (5.1) as a function of (t,x) for every fixed y. Moreover it satisfies the inequality

$$(5.4)\quad |D_t^r D_x^s \Gamma(t,x,y)| \leq A \cdot t^{-\frac{(d+2r+s)}{2}} exp\{-\alpha \frac{|x-y|^2}{t}\}, \quad x,y \in R^d, \quad t>0, \quad 2r+|s| \leq 2$$

with some positive constant A and $\alpha$.
If f is not necessarily continuous but integrable, the formula (5.3) defines a generalized solution of (5.1) for $t > 0$. In this case it satisfies the initial condition of the form

$$(5.5)\qquad \lim_{t \to 0} \|u(t,\cdot) - f\| = 0$$

Using (5.4) it is easy to verify that $u(t,\cdot) \in L^1(R^d)$ for $t \geq 0$. Thus setting

$$(5.6)\qquad P^t f(x) = u(t,x) = \int_{R^d} \Gamma(t,x,y) f(y) dy, \quad t > 0, \quad P^0 f = f,$$

we define a family of operators $P^t : L^1 \mapsto L^1$ which describes the evolution in time of solution $u(t,x)$. Using the specific "divergent" form of equation (5.1) it is easy to verify that $\{P^t\}_{t \geq 0}$ is a stochastic semigroup.

By virtue of the considerations of §§2, 3 and 4, the behavior of the solutions of the Cauchy problem (5.1), (5.2) can be stated as follows

**Theorem 5.1.** Let there exist a positive function $u_* : R^d \mapsto R_+$ satisfying the elliptic equation $Lu = 0$ ($L$ is defined by (5.1)). Suppose that there exists a family $\mathcal{A}$ of Borel sets $A \subset [0,\infty)$ having property: $[0,\infty) = \bigcup_n^\infty A_n$ for some sequence $\{A_n\} \subset \mathcal{A}$ and $u_*$ is integrable on all $A \in \mathcal{A}$. Then either $u_*$ is the unique density and for every $f \in D(R^d)$ the stochastic semigroup $\{P^t\}_{t \geq 0}$ defined by (5.6) is asymptotically stable with the limiting function $u_*$, or the stochastic semigroup $\{P^t\}_{t \geq 0}$ is sweeping on every $A \in \mathcal{A}$.

**Proof.** Assume that $u_*(x)$ is a positive solution of $Lu = 0$. It follows $u(t,x) = u_*(x)$ is a time-independent solution of (5.1). Thus, by (5.3) we have

$$u_*(x) = \int_{R^d} \Gamma(t,x,y) u_*(y) dy = \int_{R^d} \Gamma(1,x,y) u_*(y) dy.$$

Since $\Gamma(1,x,y)$ is strictly positive, then the assumptions of Corollary 3.2 are satisfied. Therefore the operator $P : L^1(R_+) \mapsto L^1(R_+)$ given by $(Pf)(x) = \int_{R^d} \Gamma(1,x,y) f(y) dy$ is either asymptotically stable or P is a sweeping operator on $\mathcal{A}$. Furthermore, the condition (5.5) assures the continuity of the semigroup $\{P^t\}_{t \geq 0}$ given by (5.6). Then by Lemma 4.1 and Lemma 4.2 we get the conclusion of Theorem 5.1. □

## 6. Applications

In order to illustrate the utility of Theorem 5.1 consider one dimensional differential equation:

$$(6.1)\qquad \frac{dx}{dt} = x(c - x^2)$$

and the corresponding stochastic differential equations:

$$(6.2)\qquad \frac{dx}{dt} = x(c - x^2) + \sigma\xi, \quad x \in R, \quad t > 0$$

and

$$\frac{dx}{dt} = x(c - x^2) + x\sigma\xi, \quad x \in R_+, \quad t > 0 \tag{6.3}$$

where $\xi$ is a (Gaussian distributed) white noise perturbation with zero mean and unit variance, $\sigma$ is a positive constant. These equations were investigated in [MLL90].

Under some standard regularity conditions (which are satisfied here), the process $x(t)$ which is the solution of the stochastic differential equation (6.2) or (6.3) has a density function $u(t,x)$ defined by:

$$Prob\{a < x(t) < b\} = \int_a^b u(t,z)dz, \quad a, b \in R.$$

It is well known that the density $u(t,x)$ satisfies the parabolic differential equation (Fokker-Planck equation).

In case (6.2), since the noise amplitude $\sigma$ is constant, equation (6.2) makes it clear that the corresponding Fokker-Planck equation is identical in Ito and Staranovich interpretations. Specifically, it takes the form:

$$\frac{\partial u}{\partial t} = \frac{1}{2}\sigma^2\frac{\partial^2 u}{\partial x^2} - \frac{\partial}{\partial x}[x(c - x^2)u]. \tag{6.4}$$

It is straightforward to show that the stationary solution of the elliptic equation:

$$\frac{1}{2}\sigma^2\frac{\partial^2 u}{\partial x^2} - \frac{\partial}{\partial x}[x(c - x^2)u] = 0$$

is given by

$$u_*(x) = K_1 \cdot exp\{\beta x^2\frac{(2c - x^2)}{4c}\}, \quad \beta = \frac{2c}{\sigma^2} \tag{6.5}$$

There exists the normalization constant $K_1$ such that $u_*(x)$ defined by (6.5) is the positive stationary density. We now turn to a consideration of the asymptotical stability of the stationary density. The Fokker-Planck (6.4) is quite regular since it is uniformly parabolic ($\sigma^2$ is a positive constant) and $x[x(c - x^2)] < 0$ for sufficiently large $x$. These properties ensure that the solutions of (6.4) are given by an integral (5.3) with a sufficiently smooth kernel. So, applying Theorem 5.1 we obtain the asymptotical stability for the stochastic semigroup generated by (6.4) for all constants $c$.

In case (6.3), it is no longer the case that the corresponding Fokker-Planck equation will be the same for the Ito and Stratonovich interpretations. Hence, assume first that we are using the Ito calculus, and replace $c$ by $c_I$ to denote this distinction. Then, the corresponding Fokker-Planck equation is

$$\frac{\partial u}{\partial t} = \frac{1}{2}\sigma^2\frac{\partial^2[x^2u]}{\partial x^2} - \frac{\partial}{\partial x}[x(c_I - x^2)u]. \tag{6.6}$$

The stationary solution $u_*(x)$ of the (Ito) Fokker-Planck equation (6.6) is given by:

$$u_*(x) = Kx^\gamma \cdot exp\{\frac{x^2}{2\sigma^2}\}, \quad \gamma = \frac{2c_I}{\sigma^2} - 2 \tag{6.7}$$

In order that $u_*(x)$ be a density, it must be integrable on $R_+$, and from (6.7) this is only possible if $\gamma > -1$ or $c_I > \frac{1}{2}\sigma^2$. Furthermore, for $\gamma \le -1$ the stationary solution $u_*$ is integrable on every set $[c, \infty)$, $c > 0$. In trying to prove that the stochastic semigroups generated by the equation (6.6) is asymptotically stable for $c_I > \frac{1}{2}\sigma^2$, we no longer can apply immediately Theorem 5.1. This is because the coefficient $\frac{\sigma^2 x^2}{2}$ vanishes at $x = 0$ and the uniform parabolicity conditions violated at $x = 0$. However, by a straightforward change of variables we may transform the Fokker-Planck equation (6.6) to circumvent this problem, and then again apply Theorem 5.1.

Define a new variable $y = \ln x$ and a new density $\tilde{u}$ by

$$\tilde{u}(t,y) = e^{2y} u(t, e^y). \tag{6.8}$$

With these changes, the Fokker-Planck equation (6.6) takes the form:

$$\frac{\partial u}{\partial t} = \frac{1}{2}\sigma^2 \frac{\partial^2 \tilde{u}}{\partial y^2} - \frac{\partial}{\partial y}[(c_I - \frac{1}{2}\sigma^2 - e^{2y})\tilde{u}]. \tag{6.9}$$

As in the case (6.4) the uniform parabolicity condition is now satisfied. Thus, using Theorem 5.1, the asymptotic stability of the equation (6.9) will be demonstrated for $c_I > \frac{\sigma^2}{2}$ which, by the change of variables (6.8), in turn implies the asymptotic stability of the stationary solution of (6.6).

## 7. References


[ArB67a] D.G.Aronson, P.Besala, Uniqueness of positive solutions of parabolic equations with unbounded coefficients, Coll.Math. 18 (1967), 125-135.

[ArB67b] D.G.Aronson, P.Besala, Parabolic Equations with Unbounded Coefficients, J.Diff. Eqs. 3 (1967), 1-14.

[Cha70] J.Chabrowski, Sur la mesure parabolique, Coll.Math. 21 (1970), 291-301.

[Eid59] S.P.Eidelnan, On the Cauchy problem for parabolic systems with increasing coefficients, Dokl.Akad. Nauk SSSR 127 (1959), 760-763 (Russian).

[Eid69] S.D.Eidelman, Parabolic Systems, North-Holland Publ. Company, Amsterdam and Wolters-Noordhoff Publ. Groningen, 1969.

[Fel65] J.Feldman, Integral kernels and invariant measures for Markov transition functions, Ann.Math.Statist., 36 (1965), 517-523.

[Fog66] S.R.Foguel, Limit theorems for Markov processes, Trans. Amer. Math. Soc. 121 (1966), 200-209.

[Fog69] S.R.Foguel, The Ergodic Theory of Markov Processes, Van Nostrand Reinchold, New York, 1969.

[Fog85] S.R.Foguel, Singular Markov operators, Houston J.Math. 11 (1985), 485-489.

[Fri64] A.Friedman, Partial Differential Equations of Parabolic Type, Prentice-Hall, Englewood Cliffs, 1964.

[Has80] R.Z.Hasminski, Stochastic Stability of Differential Equations, Sijhoff and Noordhoff, 1980.

[IKO62] A.M.Il'in, A.S.Kalashnikov and O.A.Olejnik, Second order linear equations of parabolic type, Russian Math. Surveys (Uspiehi Mat.Nauk.), 17 (1962), No.3, 1-143.

[KrS69] U.Krengel, L.Sucheston, On mixing in infinite measure space, Z.Wahr. verw. Geb. 13 (1969), 150-164.

[LaM85] A.Lasota, M.C.Mackey, Probabilistic properties of deterministic systems, Cambridge Univ. Press, 1985.

[Lin71] M.Lin, Mixing for Markov operators, Z.Wahr.verw.Geb. 19 (1971), 231-242.

[MLL90] M.C.Mackey, A.Longtin, A.Lasota, Noise induced global asymptotic stability, J.Stat. Phys. 60 (1990), 735-751.

[OrS70] D.Ornstein, L.Sucheston, An operator theorem on $L^1$ convergence to zero with applications to Markov kernels, Ann.Math.Stat. 41 (1970), 1631-1639.

[Ris84] H.Risken, The Fokker-Planck Equation, Springer Series Synergetics 18, Springer Verlag, 1984.

# GENERIC PROPERTIES OF ONE–DIMENSIONAL DYNAMICAL SYSTEMS

IVAN MIZERA

Comenius University, Bratislava

ABSTRACT. Some generic properties of continuous maps of the interval or the circle are proved, concerning global and local attractors, Ljapunov stability and pseudo-orbit shadowing.

## 1. Introduction

Let $f$ be a continuous mapping of $X$ to itself, where $X$ is the compact interval $I$ or the circle $S^1$. Let $f^n$ denote the n-th iterate of $f$. The theory of such mappings, considered as dynamical systems, has enjoyed a considerable interest in recent years. One of the areas investigated includes the question of the behaviour of generic (sometimes called typical) dynamical systems.

The most widespread notion of genericity is defined via a topology on the space of relevant dynamical systems. A natural choice for the continuous mappings of $I$ or $S^1$ to itself is the $C^0$-topology of the uniform convergence – the corresponding topological space is denoted by $C(X,X)$. We say that a property of functions from $C(X,X)$ is **generic** or holds **generically** if it holds for all $f \in \mathcal{M} \subseteq C(X,X)$, where $\mathcal{M}$ is residual in $C(X,X)$ (for the terminology see [**O**]).

In the sequel, $\mathbb{N}$ stands for the set of positive integers, $\lambda$ stands for Lebesgue measure and $d$ for the usual distance on $X$. The symbols $\operatorname{int} Y$, $\operatorname{diam} Y$, and $\partial Y$ mean the interior, the diameter and the boundary of the set $Y \subseteq X$, respectively.

Let us review some known generic properties of one-dimensional dynamical systems.

THEOREM 0. *There is a residual set $\mathcal{A} \subseteq C(X,X)$ such that for all $f \in \mathcal{A}$*

(i) *the set $B(f)$ of chain recurrent points is of zero Lebesgue measure and is nowhere dense;*
(ii) *the topological entropy $h(f) = \infty$;*
(iii) *all scrambled sets of $f$ have zero Lebesgue measure and are nowhere dense;*
(iv) *every neighbourhood of a periodic point (with a period $k$) contains points with arbitrary high periods ($nk$ for all $n \in \mathbb{N}$);*
(v) *closing lemma: the set $\Omega(f)$ of non-wandering points of $f$ is equal to the closure of $\operatorname{Per}(f)$, the set of periodic points of $f$;*
(vi) *there is no $g \in C(X,X)$ and $n$ such that $f = g^n$.*

The property (i) was proved for $\Omega(f)$ in [**ABL**] and generalized in [**Gr**] for $B(f)$; it implies that generically $f$ is not transitive and has no absolutely continuous invariant measure. The property (ii) is folklore, it implies that, generically, $f \in C(X,X)$ is chaotic in the Li-Yorke sense [**LY**]. However, (iii) indicates that this chaos is not very visible [**Mi**]. The version of (iv) by [**ABL**] was improved to the present one by [**S**]. The closing lemma (v) was proved by [**Y**] including also the $C^r$, $r > 0$ case. The proof of (vi) in [**B2**] shows that (vi) holds for a set with nowhere dense complement. The "zero Lebesgue measure" in (i) and (iii) can be improved to "zero Hausdorff dimension" [**I**].

The aim of this paper is to add some more items to this list. The proofs are postponed to Section 5; they use a general construction, which might be used also for establishing some parts of Theorem 0. The construction admits also multi-dimensional generalizations.

## 2. Attractors

Fix $f \in C(X,X)$, let $x \in X$. The smallest closed set $\omega(x)$ with the property

$$\text{for every } U \text{ open, } \omega(x) \subseteq U, \text{ there exists} \quad \text{(1)}$$
$$n_0 \in \mathbf{N} \text{ such that } f^n(x) \in U \text{ for all } n \geq n_0$$

is called the **$\omega$-limit set** of x. The set of all limit points of the forward trajectory of x coincides with $\omega(x)$.

We obtain another kind of limit set when we consider the smallest closed set $\sigma(x)$ with the property

$$\text{for every } U \text{ open, } \sigma(x) \subseteq U,$$
$$\lim_{n\to\infty} \frac{1}{n} \sum_{i=0}^{n-1} \chi_U(f^i(x)) = 1 \quad \text{(2)}$$

where $\chi_U$ is the characteristic function (indicator) of $U$. The set $\sigma(x)$ is called the **$\sigma$-limit set** of x (or, according to [**ŠKSF**], the **statistically limit set** of $x$, or also the **minimal centre of attraction** of $x$).

Recall the following folklorical properties of limit sets

$$f(\omega(x)) = \omega(x)$$
$$f(\sigma(x)) = \sigma(x) \quad \text{(3)}$$
$$\sigma(x) \subseteq \omega(x)$$

The last inclusion can be sharp; however, there is an important case when the equality occurs. An $\omega$-limit set $\omega(x)$ is called a **solenoid** if it can be written in the form

$$\omega(x) = \bigcap_{i=1}^{\infty} \bigcup_{j=1}^{k_i} I_{ij}$$

where $I_{ij}$, $i = 1, 2, \ldots$, $j = 1, 2, \ldots, k_i$, are segments such that

$$\bigcup_{j=1}^{k} I_{ij} \subseteq \bigcup_{j=1}^{k_{i+1}} I_{i+1,j} \quad \text{for all } i, \quad \text{(4)}$$

$$f(I_{ij}) \subseteq I_{i,j+1 \pmod{k_i}} \quad \text{for all } i, j \quad \text{(5)}$$

and

$$\lim_{i\to\infty} \max_j \operatorname{diam} I_{ij} = 0. \quad \text{(6)}$$

When the sequence $k_1, k_2, \ldots$ is bounded (i.e. eventually constant), the solenoid is "degenerated" – it is a periodic orbit.

**Lemma** 1. *If $\omega(x)$ is a solenoid, then $\sigma(x) = \omega(x)$.*

**Proof**. The statement holds for $\omega(x)$ being a periodic orbit. Assume that $\sigma(x)$ is strictly smaller than $\omega(x)$. Pick $y \in \omega(x)$, $y \notin \sigma(x)$. Since $\sigma(x)$ is closed, there exist open $U$ and $V$ such that $U \cap V = \emptyset$, $\omega(x) \subseteq U$, $y \in V$. By (2)

$$\lim_{n\to\infty} \frac{1}{n} \sum_{i=0}^{n-1} \chi_V(f^i(x)) = 0 \tag{7}$$

Pick $i$ such that, according to (6), $\max_j \operatorname{diam} I_{ij} < \frac{1}{2} \operatorname{diam} V$. Let $y \in I_{ij}$; since $\omega(x)$ is not periodic, there exists $m$ such that $f^m(x) \in I_{ij}$. But then from (5) it follows that $f^{m+k_i}(x), f^{m+2k_i}(x), \cdots \in I_{ij}$ and since $I_{ij} \subseteq V$, we have contradiction with (7).

Now we are ready to define global attractors. Following the approach of Milnor [M], a global attractor is the smallest closed subset of $X$, containing the limit sets of almost all $x \in X$. We can take either $\omega$- or $\sigma$-limit sets and "almost all" can be understood in the sense of category or measure. Let $A \subseteq X$. We use the notation

$$\varrho_\omega(A) = \{x : \omega(x) \subseteq A\}$$
$$\varrho_\sigma(A) = \{x : \sigma(x) \subseteq A\}$$

We obtain four types of global attractors:

- $\Lambda_\omega(f)$ – the smallest closed set such that $\varrho_\omega(\Lambda_\omega(x))$ is of full measure (**likely limit set** in the sense of [M]);
- $\Gamma_\omega(f)$ – the smallest closed set such that $\varrho_\omega(\Gamma_\omega(x))$ is residual (**generic limit set** in the sense of [M]);
- $\Lambda_\sigma(f)$ – the smallest closed set such that $\varrho_\sigma(\Lambda_\sigma(x))$ is of full measure;
- $\Gamma_\sigma(f)$ – the smallest closed set such that $\varrho_\sigma(\Gamma_\sigma(x))$ is residual.

The last two types were introduced in [**ŠKSF**] and the question of their equality to their $\omega$-limit counterparts in the generic case was raised; in [M] the question whether generically $\Lambda_\omega(f) = \Gamma_\omega(f)$ was considered. The partial answer (for $C^0$ case) is given by

**Theorem** 1. *There is a residual set $\mathcal{B} \subseteq C(X, X)$ such that for every $f \in \mathcal{B}$ there exists a set $Z(f)$ with the following properties:*

(i) *$Z(f)$ is residual and $X \setminus Z(f)$ is of zero Hausdorff dimension;*
(ii) *for every $x \in Z(f)$, $\omega(x)$ is a solenoid;*
(iii) *the set $\varrho_\omega(\omega(x))$ is nowhere dense and of zero measure for every $x \in Z(f)$;*
(iv) *the system $\mathcal{Z}(f) = \{\omega(x) : x \in Z(f)\}$ has the power of continuum;*
(v) *the set $A(f) = \cup\mathcal{Z}(f)$ is of Hausdorff dimension zero and nowhere dense in $X$;*
(vi) *for every relatively open subset $Y$ of $A(f)$ the set $\{x \in X : \omega(x) \cap Y \neq \emptyset\}$ is of the second Baire category and with positive measure.*

**Proof**. Section 5.

(Recall that $E$ is a relatively open subset of $F$ if for every $x \in E$, there exist an $\varepsilon$ such that for all $y \in X$ with $d(x, y) < \varepsilon$ is $y \in E$. If $G \subset F$ are both closed, then $F \setminus G$ is relatively open in $F$.)

COROLLARY 1. *Generically,* $\Lambda_\omega(f) = \Gamma_\omega(f) = \Lambda_\sigma(f) = \Gamma_\sigma(f)$.

PROOF. All four sets are equal to $A(f)$. To see this, note first that $\Lambda_\omega(f)$ and $\Gamma_\omega(f)$ are contained in $A(f)$. From (vi) then follows the equality. The rest of the statement holds due to (ii) and Lemma 1.

Now we turn to local attractors. A **local attractor** or, simply, an **attractor** (again in the sense of [M]) is a closed set $L \subseteq X$ such that

(8) $\varrho_\omega(L)$ is of positive measure

and

(9) there is no strictly smaller $L' \subset L$ such that $\varrho_\omega(L) \setminus \varrho_\omega(L')$ is of zero measure.

An attractor is **minimal** if it does not contain any proper subset which is an attractor.

In [M] the hope was expressed that, at least in the generic case, each attractor contains a minimal attractor. This is not true for $C^0$ case.

THEOREM 2. *There is a residual set* $\mathcal{C} \subseteq C(X,X)$ *such that every* $f \in \mathcal{C}$ *has no minimal attractors.*

PROOF. Section 5.

## 3. Ljapunov stability

A point $x \in X$ is called **Ljapunov stable** if for every $\varepsilon > 0$ there exists a $\delta > 0$ such that $d(x,y) < \delta$ implies that $d(f^n(x), f^n(y)) < \varepsilon$ for all $n = 0,1,2,\ldots$. If $x$ is not Ljapunov stable, there exists $\varepsilon > 0$ such that for every $\delta > 0$ there exists $n$ and $y$ with $d(x,y) < \delta$ such that $d(f^n(x), f^n(y)) > \varepsilon$. Such a point is called $\varepsilon$**-sensitive** [B1]. The set of all $\varepsilon$-sensitive points of $f$ is denoted by $S_\varepsilon(f)$; $S_{\varepsilon_1}(f) \subseteq S_{\varepsilon_2}(f)$ if $\varepsilon_1 \geq \varepsilon_2$ and

$$S(f) = \bigcup_{\varepsilon > 0} S_\varepsilon(f)$$

is the set of all **sensitive** or **Ljapunov unstable** points.

A function $f \in C(X,X)$ is called Ljapunov stable if almost all $x \in X$ (this can be understood in the sense of category or measure) are Ljapunov stable with respect to $f$. The definition, together with the question about the size of the set of all Ljapunov stable functions in $C(X,X)$, comes from [ŠMR].

THEOREM 3. *There is a residual set* $\mathcal{D} \subseteq C(X,X)$ *such that for every* $f \in \mathcal{D}$ *the set* $S(f)$ *is of zero Hausdorff dimension and of the first Baire category.*

PROOF. Section 5.

COROLLARY 2. *Generically, all* $f \in C(X,X)$ *are Ljapunov stable (either in the sense of category or measure).*

A function $f \in C(X,X)$ is said to have **sensitive dependence on initial conditions** (in the sense of Guckenheimer [G]) if $S_\varepsilon(f)$ is of positive measure for some $\varepsilon > 0$.

COROLLARY 3. *Generically, there is no sensitive dependence on initial conditions.*

## 4. Pseudo-orbit shadowing

A sequence $\{z_n\}_{n=0}^{\infty}$ is called a **$\delta$-pseudo-orbit** if $d(z_{n+1}, f(z_n)) < \delta$ for all $n = 0, 1, 2, \dots$ . This pseudo-orbit is **$\varepsilon$-shadowed** by an actual orbit $\{f^n(x)\}_{n=0}^{\infty}$ if $d(z_n, f^n(x)) < \varepsilon$ for all $n$. A function $f \in C(X, X)$ has **pseudo-orbit shadowing property** (or, in short, **shadowing property**) if for every $\varepsilon > 0$ there exists a $\delta > 0$ such that every $\delta$-pseudo-orbit is $\varepsilon$-shadowed by some actual orbit.

The results in [**CKY**], [**C1**], [**C2**], [**GK**] indicate that many $f \in C(X, X)$ have this property.

THEOREM 4. *There exists a residual set $\mathcal{E} \subseteq C(X, X)$ such that every $f \in \mathcal{E}$ has the pseudo-orbit shadowing property.*

PROOF. Section 5.

## 5. The basic construction

Fix $m \in M$. We introduce the following notation:

$$M = 2^m, \quad \eta = 2^{-2m(m+1)};$$

$$a_i = \frac{i}{M}, \quad i = 0, 1, 2, \dots, M;$$

$$c_i = \frac{1}{2}(a_{i-1} + a_i), \quad i = 1, 2, \dots, M;$$

$$E^i_m = [a_{i-1} + \eta, a_i - \eta], \ F^i_m = [c_i - \eta, c_i + \eta], \quad i = 1, 2, \dots, M;$$

$$H^i_m = (a_i - \eta, a_i + \eta), \ i = 1, 2, \dots, M-1,$$

$$H^0_m = [0, \eta), \ H^M_m = (1 - \eta, 1];$$

$$E_m = \bigcup_{i=1}^{M} E^i_m, \ F_m = \bigcup_{i=1}^{M} F^i_m, \ H_m = \bigcup_{i=0}^{M} H^i_m;$$

$$L_m = \bigcup_{i\geq 0} E^{2i+1}_m, \ G_m = \bigcup_{i\geq 0} F^{2i+1}_m;$$

$$R_m = \bigcup_{i\geq 1} E^{2i}_m, \ D_m = \bigcup_{i\geq 1} F^{2i}_m;$$

$$C_m = \{c_1, c_2, \dots, c_M\}.$$

Note that

(10) the intervals $E^i_m$ and $H^i_m$ form all together a partition of $I$;

(11) $F^i_m \subset E^i_m$, $D^i_m \subset R^i_m$, $G^i_m \subset L^i_m$;

(12) $\sum_i (\operatorname{diam} H^i_m)^{\frac{1}{m}} < 2^{-m}$, in particular $\operatorname{diam} H^i_m < 2^{-m}$ for all $i$;

(13) $\sum_i (\operatorname{diam} F^i_m)^{\frac{1}{m}} < 2^{-m}$, $\operatorname{diam} F^i_m < 2^{-m}$ for all $i$;

(14) if $n \geq 2m(m+1)$, then, for every $i$, the intersections $F^i_m \cap G_n$, $F^i_m \cap D_n$ are nonempty (hence so is $F^i_m \cap F_n$) and consist of precisely those components of $G_n$ or $D_n$ which have nonempty intersections with $F^i_m$;

(15) $L_m \cap R_m = D_m \cap G_m = \emptyset$;

(16) if $\{Y_j\}_{j=1}^{\infty}$ is a sequence such that $Y_j = L_{i_j}$ or $R_{i_j}$, where $i_j \in \mathbb{N}$, $i_1 < i_2 < \dots$, then both $\bigcap_{j=1}^{\infty} Y_j$ and $\bigcap_{j=1}^{\infty}(X \setminus Y_j)$ are nowhere dense and of measure zero.

Let $f \in C(X,X)$. If $X = I$, we can consider $f$ as a mapping from $[0,1]$ to iself. If $X = S^1$, we can do the same, assuming that $f(0) = f(1)$. In this way we can understand the notation introduced above.

A set $\mathcal{A}_m$ will be the set of all $f \in C(X,X)$ such that

(17) $$f(L_m) \subset \operatorname{int} G_m, \quad f(R_m) \subset \operatorname{int} D_m$$

(hence $f(E_m) \subset \operatorname{int} F_m$) and, for all $i = 0, 1, \dots, M$,

(18) $$\partial(f(H_m^i)) \subset \operatorname{int} E_m$$

Let
$$\mathcal{M} = \bigcap_{n=1}^{\infty} \bigcup_{m=n}^{\infty} \mathcal{A}_m$$

LEMMA 2. *$\mathcal{M}$ is residual in $C(X,X)$.*

PROOF. It is sufficient to prove that for all $n \in \mathbb{N}$, the set

$$\mathcal{B}_n = \bigcup_{m=n}^{\infty} \mathcal{A}_n$$

is open and dense in $C(X,X)$. From the stability of (17) and (18) under small perturbations of $f$ it follows that $\mathcal{A}_m$ is open for every $m$, hence $\mathcal{B}_n$ is, too. Let $f \in C(X,X)$ and $\varepsilon > 0$ be arbitrary and fixed. We shall construct a $g \in \mathcal{A}_m$, $m$ sufficiently large, such that

(19) $$d(f(x), g(x)) < \varepsilon, \quad \text{for all } x \in X.$$

Let $m \geq n$ be such that $2^{-m} < \frac{\varepsilon}{8}$ and

(20) $$d(f(x), f(y)) < \tfrac{\varepsilon}{8} \quad \text{whenever } d(x,y) < 2^{-m}.$$

(We use the uniform continuity of $f$.) A function $g$ is defined in the following way:

(21) for $x = c_i \in C_m$, $g(c_i)$ is the nearest to $f(c_i)$ element of
- $D_m \cap C_m$ if $c_i \in D_m$;
- $G_m \cap C_m$ if $c_i \in G_m$;

(22) for $x \in E_m^i$, $g(x) = g(c_i)$;

(23) $g$ is linear on every $H_m^i$ and continuous on $X$ (the steps are linearly connected);

(24) if $X = I$, then $g(0)$ and $g(1)$ are the nearest to $f(0)$, $f(1)$ elements of $C_m$, respectively.

The validity of the properties (17) and (18) can be verified by a straightforward check. Now we shall prove (19). Note that according to (21)

(25) $$d(f(c_i), g(c_i)) < \tfrac{\varepsilon}{4}$$

for every $c_i \in C_m$. From (20) we have for $x \in E_m^i$

$$d(f(x), f(c_i)) < \tfrac{\varepsilon}{8} \tag{26}$$

hence from (26), (25) and (22) we obtain that

$$d(f(x), g(x)) < \varepsilon \tag{27}$$

for every $x \in E_m$. Now, consider $x \in H_m^i$. Let

$$\begin{array}{lll} c = 0, & c' = c_1, & \text{if } X = I \text{ and } i = 0; \\ c = c_{2^m}, & c' = 1, & \text{if } X = I \text{ and } i = 2^m; \\ c = c_{2^m}, & c' = c_1, & \text{if } X = S^1 \text{ and } i = 0 \text{ or } 2^m; \\ c = c_i, & c' = c_{i+1} & \text{otherwise.} \end{array}$$

Due to (20) we have

$$d(f(c), f(c')) < \tfrac{\varepsilon}{8} \tag{28}$$

From this and from (25) and (24) it follows that

$$d(g(c), g(c')) < \tfrac{5\varepsilon}{8} \tag{29}$$

Since $g(x)$ lies in the convex hull of $g(c)$ and $g(c')$, we obtain, using (26) (valid also for $x \in H_m^i$), (25) and (29), that (27) holds for every $x \in H_m$. All this implies (19).

PROOF OF THEOREM 1. Put $\mathcal{B} = \mathcal{M}$. Let $f \in \mathcal{B}$. There is an increasing sequence $n_j$ such that $f \in \mathcal{A}_{n_j}$ for all $j \in \mathbb{N}$. Let

$$Z(f) = \bigcap_{k=1}^{\infty} \bigcup_{j=k}^{\infty} E_{n_j}.$$

Let $Y = X \setminus Z(f)$. We have

$$Y = \bigcup_{k=1}^{\infty} Y_k$$

where

$$Y_k = \bigcap_{j=k}^{\infty} H_{n_j}.$$

Fix $\alpha, \varepsilon > 0$. Pick $k$ such that $\frac{1}{n_k} < \alpha$ and $2^{-n_k} < \varepsilon$. Since $Y_k \subseteq H_{n_k}$, (12) implies that $Y_k$ is of dimension zero.

Now fix a segment $J$. Choose $k$ such that $\frac{1}{n_k} < \operatorname{diam} J$. Since $\operatorname{diam} H_{n_k}^i < \frac{1}{n_k}$ for all $i$, a part of $J$ is not contained in $H_{n_k}^i$ and hence also not in $Y_k$, thus $Y_k$ is nowhere dense.

We have proved that $Y$ is of zero dimension and of the first category – this yields (i).

Let $x \in Z(f)$. We can pick a subsequence $\{\ell_k\}_{k=1}^{\infty}$ of $\{n_k\}_{j=1}^{\infty}$ such that $x \in E_{\ell_k}$ and $\ell_{k+1} \geq 2\ell_k(\ell_k+1)$. Note that by (17), the whole trajectory $f^m(x)$, $m \geq 1$, lies in $F_{\ell_k}$, $k \in \mathbb{N}$. We take the components of $F_{\ell_k}$, which are visited by the trajectory of $x$ infinitely often and denote them by $I_{k_j}$, choosing the right order to match (5) – note that these components are mapped cyclically one into another by $f$. Due to (14) we have (4); (13) implies (6) and also

$$\omega(x) = \bigcap_{k=1}^{\infty} \bigcup_{j} I_{k_j}.$$

Hence $\omega(x)$ is a solenoid, proving (ii).

Note that $\omega(x) \subseteq \bigcap_{k=1}^{\infty} Y_k$, where $Y_k$ is either $D_{\ell_k}$ or $G_{\ell_k}$; if $\omega(x) \subseteq D_{\ell_k}$ $(G_{\ell_k})$, then $\varrho_\omega(\omega(x)) \subseteq X \setminus L_{\ell_k}$ $(X \setminus R_{\ell_k})$. From (16) we obtain (iii).

From (5) and (6) it follows that if $\omega(x)$ is a solenoid, then every $y \in \omega(x)$ has a dense trajectory in $\omega(x)$, so that two solenoids are either equal or disjoint. Hence $\mathcal{A}(f)$ has power less than or equal to the power of the continuum. Now, consider the set

$$Z_0 = \bigcap_{k=1}^{\infty} E_{n_k} \subset Z(f).$$

If $z \in Z_0$, there is a unique sequence $Y_k$ such that

$$\{z\} = \bigcap_{k=1}^{\infty} Y_k \tag{30}$$

and $Y_k = L_{n_k}$ or $R_{n_k}$. This yields that $\omega(z_1) \neq \omega(z_2)$ for $z_1, z_2 \in Z_0$, $z_1 \neq z_2$. Since, conversely, for every sequence appearing in (30) there exists a unique $z \in Z_0$ satisfying (30), we obtain (iv).

Consider $A(f) = \cup \mathcal{Z}(f)$; we have

$$A(f) \subseteq \bigcap_{j=1}^{\infty} \bigcup_{k=j}^{\infty} F_{n_k}. \tag{31}$$

Fix $\alpha, \varepsilon > 0$. Pick $j$ such that $\frac{1}{n_j} < \alpha$ and $2^{-n_j+1} < \varepsilon$. Since

$$A(f) \subseteq \bigcup_{k=j}^{\infty} F_{n_k} \subseteq \bigcup_{k=j}^{\infty} \bigcup_{i} F_{n_k}^i$$

and due to (13) is

$$\sum_{k=j}^{\infty} \sum_{i} (\operatorname{diam} F_{n_k}^i)^\alpha \leq \sum_{k=j}^{\infty} 2^{-n_k} < \varepsilon,$$

hence $A(f)$ is of dimension zero.

To see that $A(f)$ is nowhere dense, consider a segment $J$. For $k$ large enough there exists a component $E^i_{n_k}$ of $E_{n_k}$ such that $E^i_{n_k} \subseteq J$. But then a set $E^i_{n_k} \setminus F^i_{n_k}$, consisting of two nondegenerated intervals, does not contain any element of $A(f)$; this proves (v).

Finally, let $Y$ be a nonempty relatively open subset of $A(f)$. Pick $x \in Y$; there exists a segment $J$ such that $x \in J$ and $J \cap A(f) \subseteq Y$. By (31), there is a component $F^i_{n_k}$ of $F_{n_k}$ such that $x \in F_{n_k} \subseteq J$. Since $A(f) \cap F^i_{n_k}$ is nonempty, there exists $\ell$ such that $f^\ell(F_{n_k}) \subseteq F_{n_k}$. But then for all $x \in F^i_{n_k}$, $\omega(x) \cap F^i_{n_k} \neq \emptyset$ and, since $\varrho_\omega(A(f))$ is of full measure and residual, there exists a set $Y' = \varrho_\omega(A(f)) \cap F^i_{n_k}$ of the second category and with positive measure such that for all $x \in Y', \omega(x) \cap Y \neq \emptyset$.

The proof of (vi) and hence of Theorem 1 is completed.

Proof of Theorem 2. Put $\mathcal{C} = \mathcal{M}$ (we use the same notation as above). Let $f \in \mathcal{C}$ and let $M \subseteq A(f)$ be a minimal attractor. We claim that there exists $k$ such that

$$\lambda(\varrho_\omega(M) \cap L_{n_k}) > 0 \quad \text{and} \quad \lambda(\varrho_\omega(M) \cap R_{n_k}) > 0. \tag{32}$$

Assume the contrary. Let $Y'_k$ be a sequence such that $X \setminus Y'_k$ is equal to $L_{n_k}$ or $R_{n_k}$ and

$$\lambda(\varrho_\omega(M) \cap Y'_k) = \lambda(\varrho_\omega(M)).$$

Then

$$\lambda(\varrho_\omega(M) \cap \bigcap_{k=1}^{\infty} Y'_k) = \lambda(\varrho_\omega(M) > 0,$$

the last inequality due to (8). This is in contradiction with (16), so we can pick

$$M_L = \bigcup\{\omega(x) : x \in \varrho_\omega(M) \cap L_{n_k}\}$$

and

$$M_R = \bigcup\{\omega(x) : x \in \varrho_\omega(M) \cap R_{n_k}\}.$$

The sets $M_L$ and $M_R$ are disjoint, (32) implies that $\lambda(\varrho_\omega(M_L)) > 0$, $\lambda(\varrho_\omega(M_R)) > 0$, hence (see [M]) they contain two attractors strictly smaller than $M$. This proves the non-existence of minimal attractors.

Proof of Theorem 3. Again, put $\mathcal{D} = \mathcal{M}$. Let $f \in \mathcal{D}$. We shall show that

$$S(f) \subseteq X \setminus \bigcap_{j=1}^{\infty} \bigcup_{k=j}^{\infty} \operatorname{int} E_{n_k}.$$

The proof that $S(f)$ is zero-dimensional and of the first category is the same as the proof of Theorem 1(i). So we only have to show that every $x$ belonging to infinitely many $\operatorname{int} E_{\ell_k}$ is Ljapunov stable. Fix $\varepsilon > 0$ and pick $k$ such that $\operatorname{diam} F_{\ell_k} < \varepsilon$. Pick $\delta$ such that for every $y$ with $d(x,y) < \delta$, $y$ lies in the same component of $\operatorname{int} E_{\ell_k}$ as $x$. For such a $y$, the trajectories of $x$ and $y$ travel together in the same components of $F_{\ell_k}$, hence $d(f^n(x), f^n(y)) < \varepsilon$ for all $n = 1, 2, \ldots$, and the proof is completed.

Proof of Theorem 4. Let $\mathcal{E} = \mathcal{M}$, let $f \in \mathcal{E}$. Fix $\varepsilon > 0$ and pick $k$ such that

$$\operatorname{diam} E^i_{n_k} < \varepsilon. \tag{33}$$

Note that (33) implies

(34) $$\operatorname{diam} F^i_{n_k} < \varepsilon, \quad \operatorname{diam} H^i_{n_k} < \varepsilon.$$

Let $Y$ be a set, $\delta > 0$. We introduce a notation

$$Y_\delta = \{y \in X : d(x,y) < \delta \text{ for some } x \in Y\}.$$

Take $\delta < \varepsilon$ such that for all i

$$(F^i_{n_k})_\delta \subseteq E^i_{n_k}$$

(say, $\delta < 2^{-n_k-1}$) and

(35) $$\delta < \min\{d(x,y) : x \in \bigcup_{i=0}^{M} \partial(f(H^i_{n_k})) \cup \partial E_{n_k}, y \in \partial E_{n_k}\}$$

(due to (18), $\delta$ satisfying (35) can be taken positive).

Let $\{z_j\}_{j=0}^\infty$ be a $\delta$-pseudo-orbit. If $z_0 \in E_{n_k}$, then for every $j$ there is $i$ such that $z_j \in (F^i_{n_k})_\delta$ and (34) implies that $\{z_j\}_{j=0}^\infty$ is $\varepsilon$-shadowed by $\{f^j(z_0)\}_{j=0}^\infty$. If $z_0 \in H_{n_k}$, then two possibilities can occur.

If $z_j \in H_{n_k}$ for all $j$, say, $z_j \in H^i_{n_k}$, then if $z_{j+1} \in H^\ell_{n_k}$ we have

(36) $$f(H^i_{n_k}) \cap H^\ell_{n_k} \neq \emptyset$$

due to (35) (the components of $H_{n_k}$ met by $f(H^i_{n_k})$ are the same as those met by $(f(H^i_{n_k}))_\delta$). From (18) and (36) it follows that

$$f(H^i_{n_k}) \supseteq H^\ell_{n_k}.$$

Denoting by $Y_j$ the component of $H_{n_k}$ such that $z_j \in Y_j$, we obtain (by common argument, see e.g. [**BGMY**]) that there is $y \in H_{n_k}$ such that $f^i(y) \in Y_j$ and hence, by (33), $\{z_j\}_{j=0}^\infty$ is $\varepsilon$-shadowed by $\{f^j(y)\}_{j=0}^\infty$.

The last possibility is that $z_j \in H_{n_k}$, for $j \leq j_0$, $z_{j_0} \in H^i_{n_k}$, $z_{j_0+1} \in E^\ell_{n_k}$. If $f(z_{j_0}) \in E_{n_k}$ then $f(z_{j_0}) \in E^\ell_{n_k}$ by (35) and

(37) $$f(H^i_{n_k}) \cap E^\ell_{n_k} \neq \emptyset.$$

If $f(z_{j_0}) \in H_{n_k}$, then a component $H^m_{n_k}$ of $H_{n_k}$, such that $f(z_{j_0}) \in H^m_{n_k}$, must be adjacent to $E^\ell_{n_k}$ and hence we have again (37) due to (18). Again by common argument we conclude that there exist $y \in H_{n_k}$ such that $f^i(y) \in Y_j$, $j \leq j_0$, where $Y_j$ are the components of $H_{n_k}$ containing $z_j$, $f^{j_0+1}(y) \in E^\ell_{n_k}$ and, obviously, $f^j(y) \in F_{n_k}$ for $j \geq j_0 + 2$, travelling in the same components as $z_j$. Hence, due to (33) and (34), $\{f^j(y)\}_{j=0}^\infty$ $\varepsilon$-shadows $\{z_j\}_{j=0}^\infty$ .

## REFERENCES


[ABL] S. Agronsky, A. Bruckner and M. Laczkovich, *Dynamics of typical continuous functions*, J. London Math. Soc. **40(2)** (1990).

[BGMY] L. Block, J. Guckenheimer, M. Misiurewicz and L. S. Young, *Periodic orbits and topological entropy for one-dimensional maps*, Global Theory of Dynamical Systems, Lecture Notes in Math., vol. 819, Springer, New York – Heidelberg – Berlin.

[B1] A. M. Blokh, *On sensitive mappings of the interval*, Dokl. Akad. Nauk USSR **37(224)2** (1982), 189–190 (in Russian); Russ. Math. Surv. **37(2)** (1982), 203–204 (English translation).

[B2] A. M. Blokh, *The set of all iterates is nowhere dense in* $C([0,1],[0,1])$, preprint.

[C1] L. Chen, *Shadowing property for nondegenerate zero entropy piecewise monotone maps*, Preprint 1990/9 Inst. for Math. Sciences, SUNY at Stony Brook.

[C2] L. Chen, *Linking and the shadowing property for piecewise monotone maps*, preprint.

[CKY] E. Coven, I.Kan and J. A. Yorke, *Pseudo-orbit shadowing in the family of tent maps*, Trans. Amer. Math. Soc. **308** (1988), 227–241.

[GK] T. Gedeon and M. Kuchta, *Shadowing property of continuous maps*, preprint.

[Gr] N. Grznárová, *Typical continuous function has the set of chain recurrent points of zero Lebesgue measure*, Acta Math. Univ. Comen. (to appear).

[G] J. Guckenheimer, *Sensitive dependence on initial conditions for one-dimensional maps*, Comm. Math. Phys. **70** (1979), 133–160.

[I] A. Iwanik, personal communication.

[LY] T. Y. Li and J. Yorke, *Period three implies chaos*, Amer. Math. Monthly **82** (1975), 985–992.

[M] J. Milnor, *On the concept of attractor*, Comm. Math. Phys. **99** (1985), 177–195; *Correction and remarks*, Comm. Math. Phys. **102** (1985), 517–519.

[Mi] I. Mizera, *Continuous chaotic functions of an interval have generically small scrambled sets*, Bull. Austral. Math. Soc. **37** (1988), 89–92.

[O] J. C. Oxtoby, *Measure and Category*, Springer, New York – Heidelberg – Berlin, 1971.

[S] K. Simon, *On the periodic points of a typical continuous function*, Proc. Amer. Math. Soc. **105** (1989), 244–249.

[ŠMR] A. N. Šarkovskii, Yu. N. Majstrenko and E. Yu. Romanenko, *Difference equations and their applications*, Naukova dumka, Kiev, 1986 (in Russian).

[ŠKSF] A. N. Šarkovskii, S. F. Koljada, A. G. Sivak and E. Yu. Romanenko, *Dynamics of one-dimensional mappings*, Naukova dumka, Kiev, 1989 (in Russian).

[Y] L. S. Young, *A closing lemma on the interval*, Invent. Math. **54** (1979), 179–187.



DEPARTMENT OF MATHEMATICS, COMENIUS UNIVERSITY, MLYNSKÁ DOLINA, CS–84215, BRATISLAVA, CZECHOSLOVAKIA

# Hölder Continuity Of The Holonomy Maps For Hyperbolic Basic Sets I

J. Schmeling & Ra. Siegmund-Schultze
Karl-Weierstraß-Institut fur Mathematik
Mohrenstr.39, O-1086 Berlin, Germany

## 0. Introduction

In this note we consider -from a geometric point of view- the Hoelder continuity of the holonomy map for a hyperbolic basic set. This is the map between two near-by pieces of stable manifolds which is obtained by sliding along the unstable manifolds. As it seems to be well-known (cf. [1], [2] for a corresponding result about the stable and unstable splitting of the tangent space), this map is Hoelder continuous, but there are no efficient estimates on the Hoelder exponent. It is the aim of this paper to fill this gap and to derive some corollaries concerning the local Hausdorff dimension of the basic set.
In a forthcoming paper we show that the exponent which is derived in this note is generically the best one in an open set of $C^2$-diffeomorphisms of the 3-dimensional sphere.
Our estimate of the Hoelder exponent is taken from the first author's unpublished thesis ([3]).

Consider a compact Riemannian manifold M and a $C^2$-diffeomorphism $f$ of M onto itself. Assume that there is a *hyperbolic basic set* $\Lambda \subset M$ connected with $f$, i.e.

1. $\Lambda$ is *f-invariant* ($f(\Lambda)=\Lambda$)
2. there is a *f-invariant splitting*

$$T_\Lambda(M) = E^s \oplus E^u$$

of the restriction $T_\Lambda(M)$ of the tangent bundle $T(M)$ of M to $\Lambda$ as a direct sum of two continuous subbundles $E^s$, $E^u$, such that for certain constants $a,b>0$ , $0<\alpha<1<\beta$ the inequalities

$$\| Df^k|_{E^s_p} \| \le a\cdot\alpha^k$$

$$\| Df^k|_{E^u_p} \| \geq b\cdot\beta^k \quad ,$$

hold for all $p\in\Lambda$, where $E^s_p$ and $E^u_p$ are the fibres of $E^s$ and $E^u$ at $p$.

3. $\Lambda$ has a neighbourhood $\mathcal{U}$ in M such that

$$\Lambda = \bigcap_{k=-\infty}^{\infty} f^k(\mathcal{U})$$

4. $\Lambda$ is minimal in the sense that it contains a dense orbit
5. $\Lambda$ is maximal in the sense that it is the closure of the periodic points in $\mathcal{U}$

Without any loss of generality we assume that M is embedded into some $\mathbb{R}^d$ and that the natural Riemannian metric $d$ on M is *adapted* to $f$, i.e. we assume that there exists a positive constant $\alpha<1$ such that for the *invariant splitting*

$$T_pM = E^u_p \oplus E^s_p \qquad , \; p \in \Lambda,$$

the following relations are fulfilled

$$\begin{aligned} &\| Df|_{E^s_p} \| < \alpha \\ &\| (Df|_{E^u_p})^{-1} \|^{-1} < \alpha \qquad , \; p \in \Lambda. \end{aligned} \tag{1}$$

For any point $p \in \Lambda$ we consider the *stable* and *unstable manifolds*

$$W^s_p = \{x\in M: \; d(f^n p, f^n x) \xrightarrow[n\to\infty]{} 0\},$$

$$W^u_p = \{x\in M: \; d(f^{-n} p, f^{-n} x) \xrightarrow[n\to\infty]{} 0\}.$$

They are $C^2$-immersions of $E^s_p$ and $E^u_p$ in M, respectively, see [4], [5].

We choose some $\delta>0$ and consider the *local stable* and *unstable manifolds* $W^s_p(\delta)$ and $W^u_p(\delta)$, which are defined as $\delta$-neighbourhoods of $p$ in $W^s_p$ and $W^u_p$ with respect to the natural Riemannian metrics in $W^s_p$ and $W^u_p$.

From the theorem about the *local product structure* (see [6]) we have the existence of two constants $\beta,\beta'>0$ such that for any two points $x,y\in\Lambda$ with $d(x,y)<\beta$ the sets $W^u_y(\beta')$ and $W^s_x(\beta')$ have a unique intersection point that will be denoted by $[x,y]$. This point belongs to $\Lambda$.

For $p\in\Lambda$ and $k\in\mathbb{N}$ define

$$\lambda_k(p) = \| Df^k|_{E^s_p} \|$$

$$\mu_k(p) = \| (Df^k|_{E^s_p})^{-1}\|^{-1}$$

$$\eta_k(p) = \| (Df^k|_{E^u_p})^{-1}\|^{-1}.$$

From *(1)* we infer

$$\mu_k(p)\le\lambda_k(p)<\alpha^k<1<\alpha^{-k}<\eta_k(p). \qquad (1a)$$

Let

$$\lambda(p)=\lambda_1(p),\ \mu(p)=\mu_1(p),\ \eta(p)=\eta_1(p).$$

We define

$$\varkappa_o = \sup_{k\ge1}\ \liminf_{n\to\infty}\ \inf_{p\in\Lambda} \frac{\sum_{i=0}^{n}[\log\lambda_k(f^{ki}p) - \log\eta_k(f^{ki}p)]}{\sum_{i=0}^{n}\log\mu_k(f^{ki}p)}.$$

Remark : The number $\varkappa_o$ does not depend on the Riemannian metric on M, in particular it does not depend on the embedding of M into $\mathbb{R}^d$. In fact, if we omit in the definition of $\varkappa_o$ the supremum over $k$ and denote the value which is obtained in that way by $\varkappa(k)$, we easily realize that, for each $k_o$, the sequence $[\varkappa(k_o2^k)]_{k=0,1,2,...}$ is non-decreasing, such that we might have defined $\varkappa_o = \limsup_{k\to\infty}\varkappa(k)$ as well. Now from *(1a)* we conclude that a change to an equivalent Riemannian metric cannot change $\varkappa_o$ since this results only in a modification of $\mu_k(p)$, $\lambda_k(p)$ and $\eta_k(p)$ by some factors which are uniformly bounded away from 0 and $\infty$.

The result of this note is

Theorem *For any* $\varkappa<\min(1,\varkappa_0)$ *there is a positive constant* $C_\varkappa$ *such that for any two points* $x,y\in\Lambda$ *with distance* $d(x,y)<\beta$ *we have*

$$d(y,[y,x]) \le C_\varkappa (d(x,[x,y]))^\varkappa . \qquad (2)$$

The proof of this theorem is rather technical in detail. So we first give a heuristic outline.

It is not hard to see that we may restrict our attention to the case where, for some given and sufficiently small $\varepsilon$, the points $x, y$ and the corresponding local stable and unstable manifolds are contained in an $\varepsilon$-neighbourhood of $x$. Of course we may assume that $x'$ is very close to $x$, even much closer than the distance $\varepsilon$.

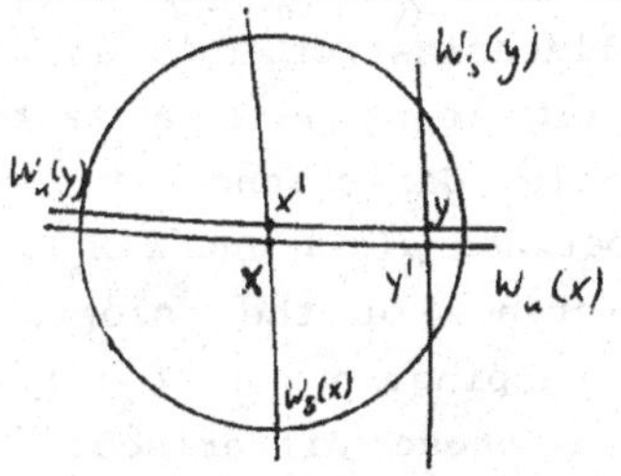


Here we wrote $x'$ for $[x,y]$ and $y'$ for $[y,x]$.

So we may trace back a long part of the negative semi-orbit of $x$, $x'$ until $\mathrm{dist}(f^{-n}(x), f^{-n}(x'))$ reaches the level $\varepsilon$ for the first time. Denote the corresponding number of steps by $n_0$.

Consider now all the pre-images $f^{-n}(x), f^{-n}(y), f^{-n}(x'), f^{-n}(y')$ for n=0,1,2,... and let us denote these points by $x_n, y_n, x_n', y_n'$. So $y_{n_0}'$ is very close to $x_{n_0}$ and $y_{n_0}$ is very close to $x_{n_0}'$, due to the exponential decay of distances of points on the unstable manifold for negative iterates of $f$.

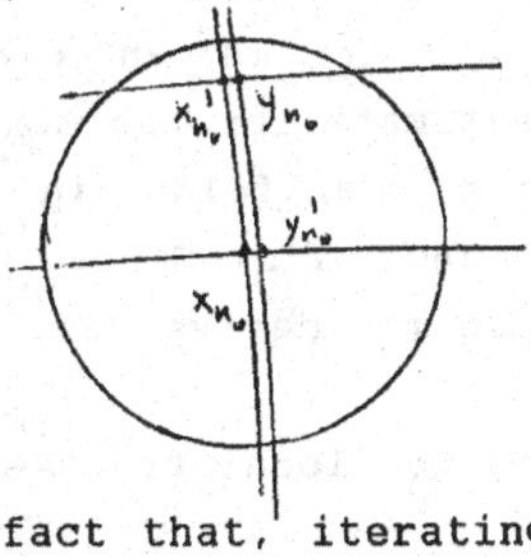


What is important is the fact that, iterating $f$, on the way from $x_{n_0}$, $y_{n_0}$ to $x$, $y$ the corresponding local stable and unstable manifolds of $x_n$ and $y_n$ remain in an $\varepsilon$-neighbourhood of $x_n$. We interpret these manifolds as graphs of some mappings $\varkappa_n(.)$, $\psi_n(.)$, introducing suitable local coordinate systems.

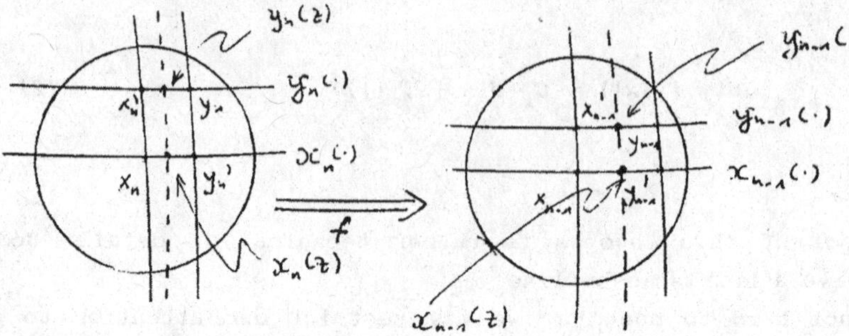

Now observe that, for small $\varepsilon$, the action of $f$ on an $\varepsilon$-neighbourhood of $x_n$ can be approximated with an arbitrary accuracy by the action of $Df_{x_n}$ (in local coordinates) on this neighbourhood. This linearization acts as a pure expansion in the unstable directions (at least with a factor $\eta(x_n) > \alpha^{-1} > 1$), whereas in the stable directions it is a contraction with contraction factors between $\mu(x_n)$ and $\lambda(x_n)$.
We consider the difference of the "slopes" (to be defined in a suitable way) of the mappings $\alpha_n(.)$, $y_n(.)$ at some position $z$ and denote the maximum of these differences (for $z$ varying in the $\varepsilon$-neighbourhood of $x_n$ on the unstable manifold of $x_n$) by $s_n$. Taking into account the combined effect of stretching and contracting, we come to the conclusion that it should be possible to estimate $s_{n-1}$ by something like

$$s_{n-1} \le s_n \frac{\lambda(x_n)}{\eta(x_n)} + C\cdot \mathrm{dist}(x_n', x_n) \,, \qquad 0 \le n \le n_0 .$$

With this key estimate being true we find an expression for $s_0$, which is approximately the maximum angle between the local unstable manifolds through $x$ and $x'$ in an $\varepsilon$-neighbourhood of $x$. This estimate leads to an estimate for the maximum distance (in the same neighbourhood) of these manifolds in terms of $\lambda, \mu, \eta$ and the distance between $x$ and $x'$. Since $y$ and $y'$ are in the $\varepsilon$-neighbourhood of $x$, we may derive *(2)*.

Applying this result to the local transverse Hausdorff dimension (see [7])

$\dim_{tr}(\Lambda, p) =$
$\quad = \inf \{ \dim_H(\mathcal{U} \cap \Lambda) : \mathcal{U}$ is a compact neighbourhood of $p$ in $W^s_p \}$

we get

<u>Corollary 1</u> *For any two points $p, q$ in $\Lambda$ the local transverse*

*Hausdorff dimensions fulfil the inequalities*

$$\varkappa_o \le \frac{\dim_{tr}(\Lambda,p)}{\dim_{tr}(\Lambda,p)} \le 1/\varkappa_o \ .$$

Corollary 2 *The local transverse Hausdorff dimension of* $\Lambda$ *is constant if* $\varkappa_o$ *is not less than* 1.

Proof of corollary 1

Let $\varepsilon$ be a positive number. Choose a radius $\rho$ such that $W^s_q(\rho)\cap\Lambda$ has a Hausdorff dimension which is not larger than $\dim_{tr}(\Lambda,q)+\varepsilon$. Let even $\rho$ be so small that for any point $x\in\Lambda$ the restriction of $f$ to $W^s_x(\rho)$ is a contraction which shortens all distances at least with a factor $\alpha'<1$ and assume the same to be true for $f^{-1}$ with respect to $W^u_x(\rho)$. This choice of $\rho$ is possible since the Riemannian metric is adapted and since $\Lambda$ is compact.

From the existence of a dense orbit in $\Lambda$ we derive that we find points $r$ and $r'$ in $\Lambda$ which are arbitrarily close to $p$ and $q$, respectively, and such that $r'$ is from the positive semi-orbit of $r$. We choose $r'$ so close to $q$ that $[q,r']=:a'$ is defined and belongs to $W^u_{r'}(\rho)\cap W^s_q(\rho/2)$. Now from the fact that $f^{-1}$ is contractive in the sense mentioned above on the local unstable manifolds with radius $\rho$ we derive that we can choose not only $r$ as close to $p$ as desired, but also $a:=f^{-k}(a')$, where $k\ge 0$, $f^k(r)=r'$. So in view of the theorem we can manage $a$ to fulfil the relation

$$\dim_H(W^s_a(\rho/2)\cap\Lambda) \ \ge\ \varkappa_o\cdot\dim_{tr}(\Lambda,p)-\varepsilon .$$

Now, since $f$ preserves the Hausdorff dimension and since it is contractive on local stable manifolds with radius $\rho$, we have

$$\dim_H(W^s_{a'}(\rho/2)\cap\Lambda) \ \ge\ \varkappa_o\cdot\dim_{tr}(\Lambda,p)-\varepsilon ,$$

such that from $a'\in W^s_q(\rho/2)\cap\Lambda$ we have

$$\dim_H(W^s_q(\rho)\cap\Lambda) \ \ge\ \varkappa_o\cdot\dim_{tr}(\Lambda,p)-\varepsilon .$$

Now, since we chose $\rho$ small enough, we get

$\dim_{tr}(\Lambda,q) \ \ge\ \varkappa_o\cdot\dim_{tr}(\Lambda,p)-2\varepsilon$. This is all we had to prove.

## 1. Proof of the Theorem

1. We introduce the following assumption, which later can be deleted without greater trouble, and which depends on a parameter $\varepsilon<\beta$.

*For any two of the four points $x,x',y,y'$ their distance is less than $\varepsilon$.* $(A_\varepsilon)$

Denote by $n_o$ the largest integer with $x_n,y_n,x'_n,y'_n$ fulfilling the condition $(A_\varepsilon)$ for all $n$ with $0\le n\le n_o$. This number is finite, supposed that $\varepsilon$ is sufficiently small and that $x\neq x'$. In fact, this is a consequence of the exponential growth of distances in the stable direction for negative iterates of $f$.

We denote by $m_u$ the unstable and by $m_s$ the stable dimension for points of $\Lambda$.

Let $\rho$ be a positive number such that for any $q\in M$ the orthogonal projection $\pi_q$ from $B_\rho(q)=\{q'\in M:\ d(q,q')<\rho\}$ to $T_qM$ has bounded derivatives.

For $p\in\Lambda$ we define a diffeomorphism $\Phi_p$ of $B_\rho(p)$ to some open subset $E_{p,\rho}\subset E_p:=E_p^u\times E_p^s$ by means of the set-up

$$\Phi_p(q) = (r_u(\pi_p(q)),r_s(\pi_p(q))),$$

where for any $r\in T_pM$ by $r_u$ and $r_s$ we denote the unstable and stable components of $r$, i.e. $r=r_u+r_s$, $r_u\in E_p^u$, $r_s\in E_p^s$. We equip $E_p$ with the euclidean norm, i.e. $\|r\| = (\|r_u\|^2+\|r_s\|^2)^{1/2}$.

We have

<u>Lemma 1.</u> *The family of diffeomorphisms $\{\Phi_p\}_{p\in\Lambda}$ has uniformly bounded first and second derivatives and the same is true for $\{\Phi_p^{-1}\}_{p\in\Lambda}$.*

In fact, this is an easy consequence of the continuity of the hyperbolic splitting and the compactness of $\Lambda$.

For any $p\in\Lambda$ denote by $f^{(p)}$ the map $\Phi_{f(p)}\circ f\circ\Phi_p^{-1}$, which is defined on some ball $B_{\rho'}(0)=\{r\in E_p:\ \|r\|<\rho'\}$. The positive number $\rho'$ can be chosen to be the same for all $p\in\Lambda$. Let us denote by $\mathfrak{x}_n,\mathfrak{y}_n,\mathfrak{x}_n',\mathfrak{y}_n'$ the images of $x_n, y_n, x'_n, y'_n$ with respect to $\Phi_{x_n}$. We have to choose $\varepsilon$ small enough to ensure that $y_n, x'_n, y'_n$ are in $B_\rho(x_n)$ and $\mathcal{y}_n, \mathcal{x}'_n, \mathcal{y}'_n$ are in $B_{\rho'}(\mathcal{x}_n) = B_{\rho'}(0)$. We assume here that $\rho$ was chosen small enough to ensure that for each $p \in \Lambda$ and $p' \in B_{\rho/2}(p)$ the sets $W^u_{p'}(\delta) \cap B_\rho(p)$ and $W^s_{p'}(\delta) \cap B_\rho(p)$ are $C^2$-disks. This is possible in view of the $C^2$-continuity of the families of unstable and stable local manifolds ([5]).

Denote by $\mathcal{V}^u_{\mathcal{x}_n}$ the image of $W^u_{x_n}(\delta) \cap B_\rho(x_n)$ under $\Phi_{x_n}$. In the same way define

$$\mathcal{V}^s_{\mathcal{x}_n} = \Phi_{x_n}(W^s_{x_n}(\delta) \cap B_\rho(x_n))$$

$$\mathcal{V}^u_{\mathcal{y}_n} = \Phi_{x_n}(W^s_{y_n}(\delta) \cap B_\rho(x_n))$$

$$\mathcal{V}^s_{\mathcal{y}_n} = \Phi_{x_n}(W^s_{y_n}(\delta) \cap B_\rho(x_n)).$$

For sufficiently small $\rho$ and $\varepsilon < \rho/2$ these are $C^2$-disks in $E_{x_n}$ of dimension $m_u$ or $m_s$ which lie *schlicht* over $E^u_{x_n}$ or $E^s_{x_n}$, respectively, and which intersect in $\mathcal{x}_n, \mathcal{y}_n, \mathcal{x}'_n, \mathcal{y}'_n$. In fact, the curvature of the local stable and unstable manifolds is bounded by a constant. This is again a consequence of the $C^2$-continuity theorem for the stable and unstable manifolds connected with the compact hyperbolic basic set $\Lambda$. Now from lemma 1 the same assertion follows for $\mathcal{V}^u_{\mathcal{x}_n}, \mathcal{V}^s_{\mathcal{x}_n}, \mathcal{V}^u_{\mathcal{y}_n}, \mathcal{V}^s_{\mathcal{y}_n}$. Also from lemma 1 we conclude that there is a constant $L > 0$ such that from the validity of $(A_\varepsilon)$ for the four points $x_n, y_n, x'_n, y'_n$ we get the validity of $(A_{L\varepsilon})$ for the quadrupel $\mathcal{x}_n, \mathcal{y}_n, \mathcal{x}'_n, \mathcal{y}'_n$. So let us consider the $2L\varepsilon$-neighbourhood of $\mathcal{x}_n$ and denote it by $\mathcal{U}(\mathcal{x}_n)$. Moreover, consider the cylinder

$$\mathcal{Z}(\mathcal{x}_n) = \{ r_u \in E^u_{x_n} : \| r_u \| < L\varepsilon \} \times E^s_{x_n}.$$

For sufficiently small $\varepsilon$ the intersections of $\mathcal{Z}(\mathcal{x}_n)$ with $\mathcal{V}^u_{\mathcal{x}_n}$ and $\mathcal{V}^u_{\mathcal{y}_n}$ are contained in $\mathcal{U}(\mathcal{x}_n)$ and are the graphs of two functions

$$\mathcal{x}_n(\cdot) : \{ z \in E^u_{x_n} : \| z \| < L\varepsilon \} \longrightarrow \mathcal{U}(\mathcal{x}_n)$$

$$\psi_n(\cdot) : \{z \in E^u_{x_n} : \|z\| < L\varepsilon\} \longrightarrow \mathcal{U}(x_n).$$

These functions are injective, we denote the inverse maps, which are defined on $\mathcal{V}^u_{x_n} \cap \mathcal{Z}(x_n)$ and $\mathcal{V}^u_{\psi_n} \cap \mathcal{Z}(x_n)$, respectively, by $x_n^{-1}(\cdot)$ and $\psi_n^{-1}(\cdot)$.

2. Next we intend to define the angle between two linear subspaces $\mathcal{L}_1$ and $\mathcal{L}_2$ of the same dimension $d'$ of the euclidean space $\mathbb{R}^d$. It is not hard to check that one can always find two orthogonal bases $a_1,\dots,a_{d'} \in \mathcal{L}_1$ and $b_1,\dots,b_{d'} \in \mathcal{L}_2$ with $a_i \perp b_j$ for $i \neq j$ and $(a_i,b_j) \geq 0$ for $i=1,\dots,d'$.

By the angle $\sphericalangle(\mathcal{L}_1,\mathcal{L}_2)$ we understand the maximum of the angles between $a_i$ and $b_i$ for $1 \leq i \leq d'$.

It is not hard to check that this angle is uniquely defined and is nothing but the greatest possible angle between a vector in $\mathcal{L}_1$ and its projection to $\mathcal{L}_2$.

The angle between two affine subspaces is simply defined by shifting them to the origin.

3. With this notion of an angle, let be $s_n$ the supremum of the expression

$$\sphericalangle(\mathcal{E}^u_{x_n(z)}, \mathcal{E}^u_{\psi_n(z)}) \qquad , \|z\| < L\varepsilon,$$

where $\mathcal{E}^u_{x_n(z)}$ denotes the tangent space to $\mathcal{V}^u_{x_n}$ at $x_n(z)$, $\mathcal{E}^u_{\psi_n(z)}$ the tangent space to $\mathcal{V}^u_{\psi_n}$ at $\psi_n(z)$. Here we identify the tangent spaces at all points of the euclidean space $E_{x_n}$ with $E_{x_n}$ itself, so that all the $\mathcal{E}^u_{x_n(z)}$ have a common origin.

We intend to estimate $s_0$. This will be accomplished by an argument of induction type. We have a trivial estimate

$$s_{n_0} \leq C_0,$$

where $C_0$ is some constant depending upon $\varepsilon$ and $\rho$ (it tends to zero with these parameters), but it does not depend on $x$ and $y$.

Now suppose that $s_n$ is given. We derive an estimate for $s_{n-1}$. So let $z \in E^u_{x_{n-1}}$, $\|z\| \leq L\varepsilon$. If $\varepsilon$ is small enough, we find some $z^*$ with

$$f^{(x_n)}(x_n(z^*)) = x_{n-1}(z)$$

and

$$\measuredangle(\mathcal{E}^u_{\alpha_{n-1}(z)},\mathcal{E}^u_{y_{n-1}(z)}) \le$$

$$\le \measuredangle(Df^{(x_n)}_{\alpha_n(z^*)}(\mathcal{E}^u_{\alpha_n(z^*)}),Df^{(x_n)}_{y_n(z^*)}(\mathcal{E}^u_{y_n(z^*)})) +$$

$$+ \measuredangle(Df^{(x_n)}_{y_n(z^*)}(\mathcal{E}^u_{y_n(z^*)}),\mathcal{E}^u_{y_{n-1}(z)}). \tag{2}$$

Here we made use of the following facts:

i) The angle function $\measuredangle(\cdot,\cdot)$ is a metric in the Grassmann manifold (of $m_u$-dimensional subspaces).

ii) $f(W^u_p) = W^u_{f(p)}$ , $p\in\Lambda$

iii) For sufficiently small $\varepsilon$ the graph of $\alpha_n(\cdot)$ is arbitrarily close to the "coordinate plane" $E^u_{x_n}\times\{0\}$ and the action of $f^{(x_n)}$ on this plane can be approximated as well as desired by the action of $Df_{x_n}$ on $E^u_{x_n}$, which is a pure expansion at least with the factor $\eta(x_n)>\alpha^{-1}>1$. These approximations are valid uniformly in $x$ and $n$, because $\Lambda$ is compact and in view of the uniformly bounded curvature of the local unstable manifolds. So the pre-image of $\alpha_{n-1}(z)$ is surely inside $Z(\alpha_n)$, so that $z^*$ exists.

Let us denote the first one of the two angles at the right-hand side of *(2)* by $a_1$, the second one by $a_2$. We get

$$a_1 \le \measuredangle(Df^{(x_n)}_{\alpha_n(z^*)}(\mathcal{E}^u_{\alpha_n(z^*)}),Df^{(x_n)}_{\alpha_n(z^*)}(\mathcal{E}^u_{y_n(z^*)})) +$$

$$+ \measuredangle(Df^{(x_n)}_{\alpha_n(z^*)}(\mathcal{E}^u_{y_n(z^*)}),Df^{(x_n)}_{y_n(z^*)}(\mathcal{E}^u_{y_n(z^*)})).$$

Denote the first term here by $a_{11}$, the second term by $a_{12}$.

Take any vector $v'$ from $Df^{(x_n)}_{\alpha_n(z^*)}(\mathcal{E}^u_{y_n(z^*)})$ and consider its projection $v$ to $Df^{(x_n)}_{\alpha_n(z^*)}(\mathcal{E}^u_{\alpha_n(z^*)}) = \mathcal{E}^u_{\alpha_{n-1}(z)}$. Set $w=v'-v$. Assume that $\|v\|=1$. The pre-image $\bar{v}$ of $v$ with respect to $Df^{(x_n)}_{\alpha_n(z^*)}$ has a length which can be estimated by

$$\|\bar{v}\| \le (1+\zeta_1(\varepsilon))(\eta(x_n))^{-1}\|v\| = (1+\zeta_1(\varepsilon))(\eta(x_n))^{-1}, \tag{3}$$

where $\zeta_1(\varepsilon)$ is some function of $\varepsilon$ tending towards 0 as $\varepsilon\to0$ and

which does not depend on $x,y,n$.
This is due to the fact that, depending on $\varepsilon$, we are arbitrarily close to the situation where $Df^{-1}_{x_n}$ is applied to a vector in the unstable tangent space. The same kind of argument applies to $w$. Since $w$ is perpendicular to $\mathcal{E}^u_{x_{n-1}(z)}$, we infer that $w/\|w\|$ is arbitrarily close to a unit vector in the stable tangent space at $x_{n-1}$ and, depending on $\varepsilon$, we are arbitrarily close to the situation, where $Df^{-1}_{x_n}$ is applied to a stable direction. So for the pre-image $\bar{w}$ of $w$ we have

$$\|\bar{w}\| \geq (1-\zeta_2(\varepsilon))(\lambda(x_n))^{-1}\|w\|. \tag{4}$$

Finally, since $Df$ conserves the stable and unstable directions, the deviation of $\bar{w}/\|\bar{w}\|$ from a unit vector in the stable tangent space at $x_n$ is arbitrarily small. This means that $\bar{v}'=\bar{v}+\bar{w}$, which belongs to $\mathcal{E}^u_{y_n(z^*)}$, has a projection to $\mathcal{E}^u_{x_n(z^*)}$ which is arbitrarily close to $\bar{v}$ as $\varepsilon$ becomes small. Note that $\|\bar{v}\|$ is bounded from below in accordance with *(3)*. So we may conclude that for some function $\zeta(\varepsilon)$ which goes to 0 with $\varepsilon$ we have

$$\sphericalangle(\mathcal{E}^u_{x_n(z^*)},\mathcal{E}^u_{y_n(z^*)}) \geq$$
$$\geq (1+\zeta(\varepsilon))^{-1}\,\frac{\eta(x_n)}{\lambda(x_n)}\cdot\sphericalangle(Df^{(x_n)}_{x_n(z^*)}(\mathcal{E}^u_{x_n(z^*)}),Df^{(x_n)}_{x_n(z^*)}(\mathcal{E}^u_{y_n(z^*)})),$$

so that we obtain

$$a_{11} \leq (1+\zeta(\varepsilon))\,\frac{\lambda(x_n)}{\eta(x_n)}\,s_n. \tag{5}$$

For $a_{12}$ we get for some absolute constants $C,C'$

$$a_{12} \leq C'\|Df^{(x_n)}_{x_n(z^*)}-Df^{(x_n)}_{y_n(z^*)}\| \leq$$
$$\leq C\|x_n(z^*)-y_n(z^*)\|. \tag{6}$$

Here we made use of the fact that the numbers $\mu(p)$ are bounded away from zero on the compact set $\Lambda$ and that $Df$ is a Lipschitz function on M (this in connection with lemma 1).

Now we intend to estimate $\alpha_2$. We get for some constants $C_1$ and $C_2$ and for small $\varepsilon$

$$\alpha_2 \le C_1 \| f^{(x_n)}(y_n(z^*)) - y_{n-1}(z) \| \le$$
$$\le C_2 \| y_n(z^*) - (f^{(x_n)})^{-1}(y_{n-1}(z)) \| .$$

Now take into account that $y_n(z^*)$ is the intersection of the unstable manifold $\mathcal{V}^u_{y_n}$ with the affine subspace of dimension $m_s$ which goes through $x_n(z^*)$ and which is perpendicular to $E^u_{x_n}$ (i.e. parallel to $E^s_{x_n}$).

At the other hand, $(f^{(x_n)})^{-1}(y_{n-1}(z))$ is the intersection of $\mathcal{V}^u_{y_n}$ with the pre-image of the affine subspace which goes through $x_{n-1}(z)$ and is perpendicular to $E^u_{x_{n-1}}$. From this we derive that for some $C_3$, which does not depend upon $x, y, n, \varepsilon$, we have

$$\alpha_2 \le C_3 \| y_n(z^*) - x_n(z^*) \|$$

(we could even have written $C_3\varepsilon$ instead of $C_3$).
So we come to the conclusion that for some $C_4$

$$\sphericalangle(\mathcal{E}^u_{x_{n-1}(z)}, \mathcal{E}^u_{y_{n-1}(z)}) \le s_n(1+\zeta(\varepsilon))\frac{\lambda(x_n)}{\eta(x_n)} + C_4 \| y_n(z^*) - x_n(z^*) \| , \qquad (6)$$
$$\|z\| \le L\varepsilon .$$

Taking into consideration that for small $\varepsilon$ the value

$$\sup_{\substack{n,z \\ \|z\| \le L\varepsilon}} \sphericalangle(E^u_{x_n}, \mathcal{E}^u_{x_n(z)})$$

can be made arbitrarily small (we need only that it is less than, say, $\pi/4$), we can conclude that for some $K$ (not depending on $x, y, n, \varepsilon$)

$$\| y_n(z^*) - x_n(z^*) \| \le \| y_n(0) - x_n(0) \| + s_n K\varepsilon .$$

So altogether we have for some function $\zeta_3(\varepsilon)$ (tending to zero as $\varepsilon$ does so) and for some constant $K_1$ (that does not depend on $x, y, n, \varepsilon$):

$$s_{n-1} \le s_n (1+\zeta_3(\varepsilon)) \frac{\lambda(x_n)}{\eta(x_n)} + K_1 \| y_n(0) - x_n(0) \| . \tag{7}$$

4. Define for $0 \le n \le n_0$ the expression

$$\varphi(n) = \prod_{i=1}^{n} \frac{\lambda(x_i)}{\eta(x_i)} .$$

Set $\varphi(-1)=1$.
From *(7)* we get

$$s_0 \le C_0 \varphi(n_0)(1+\zeta_3(\varepsilon))^{n_0} + \sum_{j=0}^{n_0-1} K_1 \varphi(j)(1+\zeta_3(\varepsilon))^j \| y_{j+1}(0) - x_{j+1}(0) \| .$$

So for each $z$ with $\|z\| \le L\varepsilon$ we get

$$\| y_0(z) - x_0(z) \| \le \| y_0(0) - x_0(0) \| +$$
$$+ C_0 K\varepsilon \varphi(n_0)(1+\zeta_3(\varepsilon))^{n_0} + \sum_{j=0}^{n_0-1} K_1 K\varepsilon \varphi(j)(1+\zeta_3(\varepsilon))^j \| y_{j+1}(0) - x_{j+1}(0) \| .$$

so that for some constant $R$ (not depending on $x, y, \varepsilon$) we get

$$d(y,y') \le d(x,x') +$$
$$R\varepsilon \Big( \varphi(n_0)(1+\zeta_3(\varepsilon))^{n_0} + \sum_{j=0}^{n_0-1} \varphi(j)(1+\zeta_3(\varepsilon))^j \cdot d(x_{j+1}, x'_{j+1}) \Big).$$

Here we made use of the fact that the stable manifolds $V^s_{x_n}$ and $V^s_{y_n}$ approximate $E^s_{x_n}$ as well as desired for small $\rho$. Moreover we applied lemma 1.
Taking into account that $n_0$ is the last index with the property that $x_n, y_n, x'_n, y'_n$ fulfil $(A_\varepsilon)$, $0 \le n \le n_0$, we easily derive that there is some constant $R'$ not depending on $x, y, \varepsilon$, such that

$$d(y,y') \le R' \sum_{j=0}^{n_0+1} \varphi(j-1)(1+\zeta_3(\varepsilon))^j \cdot d(x_j, x'_j) . \tag{8}$$

In fact, since we know that $x_{n_0}, y_{n_0}, x'_{n_0}, y'_{n_0}$ fulfil $(A_\varepsilon)$, the distances $d(x_{n_0}, y'_{n_0})$ and $d(x'_{n_0}, y_{n_0})$ are less than $\varepsilon$. But since $x_{n_0}, y'_{n_0}$ are on the same local unstable manifold, we may conclude (for small $\varepsilon$), that $d(x_{n_0+1},y'_{n_0+1})$ is less than

$$\eta(x_{n_0+1})^{-1}(1+\zeta_4(\varepsilon))\varepsilon,$$

where $\zeta_4$ tends to zero as $\varepsilon$ does so. The same holds for the pair $(x'_{n_0},y_{n_0})$. So the quadrupel $(x_{n_0+1},y_{n_0+1},x'_{n_0+1},y'_{n_0+1})$ cannot fail to fulfil $(A_\varepsilon)$ unless $d(x_{n_0+1},x'_{n_0+1}) > (1-\eta(x_{n_0+1})^{-1}(1+\zeta_4(\varepsilon)))\varepsilon$ or $d(y_{n_0+1},y'_{n_0+1}) > (1-\eta(x_{n_0+1})^{-1}(1+\zeta_4(\varepsilon)))\varepsilon$. Finally we use the fact that for small $\rho$ the $C^2$-disks $W^u_{x_{n_0+1}}(\delta)\cap B_\rho(x_n)$ and $W^u_{y_{n_0+1}}(\delta)\cap B_\rho(x_n)$ as well as $W^s_{x_{n_0+1}}(\delta)\cap B_\rho(x_n)$ and $W^s_{y_{n_0+1}}(\delta)\cap B_\rho(x_n)$ are flat and parallel with an arbitrary accuracy and that the angle between the stable and unstable direction is bounded from below on $\Lambda$. So we may conclude that for some $\zeta_5(\varepsilon)$ tending to 0 with $\varepsilon$ we have

$$d(x_{n_0+1},x'_{n_0+1}) > (1-\eta(x_{n_0+1})^{-1}(1+\zeta_5(\varepsilon)))\varepsilon, \tag{9}$$

so that *(8)* is verified.
So for some positive $\zeta_6(\varepsilon)$ (with the properties as before) we have

$$\log d(y,y') \le \log R' + \log n_0 + \max_{0\le j\le n_0+1} [\, \log\varphi(j-1) + j\zeta_6(\varepsilon) + \log d(x_j,x'_j)].$$

Hence for small $\varepsilon$ we get

$$\frac{\log d(y,y')}{\log d(x,x')} \ge \tag{10}$$
$$\ge \frac{\log R' + \log n_0}{\log d(x,x')} + \min_{0\le j\le n_0+1} \frac{\log\varphi(j-1) + j\zeta_6(\varepsilon) + \log d(x_j,x'_j)}{\log d(x,x')}$$

Let us denote the expression that has to be minimized by $a(j)$. Further set $\sigma_j := (d(x_j,x'_j))^{-1} d(x_{j-1},x'_{j-1})$. Then we have

$$a(j) = \frac{\sum_{i=1}^{j-1} (-\log \frac{\lambda(x_i)}{\eta(x_i)}) - j\zeta_6(\varepsilon) + \sum_{i=j+1}^{n_0+1} (-\log \sigma_i) - \log d(x_{n_0+1},x'_{n_0+1})}{\sum_{i=1}^{n_0+1} (-\log \sigma_i) - \log d(x_{n_0+1},x'_{n_0+1})}$$

Now choose some $\varkappa<\min(1, \liminf_{n\to\infty} \inf_{p\in\Lambda} \frac{\sum_{i=0}^{n} [\log \lambda(f^i p) - \log \eta(f^i p)]}{\sum_{i=0}^{n} \log \mu(f^i p)})$.

Denote the expression under the $\liminf_{n\to\infty} \inf_{p\in\Lambda}$ by $a(p,n)$. Due to the choice of $\varkappa$ there can be found some $\tau>0$ and some $N$ such that for any $p\in\Lambda$ and $n>N$ we have

$$\varkappa < \min(1, a(p,n)) - \tau.$$

Now observe that for sufficiently small $\varepsilon$ we have for some $\zeta_7(\varepsilon)$ tending to zero as $\varepsilon$ does so

$$-\log\mu(x_j) + \zeta_7(\varepsilon) \geq -\log\sigma_j \geq -\log\lambda(x_j) - \zeta_7(\varepsilon), \qquad 0\leq j\leq n_0+1.$$

Let $\mu=\min_{p\in\Lambda}\mu(p)$, $\lambda= \max_{p\in\Lambda}\lambda(p)$. We have $0<\mu\leq\lambda<\alpha<1$. ($\Lambda$ is a compact.) We get from $d(x_{n_0}, x'_{n_0})<\varepsilon$ the estimate

$$n_0 \leq \frac{-\log\varepsilon + \log d(x,x')}{\log\lambda + \zeta_7(\varepsilon)}.$$

So for fixed $\varepsilon$ such that $\log\lambda + \zeta_7(\varepsilon)<0$ we obtain that the first term at the right hand side of *(10)* can be made less than $\tau/4$, supposed $d(x,x')$ is small enough. Taking into account the estimate *(9)* we easily derive that the difference between $a(j)$ and

$$a_1(j) = \frac{\sum_{i=1}^{j} (-\log \frac{\lambda(x_i)}{\eta(x_i)}) - j\zeta_6(\varepsilon) + \sum_{i=j+1}^{n_0+1} (-\log \sigma_i)}{\sum_{i=1}^{n_0+1} (-\log \sigma_i)}$$

less than $\tau/4$ (uniformly in $j$), supposed $d(x,x')$ is small enough. If $\varepsilon$ is small enough, the difference between $a_1(j)$ and

$$a_2(j) = \frac{\sum_{i=1}^{j} (-\log \frac{\lambda(x_i)}{\eta(x_i)}) + \sum_{i=j+1}^{n_0+1} (-\log \sigma_i)}{\sum_{i=1}^{n_0+1} (-\log \sigma_i)}$$

is less than $\tau/4$ (uniformly in $j$). So for sufficiently small $\varepsilon$ and (assuming $(A_\varepsilon)$) for sufficiently small $d(x,x')$ we have

$$\frac{\log d(y,y')}{\log d(x,x')} > \min_{0\le j\le n_0+1} \frac{\sum_{i=1}^{j}\left(-\log \frac{\lambda(x_i)}{\eta(x_i)}\right) + \sum_{i=j+1}^{n_0+1}(-\log \sigma_i)}{\sum_{i=1}^{n_0+1}(-\log \sigma_i)} - 3/4\cdot\tau.$$

Now make $d(x,x')$ so small, i.e. make $n_0$ so large that

$$\frac{\sum_{i=N+1}^{n_0+1}(-\log \sigma_i)}{\sum_{i=1}^{n_0+1}(-\log \sigma_i)} > 1-\tau/4.$$

(For this it is sufficient to have $\dfrac{N\,(-\log\mu+\zeta_7(\varepsilon))}{n_0\,(-\log\lambda-\zeta_7(\varepsilon))} < \tau/4$.)

If this is fulfilled, we have

$$\frac{\log d(y,y')}{\log d(x,x')} > \min\left(1-\tau,\ \min_{N<j\le n_0+1} \frac{\sum_{i=1}^{j}\left(-\log \frac{\lambda(x_i)}{\eta(x_i)}\right) + \sum_{i=j+1}^{n_0+1}(-\log \sigma_i)}{\sum_{i=1}^{n_0+1}(-\log \sigma_i)} - 3/4\cdot\tau\right).$$

Now for those $j$ with $N<j\le n_0+1$ and such that

$$\sum_{i=1}^{j}\left(-\log \frac{\lambda(x_i)}{\eta(x_i)}\right) \ge \sum_{i=1}^{j}(-\log \sigma_i)$$

the expression $a_2(j)$ is not less than one, so that these $j$ do not affect the minimum. For the other $j$ with $N<j\le n_0+1$ we may estimate

$$a_2(j) \geq \frac{\sum_{i=1}^{j} (-\log \frac{\lambda(x_i)}{\eta(x_i)})}{\sum_{i=1}^{j} (-\log \sigma_i)} \geq \frac{\sum_{i=1}^{j} (-\log \frac{\lambda(x_i)}{\eta(x_i)})}{\sum_{i=1}^{j} (-\log\mu(x_i)) + j\zeta_7(\varepsilon)}.$$

Making $\varepsilon$ small enough, we finally get

$$\frac{\log d(y,y)}{\log d(x,x')} > \min(1, \min_{N<j} a(x_j, j)) - \tau > \varkappa,$$

i.e. we have shown

$$d(y,y') \leq (d(x,x'))^{\varkappa}$$

holds for sufficiently small $d(x,x')$. Then, of course, we find some $C_\varkappa$ such that

$$d(y,y') \leq C_\varkappa(d(x,x'))^{\varkappa}$$

is fulfilled as soon as $x,y,x',y'$ fulfil $(A_\varepsilon)$.

So we are through except for the deletion of the condition $(A_\varepsilon)$. First consider the case $d(x,y')<\varepsilon/2$. From the continuity theorem for the stable and unstable manifolds we easily derive the equicontinuity of the family of holonomy maps on the compact $\Lambda$ which means that we find some $\varepsilon'>0$ depending on $\varepsilon$ such that $(A_\varepsilon)$ cannot be violated unless $d(x,x')>\varepsilon'$. But then *(2)* is trivially fulfilled for some suitable $C_\varkappa$ because $\Lambda$ is compact.

Now assume $d(x,y')\geq\varepsilon/2$. Since $y'\in W^u_x(\beta')$, we find some $k_o$ such that $d(f^{-k_o}(x), f^{-k_o}(y'))<\varepsilon/2$. It is easy to conclude from the compactness of $\Lambda$ and from the continuity theorem for the unstable manifolds that we can choose $k_o$ independently of what $x$ and $y'$ are, it depends just on $\beta'$ and $\varepsilon$.

Next we choose some $\varepsilon_1<\varepsilon/2$ such that whenever $p,q\in\Lambda$ and $q\in W^s_p(\varepsilon_1)$, so $f^{-k_o}(q)\in W^s_{f^{-k_o}(p)}(\varepsilon/2)$. Finally, using again the equicontinuity of the family of holonomy maps on $\Lambda$, we find some $\varepsilon_2<\varepsilon_1$ such that whenever $p,q\in\Lambda$, $d(p,q)<\beta$ and $[p,q]\in W^s_p(\varepsilon_2)$, then we have $[q,p]\in W^s_q(\varepsilon_1)$.

Now obviously we may restrict our attention to the case $x'\in W^s_x(\varepsilon_2)\subset W^s_x(\varepsilon_1)$. The we have $y'\in W^s_y(\varepsilon_1)$. Hence

$$f^{-k_o}(x')\in W^s_{f^{-k_o}(x)}(\varepsilon/2)\subset W^s_{f^{-k_o}(x)}(\beta')$$

and

$$f^{-k_o}(y')\in W^s_{f^{-k_o}(y)}(\varepsilon/2)\subset W^s_{f^{-k_o}(y)}(\beta').$$

Hence $d(f^{-k_o}(x),f^{-k_o}(y))<\varepsilon<\beta$, so that $f^{-k_o}(x')=[f^{-k_o}(x),f^{-k_o}(y)]$ and $f^{-k_o}(y')=[f^{-k_o}(y),f^{-k_o}(x)]$. The quadrupel $(f^{-k_o}(x),f^{-k_o}(y),f^{-k_o}(x'),f^{-k_o}(y'))$ fulfils $(A_\varepsilon)$, so that *(2)* is valid for these points, with some constant $C_*$. Since $f^{k_o}$ is a fixed diffeomorphism, we have derived *(2)* in the general case. Of course, we have to enlarge $C_*$ depending on the Lipschitz constant for $f^{k_o}$ on $\Lambda$. The proof of the theorem is complete.

## References


[1] M. Brin, Ya. Pesin, Partially hyperbolic dynamical systems, *Izv.Acad.Nauk SSSR ser., mat.* No.1 (1974), 170-212

[2] M. Brin, Yu. Kifer, Dynamics of Markov chains and stable manifolds for random diffeomorphisms, *Ergodic Theory and Dyn. Syst.* 7 (1987), 351-374

[3] J. Schmeling, Holderexponenten fur die Holonomieabbildung von eindimensionalen hyperbolischen Attraktoren und der Zusammenhang zur lokalen Hausdorffdimension, *Thesis* 1990

[4] S. Smale, Differentiable Dynamical Systems, *Bull. Amer. Math. Soc.*73 (1967), 747-817

[5] M. W. Hirsch, C. C. Pugh, Stable Manifolds and Hyperbolic Sets, *Proc. Symp. Pure Math.*, Vol. XIV (1970), 133-163

[6] A. B. Katok, Local'nye svojstva giperboliceskych mnoshestv, in: *Differentiable dynamics* (by Z. Nitecki), russ. edition, Mir publishers, Moscow 1975, 214-232

[7] H.-G. Bothe, Some remarks concerning the fractal dimensions of expanding attractors, *Report K.-Weierstraß-Inst. f. Math. AdW der DDR*, Berlin 1988

# PECULIAR SUBMEASURES ON FINITE ALGEBRAS

Jan Sipos
Stavebna fakulta, Radlinskeho 11,
813-68 Bratislava, Czechoslovakia

IF $\mu$ is a finite submeasure on a measurable space $(X,\mathbb{A})$ then it may happen that $\nu(X) < \mu(X)$ for any finitely addititve nonnegative measure $\nu$ on $\mathbb{A}$ which is dominated by $\mu$.

The number

$$\bar{\mu} = (\mu(X) - \sup \{ \nu(X) \mid \nu \text{ is a measure dominated by } \mu\}) / \mu(X)$$

expresses, in a way, the degree of how peculiar the submeasure $\mu$ is. It was shown in [H-CH] that $\bar{\mu}$ can even be one (in this case $\mu$ is called a pathological submeasure). On the other hand, it is easy to see that if $\mathbb{A}$ is a finite algebra then $\bar{\mu}$ must be less than one.

The purpose of this article is to give some properties and estimates for the number

$$\rho_k = \sup\{\bar{\mu} \mid \mu \text{ is a normalized submesure on } (X,\mathbb{P}(X)) \text{ and } \operatorname{card}(X)=k\}$$

which can be considered as the highest degree of how peculiar a submeasure on a finite algebra having exactly $k$ atoms can be.

Our main result is that

$$\frac{1}{2}\,\frac{1}{\log k + \log 2} \le 1 - \rho_k \le \frac{\log 4}{\log k}.$$

First we fix some terminology. Let $(X,\mathbb{A})$ be a measurable space. A real valued set function $\mu$ defined on $\mathbb{A}$ is called a submeasure iff it is increasing, subadditive and vanishes at empty set. All measures considered in the present paper are finitely additive.

Let $X_k = \{x_1, x_2, \dots, x_k\}$, and let $\mu$ be a normalized submeasure on $\mathbb{A}_k = \mathbb{P}(X_k)$. We put

$$\underline{\mu} = \sup \nu(X_k)$$

where the supremum is taken over all measures on $\mathbb{A}_k$ which are dominated by $\mu$ (shortly $\nu \le \mu$). Further we put

$$q_k = \inf \underline{\mu}$$

where the infimum is taken over all normalized submeasures on $\mathbb{A}_k$.

Finally we put

$$\rho_k = 1 - q_k.$$

Let us start with example.

1. *Example.* Let $X = \{x_1, x_2, x_3\}$ and let $\mathbb{A} = \mathbb{P}(X)$.

*a)* Put

$$\mu(A) = \begin{cases} 0 & \text{if} \quad A = \emptyset \\ 1/2 & \text{if} \quad \emptyset \neq A \neq X \\ 1 & \text{if} \quad A = X\,. \end{cases}$$

It is easy to check that $\mu$ is a submeasure on $\mathbb{A}$. Let $\nu$ be a measure on $\mathbb{A}$ dominated by $\mu$. Then

$$\begin{aligned} 2.\nu(X) &= \nu(\{x_1, x_2\}) + \nu(\{x_1, x_3\}) + \nu(\{x_2, x_3\}) \\ &\leq \frac{1}{2} + \frac{1}{2} + \frac{1}{2} \\ &= \frac{3}{2}\,, \end{aligned}$$

and so $\underline{\mu} \leq \frac{3}{4}$ which showes that $q_3 \leq \frac{3}{4}$.

*b)* Let $\mu$ be a normalized submeasure on $\mathbb{A}$. We are going to show that $\underline{\mu} \geq \frac{3}{4}$. Put

$$t = \frac{\mu(\{x_1, x_2\}) + \mu(\{x_1, x_3\}) + \mu(\{x_2, x_3\})}{2}\,.$$

Via subadditivity of $\mu$ we get

$$\begin{aligned} \mu(\{x_1, x_2\}) + \mu(\{x_1, x_3\}) &\geq 1 \\ \mu(\{x_1, x_2\}) + \mu(\{x_2, x_3\}) &\geq 1 \\ \mu(\{x_1, x_3\}) + \mu(\{x_2, x_3\}) &\geq 1 \end{aligned}$$

If we add the last three inequalities and divide the result by two we get

$$t \geq \frac{3}{4}\,.$$

Now let $\nu$ be the measure on $\mathbb{A}$ defined by

$$\nu(\{x\}) = \min\{1, t\} - (\min\{1, \frac{1}{t}\}).\mu(X - \{x\})$$

for $x \in X$. It is easy to see that $\nu$ is dominated by $\mu$ and so

$$\underline{\mu} \geq \nu(X) = \min\{1, t\}.$$

We have just obtained that $\frac{3}{4} \leq q_3$ which together with *a)* implies that $q_3 = \frac{3}{4}$ and $p_3 = \frac{1}{4}$.

Intuively it seems to be clear that if we want to 'press' the number $\underline{\mu}$ down then we must take 'large' sets with 'small' $\mu$-value on them so that they are covering the space $X$ as many time as possible. Really the following theorem holds true.

**2. Theorem.** *Let* $\mu$ *be a normalized submeasure on* $(X_k, \mathbb{A}_k)$. *Then*

$$\underline{\mu} = \inf \left\{ \frac{1}{r} \sum_{i=1}^{n} \mu(A_i) \mid r.\chi_{X_k} \le \sum_{i=1}^{n} \chi_{A_i} \right\}.$$

For the proof we need some auxiliary results. Let us introduce a new notion at first. We say that a submeasure on $(X, \mathbb{A})$ is an *m-submeasure* (measure like submeasure, because it has some measure like properties) if

$$r.\chi_A \le \sum_{i=1}^{n} \chi_{A_i} \quad \text{implies} \quad r.\mu(A) \le \sum_{i=1}^{n} \mu(A_i).$$

$A$ and $A_i$ being from $\mathbb{A}$.

**3. Theorem.** *Let* $\mu$ *be an m-submeasure on* $(X_k, \mathbb{A}_k)$. *Then*

$$\mu = \max_{\nu \le \mu} \nu .$$

*Proof*. We shall give only a sketch of the proof because the theorem can be proved in more general setting using the results of [Č] and [S]. Denote by $\mathbb{F}$ the family of all $\mathbb{A}_k$ measurable functions on $X_k$. For a function $f$ in $\mathbb{F}$ put

$$\|f\| = \inf \left\{ \frac{1}{r} \sum_{i=1}^{n} \mu(A_i) \mid r.|f| \le \sum_{i=1}^{n} \chi_{A_i} \right\}$$

where $n$ and $r$ are natural numbers and $A_i \in \mathbb{A}_k$.

It is easy to see that $\| \ \|$ is a seminorm. For a fixed $A \in \mathbb{A}_k$ put

$$\mathbb{F}_0 = \langle c.\chi_A \mid c \text{ is a real number} \rangle.$$

Let $\mathbb{F}_1, \mathbb{F}_2, \ldots, \mathbb{F}_n$ be a sequence of linear subspaces of $\mathbb{F}$ with the property

$$\mathbb{F}_1 \subset \mathbb{F}_2 \subset \ldots \subset \mathbb{F}_n = \mathbb{F}$$

and such that the codimension of $\mathbb{F}_i$ in $\mathbb{F}_{i+1}$ is 1.

For $f = c.\chi_A \in \mathbb{F}_0$ put

$$T_0(f) = c.\mu(A).$$

Then $T_0$ is a linear monotone functional on $\mathbb{F}_0$ with $T_0(f) \le \|f\|$ on $\mathbb{F}_0^+$.

Similarly as in Lemma 2 of [S] we can define a sequence $\langle T_i \rangle$ of linear monotone functionals on $\mathbb{F}_i$ with

$$T_i(f) \le \|f\|$$

on $\mathbb{F}_i^+$ and such that $T_{i+1}$ is an extension of $T_i$.

Let $T = T_n$. Put

$$\nu(E) = T(\chi_E)$$

for $E \in \mathbb{A}_k$. Then clearly $\nu$ is a measure on $\mathbb{A}_k$ which is dominated by $\mu$ and $\nu(A) = \mu(A)$. □

*Proof of Theorem 2.* Let $A$ be in $\mathbb{A}_k$. Put

$$\tau(A) = \inf\left\{ \frac{1}{r} \sum_{i=1}^{n} \mu(A_i) \;\middle|\; r.\chi_A \le \sum_{i=1}^{n} \chi_{A_i} \right\}.$$

It is easy to check that $\tau$ is an m-submeasure and so by Theorem 4 we get that

$$\tau(X_k) = \sup\{ \nu(X_k) \mid \nu \le \tau \}.$$

Since $\nu \le \tau$ if and only if $\nu \le \mu$ we have that Theorem 2 is true. □

The more sets the algebra $\mathbb{A}_k$, has the more possibilities to press the number $\underline{\mu}$ down. Indeed the folllowing lemma holds true.

**4. Lemma.** *The sequence* $q_k$ *is a decreasing sequence.* □

**5. Lemma.**
$$q_{\binom{2n}{n}} \le \frac{2}{n+1}.$$

*Proof.* Let $n$ be a fixed natural number. Let us consider the $(2n)\times\binom{2n}{n}$ matrix $M$ whose columns are all the $2n$-touples which can be made from $n$ zeros and $n$ ones. Let $X = \left\{ x_1, \ldots, x_{\binom{2n}{n}} \right\}$ and let $\mathbb{A} = \mathbb{P}(X)$. The $i$-th row of the matrix $M$ can be considered as the characteristic function $\chi_{E_i}$ of a subset $E_i$ of $X$, for each $i = 1,2,\ldots,2n$. Let us define a submeasure $\mu$ on $\mathbb{A}$ in the following way

$$\mu(A) = \frac{\min\left\{ m \mid A \subset \bigcup_{i=1}^{m} E_i \right\}}{n+1}.$$

Then $\mu$ is a normalized submeasure on $\mathbb{A}$ and $\mu(E_i) = \dfrac{1}{n+1}$. Since

$n.\chi_X = \sum_{i=1}^{n} \chi_{A_i}$ it follows from Theorem 3 that

$$\underline{\mu} \le \frac{2}{n+1}$$

and thus

$$q_{\binom{2n}{n}} \le \frac{2}{n+1} . \ \square$$

**6. Theorem.**

$$q_k \le \frac{\log 4}{\log k} .$$

*Proof.* From Stirling's formula we get that

$$2^n \ge \binom{2n}{n} \quad (\text{for } n \ge 3).$$

Put $k = 2^n$ ; then $n = \log k \,/\, \log 2$. Since $\{q_k\}$ is decreasing we then have

$$q_k \le \frac{2}{n} = \frac{\log 4}{\log k} . \ \square$$

If we use a slightly finer consideration we can even show that

$$q_k \le \frac{\log 4}{\log k + (1/2) \log \log k + \log 2} .$$

Let us turn now our attention to the lower bound for $q_k$.

**7. Theorem.**

$$\frac{1}{2} \cdot \frac{1}{\log k + \log 2} \le q_k .$$

*Proof.* Let $\mu$ be a submeasure on $(X, \mathbb{P}(X))$, where $\operatorname{card}(X) = k$. Let

$$r . \chi_X \le \sum_{i=1}^{n} \chi_{A_i} .$$

Put

$$m = \left[ \frac{\log 2 + \log k}{\log n - \log (n-r)} \right] + 1$$

then

$$\binom{n}{m} \ge 2k \binom{n-r}{m} .$$

The number of subsets $F \subset \{1, 2, \dots, n\}$ with $\operatorname{card}(F) = m$ and

$$\sum_{i \in F} \chi_{A_i}(x) < \chi_X(x)$$

for some $x$ in $X$ can be at most

$$k \binom{n-r}{m} .$$

Thus for at least a half of the subsets $F$ of the set $\{1, 2, \dots, n\}$ with $\operatorname{card}(F) = m$ the following holds

$$\sum_{i \in F} \chi_{A_i} \ge \chi_X .$$

And so

$$\sum_F \sum_{i \in F} \mu(A_i) \ge \frac{1}{2} \binom{n}{m}$$

(where the first sum is taken over all subsets $F \subset \{1,2,\ldots,n\}$ with card $(F) = m$ ). From which we have

$$\binom{n-1}{m-1} \sum_{i=1}^{n} \mu(A_i) \geq \frac{1}{2} \binom{n}{m}$$

which gives

$$\frac{1}{r} \sum_{i=1}^{n} \mu(A_i) \geq \frac{1}{2} \frac{\log t}{t-1} \frac{1}{\log k + \log 2}$$

(where $t = \frac{n-r}{n}$ ). Since

$$\frac{\log t}{t-1} > 1$$

we get that

$$\underline{\mu} > \frac{1}{2} \frac{1}{\log k + \log 2}$$

and so

$$q_k > \frac{1}{2} \frac{1}{\log k + \log 2} . \square$$

Using again some more finer consideration we can get that

$$\frac{1}{1 + 1 / \log k} \frac{1}{\log k + \log (1 + \log k)} < q_k .$$

**References**

[Č] Černek, P.: Integral with respect to a submeasure and the product of submeasures. Ph.D. Thesis. (In Slovak), (1982).

[H-CH] Herer, W., Christensen, J.P.R.: On the existence of pathological submeasures and the construction of exotic groups. Math. Ann. 213,203-210(1975).

[S] Sipos, J.: A note on Hahn - Banach extension theorem. To appear in Czech. Math. Journal.

# INVARIANCE PRINCIPLES AND CENTRAL LIMIT THEOREMS FOR NONADDITIVE, STATIONARY PROCESSES

Ulrich Wacker

Postal adress: Institut für Mathematische Stochastik, Universität Göttingen

Lotzestraße 13, W - 3400 Göttingen

## §1 Introduction

In this paper we shall study processes $\mathfrak{X} = ( X_{m,n} )_{0 \leq m \leq n < +\infty}$ , which are stationary in the sense that the joint distribution of the process $\mathfrak{X}$ is the same as that of the process $( X_{m+1,n+1} )_{0 \leq m < n < +\infty}$ :

$$\text{( I )} \qquad ( X_{m,n} ) \overset{\mathcal{D}}{=} ( X_{m+1,n+1} )$$

A nonadditive process $\mathfrak{X}$ is called a ***subadditive process*** if it satisfies ( I ) and the following two conditions :

( II ) The random variables $X_{m,n}$ are integrable and satisfy $\inf_{N \in \mathbb{N}} ( E( X_{0,N} ) / N ) > - \infty$ ,

$$\text{( III )} \qquad X_{m,n} \leq X_{m,k} + X_{k,n} \qquad \text{for all } 0 \leq m < k < n < + \infty.$$

$\mathfrak{X}$ is a ***superadditive process***, if $- \mathfrak{X} := ( - X_{m,n} )$ is a subadditive process. $\mathfrak{X}$ is an ***additive process*** , if $\mathfrak{X}$ is a subadditive and a superadditive process. Subsequently, we shall also admit weaker conditions.

Nonadditive processes appear in queuing theory, percolation, demography, random walks, in the study of products of random matrices and in other fields. The starting point of the ergodic theory of subadditive processes is the paper [ K 68 ] . Kingman [ K 68 ] proved the pointwise ergodic theorem for subadditive processes. Later Akcoglu-Sucheston [ A-S 78 ] generalized the Chacon - Ornstein - Theorem to the case of superadditive processes and Akcoglu-Krengel [ A-K 81 ] proved ergodic theorems for superadditive processes with multidimensional parameter. Ishitani [ I 77 ] proved a central limit theorem for subadditive processes and used this theorem to generalize the central limit theorem for products of random matrices, due to Furstenberg - Kesten [ F-K 60 ] . In this paper a more general central limit theorem and invariance principles for nonadditive processes shall be presented. These results lead to substantial generalizations of the known results for products of random matrices and for the range of random walks . The basic approach of the proof of Theorem 5 is inspired by [ K-P 80 ] , although the nonadditivity requires also very different estimates. The additive analogies of Theorem 1, Theorem 2 and Theorem 3 can be found in [ I-L 71 ] and many ideas in [ I-L 71 ] are used in the proofs of these theorems. The proof of Theorem 4 is inspired by the proof of [ K-P 80, Lemma 2.5 ]. The results in this paper and the above applications are contained in my thesis, written under the direction of U. Krengel. The applications shall be presented elsewhere.

## §2 *The central limit theorem*

The first aim in [ W 83 ] has been a central limit theorem for superadditive , independent processes. Although both conditions shall be weakened bellow, it is of interest to take a look at this situation. A nonadditive process $\mathfrak{X} = ( X_{m,n} )$ is called ***independent*** , if the random variables $( X_{m_i,n_i} )_i$ are independent , whenever $( [ m_i,n_i [ )_i$ are disjoint intervals . An additive, independent process obviously is formed by partial sums of i. i. d. random variables. In this case the central limit theorem is satisfied and the variances $V( X_{0,N} )$ tends to $\infty$ linearly with N provided $0 < V( X_{0,1} ) < \infty$. Ishitani [ 77 ] considers superadditive processes and assumes

( 2. 1 ) $\quad V( X_{0,N} ) \sim c N \qquad$ for some c with $0 < c < \infty$ ,

and

( 2. 2 ) $\quad ( E( X_{0,N} ) / N - \sup_M ( E( X_{0,M} )/M ) = o( \sqrt{N} ).$

There are examples of independent superadditive processes satisfying ( 2. 1. ) and ( 2. 2. ) with o replaced by O and in addition ( C 3 ) below, in which the the assertion of the central limit theorem fails ( s. Jain-Pruit [ J-P 73 ] ). However, condition ( 2. 1 ) can be substantially weakened. Using Kingman's decomposition theorem for superadditive processes, Ishitani proved that the following condition ( C 1 ) holds for processes satisfying ( 2. 1 ) and ( 2. 2 ) :

( C 1 ) There is an additive process $\mathfrak{Y} = ( Y_{m,n} )_{m,n}$ with $\| X_{0,N} - Y_{0,N} \|_1 = o( \sqrt{V( X_{0,N})})$ such that the joint process $( X_{m,n} , Y_{m,n} )_{0 \le m < n < \infty}$ is stationary.

It turns out that processes satisfying ( C 1 ) need not satisfy ( 2. 1 ). Recall that a positive real valued function h on $\mathbb{N}$ ( or on $\mathbb{R}^+$) is called slowly varying if

$$\lim_{N \to \infty} h( t N )/ h( N ) = 1$$

for every $t \in \mathbb{N}$ ( resp. $t \in \mathbb{R}^+$ ).

**Theorem 0** : *Let $\mathfrak{X}$ be an independent, superadditive process, which satisfies ( C 1 ) and the central limit theorem, then $V( X_{0,N} ) = N h( N )$, where h is a slowly varying function on $\mathbb{N}$.*

**Proof of Theorem 0** : Let $g_N := E( X_{0,N} )$. As $\mathfrak{X}$ is superadditive $\gamma := \lim ( g_N / N )$ exists, and $\gamma = \sup( g_N / N )$. Let k be fixed. Set

$$A_N = V( X_{0,kN} )^{1/2}$$

By superadditivity we have :

$$( 0.1 ) \qquad A_N^{-1} ( X_{0,kN} - g_{kN} ) \ge A_N^{-1} \Big( \sum_{l=1}^{k} ( X_{(l-1)N,lN} - g_N ) \Big) + A_N^{-1} ( k g_N - g_{kN} )$$

If there is a sequence $N_i$ , such that $V( X_{0,kN_i} ) / kV( X_{0,N_i} )$ tends to infinity , then the first term on the right hand side tends stochastically to 0 as $N = N_i \to \infty$. It follows from $kg_N \ge g_{kN} \ge kg\gamma$ that

$$| kg_N - g_{kN} | \le | kg_N - kN\gamma | = \| X_{0,N} - Y_{0,N} \|_1$$

Thus, ( C 1 ) implies that also the second term on the right hand side tends to 0. On the other hand, if the central limit theorem holds for $\mathfrak{X}$, the left hand side of ( 0. 1. ) tends in distribution to N( 0 , 1 ), a contradiction. So, $B_N = V( X_{0,kN} ) / ( k V( X_{0,N} ) )$ is a bounded sequence. If $\liminf B_N = 0$, look at ( 0. 1. ) with $A_N^{-1}$ replaced by $V( X_{0,N} )^{-(1/2)}$. A similiar argument as above yields a contradiction. Finally, let $N' \to \infty$ be a sequence such that $B_{N'}$

tends to a strictly positive number, which we denote by $\sigma^{-2}$. Then the left hand side in ( 0. 1 ) tends to $N(0, 1)$, the first term on the right hand side to $N(0, \sigma^2)$, and the second term on the right hand side to 0. Thus, any limit point of $B_N$ equals 1, so that $B_N$ must converge to 1. Put $h(N) = V(X_{0,N}) / N$. Then h is slowly varying on **N**.
q.e.d.

**Remark** : The following additional result can be found in [ W 83 ] : If $k_N$ is a sequence tending to $\infty$, and $\sqrt{k_N}\,( g_N - N\gamma )\, V( X_{0,N} )^{-(1/2)} \to 0$, then $h( k_N N ) / h( N ) \to 1$. The proof is partly simillar.

In view of theorem 0, we shall assume the following condition:

( C 2 ) $\quad V( X_{0,N} ) = N\, h( N )$ , where h is a slowly varying function.

**Definition** : Recall that a sequence $Z_N$, $N \in \mathbf{N}$, is called ***uniformly integrable***, if one and hence both of the following two equivalent conditions is satisfied :

( U1 ) There is a function $f : \mathbf{R}^+ \to \mathbf{R}^+$ with $\limsup( f( t ) / t ) = \infty$ and $\sup E( f( | Z_N | )) < \infty$.

( U2 ) For every $\varepsilon > 0$ exists an $k < \infty$ with $\sup E( | Z_N | \, 1_{\{|Z_N|>k\}} ) < \varepsilon$.

See Gänssler-Stute [ G-S 77 ] for a proof that ( U1 ) and ( U2 ) are equivalent. In the proofs we use ( U2 ). But it seems to bee easier in the applications to verify ( U1 ).

If $Z_N$ is a sequence of random variables with $E( Z_N ) = 0$ and $\lim V( Z_N ) = 1$, which satisfies $\lim Z_N \overset{\mathcal{D}}{=} N( 0 , 1 )$, then $( Z_N )^2$, $N \in \mathbf{N}$ is uniformly integrable. ( s. [ B 68 ] ). Thus, it is natural to impose also the following condition :

( C 3 ) $\quad V( X_{0,N} )^{-1} ( X_{0,N} - E( X_{0,N} ))^2$ , $N \in \mathbf{N}$, is uniformly integrable.

Ishitani [ I 77 ] proved a central limit theorem for mixing processes, and ( C 2 ) is a common condition for additive mixing processes. But, in the the nonadditive case , the usual $\alpha$- and $\varphi$- mixing conditions do not cover the case of independence. We will use a slightly more general definition of mixing for nonadditive processes. First we will define mixing and independence for families of σ-algebras.

**Definition** : Let $(\Omega, A, P)$ be a probability space and $\mathfrak{M} = (M_{m,n})_{0 \le m \le n < \infty}$ a family of sub-σ-algebras of A. For $-\infty \le s < t \le +\infty$ let $I_{s,t}$ be a set of disjoint intervalls $[u, v[$ with $s - 1 < u < v < t + 1$. Let $\sigma(I_{s,t}) := \sigma(M_{u,v}, [u, v[ \in I_{s,t})$. $\mathfrak{M}$ is called *α' - (φ' -) mixing*, if there is a sequence $\alpha_n$ ($\varphi_n$) of real numbers with $\lim \alpha_n = 0$ ($\lim \varphi_n = 0$) and $| P(A \cap B) - P(A)P(B) | \le \alpha_s$ ($\varphi_s P(A)$), for all $A \in \sigma(I_{0,t})$, $B \in \sigma(I_{t+s,\infty})$, all $0 < t < \infty$ and every choice of $I_{0,t}$ and $I_{t+s,\infty}$. If $\mathfrak{M}$ is a α'-mixing (φ'-mixing) family of σ-algebras with $M_{u,v} \subset M_{s,t}$ for $s \le u < v \le t$, then $\mathfrak{M}$ is called *α - (φ -) mixing*. A nonadditive process $\mathfrak{X} = (X_{m,n})_{m,n}$ is α-,α'-,φ-,φ'-mixing, if the family $\mathfrak{M} := (\sigma(X_{m,n}))_{m,n}$ is α-,α'-,φ-,φ'-mixing.

For additive, mixing processes ( C 2 ) is a usual condition. In the step from independence to mixing , the behavior of the variance changes for additive processes. For independent processes, the behavior of the variances changes in the step from additivity to nonadditivity. Then the step from independence to mixing brings no further problems.

**Theorem 1** : *Let $\mathfrak{X} = (X_{m,n})_{0 \le m < n < \infty}$ be a α' - mixing, stationary process, which satisfies ( C 1 ) , ( C 2 ) and ( C 3 ), then the central limit theorem holds :*

$$V(X_{0,N})^{-(1/2)} (X_{0,N} - E(X_{0,N})) \xrightarrow{\mathcal{D}} N(0, 1)$$

For the proof of the theorem we need some properties of slowly varying functions, which are proved in Ibragimov-Linnik [ I-L 71 ] :

( SV1 ) Every slowly varying function h has an integral representation :

$$h(x) = C(x) \exp\left( \int_{\beta}^{x} (\varepsilon(t) / t) \, dt \right), \text{ with } 0 < \beta < \infty ,$$

and

$$\lim_{x \to \infty} C(x) = C \neq 0, \qquad \lim_{x \to \infty} \varepsilon(x) = 0.$$

( SV2 ) For every $\varepsilon > 0$ the following equations are satisfied :

$$\lim_{x\to\infty} x^{\varepsilon} h(x) = \infty \quad \text{and} \quad \lim_{x\to\infty} x^{-\varepsilon} h(x) = 0.$$

**Proof of Theorem 1 :** We shall devide the proof into several lemmas . Let $\mathfrak{Y} = ( Y_{m,n} )$ be the additive process of ( C 1 ) and let $\beta_1( N ) := ( V( X_{0,N} ) )^{-(1/2)} \| X_{0,N} - Y_{0,N} \|_1$. Let h be the slowly varying function , which satisfies $V( X_{0,N} ) = N h( N )$. If $c_n$, $d_n$ are two sequences with $c_n = O( d_n )$ we write $c_n \ll d_n$.

**Lemma 1.1 :** *There are sequences of nonnegative numbers $a_i( N )$ , $1 \le i \le 4$, which go to infinity and satisfy the following conditions :*

$$a_1 ( N ) a_2 ( N ) + ( a_1 ( N ) - 1 ) a_3 ( N ) + a_4 ( N ) = N$$

$$a_1 ( N )^4 \ll a_2 ( N )$$

$$( a_2 ( N )^{-(1/2)} / a_1 ( N ) ) \ll a_3 ( N ) \ll ( a_2 ( N )^{-(1/2)} / a_1 ( N ) )$$

$$a_1 ( N ) \alpha' ( a_3 ( N ) ) \to 0$$

$$a_4( N ) \ll a_2( N ) \ll a_4( N )$$

$$\lim_{N\to\infty} a_1 ( N ) \beta_1 ( a_2 ( N ) ) = 0$$

$$\lim_{N\to\infty} \Big( \sup_{t \ge \min(a_2(N), a_4(N))} ( | \varepsilon( t ) | \log( a_1 ( N ) ) \Big) = 0$$

**Proof of Lemma 1.1 :** By ( C 1 ) and ( SV 1 ) we can choose sequences $d_1( N )$, $d_2( N )$ of positive real numbers , which satisfy the following condition for all $N \ge M$ and a large $M \ge 1024$ :

$$d_1( N ), d_2( N ) \to \infty, \; d_1( N ) . d_2( N ) = N,$$

$$\max_{n \ge [ (d_2(N)-2)^{(1/2)} / (d_1(N)-2) ]} \alpha'( n ) \le d_1( N )^{-2},$$

$$\max_{n \ge [ d_2(N)-1 ]} \beta_1 ( N ) \le d_1( N )^{-2} .$$

$$\max_{t \geq [\, d_2(N)-1 \,]} | \varepsilon( t ) | \log( d_1( N ) ) \leq 1/d_1( N ),\ d_1( N )^4 \leq d_2( N ).$$

**For N $\leq$ M we can choose $a_i( N )$, such that none of the conditions in the lemma fails. For N $\geq$ M we define:**

$$a_1( N ) := [\, d_1( N ) - 2 \,],$$
$$a_2( N ) := [\, d_2( N ) - 1 \,],$$
$$a_3( N ) := [\, ( N - a_1( N ) )\, a_2( N )^{(1/2)} / ( a_1( N ) - 1 ) \,]$$
$$a_4( N ) := N - a_1( N )\, a_2( N ) - ( a_1( N ) - 1 )\, a_3( N )$$

**q. e. d.**

**Whenever we use sequences $a_i( N )$, they will be as in lemma 1.1.**

**Lemma 1.2 :** $\lim_{N \to \infty} ( h( a_i ( N ) ) / h( N ) ) = 1$ *for i = 2,4*

**Proof of Lemma 1.2 : For large enough N and i = 2,4 we have :**

$$h( a_i( N ) ) / h( N ) = ( C( a_i( N ) ) / C( N ) ) \exp\left( \int_N^{a_i(N)} ( \varepsilon( t ) / t )\, dt \right).$$

**Lemma 1.1 implies :**

$$\left| \int_N^{a_i(N)} ( \varepsilon( t ) / t )\, dt \right| \leq \max_{t \geq a_i(N)} | \varepsilon( t ) | \int_{a_i(N)}^{N} |( 1 / t)|\, dt \leq \max_{t \geq a_i(N)} | \varepsilon( t ) | \log( N / a_i( N ) )$$

$\ll \max_{t \geq a_i(N)} | \varepsilon( t ) | \log( a_1( N ) ) \to 0$. **This together with $\lim_{x \to \infty} C( x ) = C > 0$ implies the lemma.**

**q. e. d.**

**Set $\sigma_N^2 := V( X_{0,N} )$.**

**Lemma 1.3** : *The following properties are satisfied :*

| | | |
|---|---|---|
| *( A 1 )* | $a_1(N)\,\sigma_{a_2(N)}\,\beta_1(a_2(N)) = o(\sigma_N)$ | |
| *( A 2 )* | $(a_1(N) - 1)\,\sigma_{a_3(N)} = o(\sigma_N)$ | |
| *( A 3 )* | $\sigma_{a_4(N)} = o(\sigma_N)$ | |
| *( A 4 )* | *Let* $s_N^2 := a_1(N)\,(\sigma_{a_2(N)} / \sigma_N)^2$ , | *then* $\lim_{N\to\infty} s_N^2 = 1.$ |

**Proof of lemma 1.3 : ( A 1 ): lemma 1.1 implies $\lim a_1(N)\,\beta_1(a_2(N)) = 0$. By ( C 2 ), lemma 1.1 and lemma 1.2 we obtain $\sigma_{a_2(N)} = o(\sigma_N)$. Both together imply ( A 1 ).**

**( A 2 ):** $(a_1(N)^2 a_3(N)\,h(a_3(N)))/\,N\,h(N) \le$

$(a_1(N)^2 a_3(N)\,h(a_3(N)))/(a_1(N)\,a_2(N)\,h(N)) \ll$

$(a_1(N)^2\,a_2(N)^{(1/2)}\,h(a_3(N)))/(a_1(N)^2\,a_2(N)\,h(N)) =$

$h(a_3(N))/(h(N)\,a_2(N)^{(1/2)})$.

**Lemma 1.1 implies : $\max(a_3(N), N^{(2/5)}) \ll a_2(N)^{(1/2)}$ .**

**Both together with ( SV2 ) imply : $\lim_{N\to\infty} h(a_3(N))/(h(N)\,a_2(N)^{(1/2)})$.**

**This proves ( A 2 ).**

**( A 3 ) and ( A 4 ) are immediately implied by ( C 2 ) , lemma 1.1 and lemma 1.2.**

**q. e. d.**

**For the rest of the proof we use the following definitions.**

$\zeta_{n,N} := X_{(n-1)a_2(N) + (n-1)a_3(N),\; n a_2(N) + (n-1)a_3(N)}$ , **for $1 \le n \le a_1(N)$ and $N \in \mathbb{N}$.**

$\eta_{n,N} := X_{n a_2(N) + (n-1)a_3(N),\; n a_2(N) + n a_3(N)}$ , **for $1 \le n \le a_1(N) - 1$ and $N \in \mathbb{N}$.**

$R_N := X_{N - a_4(N),\,N}$ , **for $N \in \mathbb{N}$.**

**For an integrable random variable $Z_N$ let $\tilde{Z}_N := V(X_{0,N})^{(-1/2)}(Z_N - E(Z_N))$.**

**First we prove.**

**Lemma 1.4 :** *The following ( a ) and ( b ) are equivalent.*

*( a )* $$\tilde{X}_{0,N} \xrightarrow{\mathcal{D}} N(0,1)$$

*( b )* $$\sum_{n=1}^{a_1(N)} \tilde{\zeta}_{n,N} \xrightarrow{\mathcal{D}} N(0,1)$$

**Proof of lemma 1.4 : First we show :**

$$\tilde{X}_{0,N} - \sum_{n=1}^{a_1(N)} \tilde{\zeta}_{n,N} - \sum_{n=1}^{a_1(N)-1} \tilde{\eta}_{n,N} - \tilde{R}_N \to 0 \text{ in probability.}$$

**Using the Minkovski-inequality we obtain:**

$$\left\| \sum_{n=1}^{a_1(N)} \tilde{\zeta}_{n,N} + \sum_{n=1}^{a_1(N)-1} \tilde{\eta}_{n,N} + \tilde{R}_N - \tilde{X}_{0,N} \right\|_1 \leq \| \tilde{X}_{0,N} - \tilde{Y}_{0,N} \|_1 + a_1(N) \| \tilde{\zeta}_{1,N} - \tilde{Y}_{0,a_2(N)} \|_1$$

$$+ (a_1(N) - 1) \| \tilde{\eta}_{1,N} - \tilde{Y}_{a_2(N)+1,\, a_2(N)+a_3(N)} \|_1 + \| \tilde{R}_N - \tilde{Y}_{N-a_1(N)+1,\, N} \|_1$$

**By $\| f - E(f) \|_1 \leq 2 \| f \|_1$, lemma 1.1, and lemma 1.2 the first part of the proof is finished . We complete the proof of the lemma by showing:**

$$\sum_{n=1}^{a_1(N)-1} \tilde{\eta}_{n,N} + \tilde{R}_N \to 0 \quad \text{in probability.}$$

**For any $\varepsilon > 0$ $P(|\tilde{R}_N| > \varepsilon) \leq \varepsilon^{-2} V(X_{0,N})^{-1} V(X_{0,a_4(N)})$ is satisfied , so by lemma 1.3 ( A 3 ) $\tilde{R}_N \to 0$ in probability is proved. By a simple calculation and using the Hölder-inequality we obtain:**

$$V\left( \sum_{n=1}^{a_1(N)-1} \eta_{n,N} \right) \leq V(X_{0,N})^{-1} (a_1(N) - 1)^2 \, V(X_{0,a_3(N)})^{-1}$$

**By lemma 1.3 ( A 2 ) and because $L_p$ convergence implies convergence in probability, this yields the desired result.**

**q. e. d.**

The proof of the theorem is completed by the proof of the following lemma.

**Lemma 1.5 :**

$$\sum_{n=1}^{a_1(N)} \tilde{\zeta}_{n,N} \xrightarrow{\mathcal{D}} N(0,1)$$

Proof of lemma 1.5 : For a random variable Z let $\mathcal{F}(Z)$ denote the Fourier-Transform of Z. As in [ I-L 71 ] p. 338 it can be proved, that

$$\left| \mathcal{F}\left(\sum_{n=1}^{a_1(N)} \tilde{\zeta}_{n,N}\right) - \prod_{n=1}^{a_1(N)} \mathcal{F}(\tilde{\zeta}_{n,N}) \right| \le 16\, a_1(N)\, \alpha'(a_3(N))$$

is satisfied. If in addition

( 1.5.1 ) $$\lim_{N\to\infty} \prod_{n=1}^{a_1(N)} \mathcal{F}(\tilde{\zeta}_{n,N}) = \mathcal{F}(N(0,1)),$$

then the lemma follows by lemma 1.1 and the continuity theorem of Levy. It remains to prove this equation . For every ( large enough ) N let $Z_{n,N}$ , $1 \le n \le a_1(N)$, be i.i.d. random variables with $Z_{n,N} \stackrel{\mathcal{D}}{=} \tilde{\zeta}_{1,N}$. Then ( 1.5.1 ) is equivalent with:

( 1.5.2 ) $$\sum_{n=1}^{a_1(N)} Z_{n,N} \xrightarrow{\mathcal{D}} N(0,1).$$

Let $s_N^2$ be defined as in lemma 1.3, then $s_N^{-1} Z_{n,N}$, $1 \le n \le a_1(N)$ , $N \in \mathbb{N}$ , satisfies the Lindeberg-condition. This is a consequence of ( C 3 ) and $a_1(N) \to \infty$ . By lemma 1.3 ( A 4 ) we obtain:

( 1.5.3 ) $$\left\| \sum_{n=1}^{a_1(N)} Z_{n,N} - s_N^{-1} \sum_{n=1}^{a_1(N)} Z_{n,N} \right\|_2 = \left|1 - s_N^{-1}\right| s_N = \left|1 - s_N\right| \longrightarrow 0 .$$

By the central limit theorem of Lindeberg, $s_N^{-1} \sum Z_{n,N} \xrightarrow{\mathcal{D}} N(0,1)$ is satisfied. This together with ( 1.5.3 ) implies ( 1.5.2 ) , and the Lemma is proved.

q. e. d.

In the rest of the paper we will use the following abbreviation for centering random variables their expectation : $\hat{X} = X - E(X)$. Next we prove a theorem, which can be used to verify ( C 2 ). The additive analogon in the $\varphi$ - mixing case is proved in [ I-L 71 Theorem 18. 2. 2. ] .

**Theorem 2:** *Let ( C 1 ) be satisfied. Let $V( X_{0,N} )$ tend to infinity. Let $\| X_{0,N} - Y_{0,N} \|_2 = o( \sigma_N )$ and let one of the following two conditions be satisfied.*

*( a ) $\mathfrak{X}$ is $\varphi'$ - mixing.*

*( b ) $\mathfrak{X}$ is $\alpha'$ - mixing and ( C 3 ) is satisfied.*

*Then ( C 2 ) is satisfied.*

**Proof of Theorem 2 :** Let $\beta_2( N ) := ( \| \hat{X}_{0,N} - \hat{Y}_{0,N} \|_2 ) / \sigma_N$ and $\sigma_N^2( \mathfrak{Y} ) := V( Y_{0,N} )$. First we will prove ( C 2 ) for $\mathfrak{Y}$. In [ I - L 71 p. 330 ] it is remarked that in the proof of their Theorem 18. 2. 2. only the following two conditions are used instead of $\varphi$ - mixing:

( I-L 1 ) $\sigma_N( \mathfrak{Y} ) \to \infty$ , for $N \to \infty$.

( I-L 2 ) For every $\varepsilon > 0$ there are $p, N \in \mathbb{N}$ , such that for all $n,m > N$ the following equation is satisfied: $| E( \hat{Y}_{0,n} \hat{Y}_{n+p,\, n+p+m} ) | \le \varepsilon \sigma_n ( \mathfrak{Y} ) \sigma_m( \mathfrak{Y} )$.

We shall prove , that $\mathfrak{Y}$ satisfies these conditions. $\beta_2( N ) \to 0$ implies $\sigma_N \sim \sigma_N( \mathfrak{Y} )$ and so ( I-L 1 ) is proved. Now we prove ( I-L 2 ) :

$$| E( \hat{Y}_{0,n} \hat{Y}_{n+p,n+p+m} ) |$$

$$\le | E( [ \hat{X}_{0,n} - ( \hat{X}_{0,n} - \hat{Y}_{0,n} )] [ \hat{X}_{n+p,n+p+m} - ( \hat{X}_{n+p,n+p+m} - \hat{Y}_{n+p,n+p+m} ) ] )$$

$$\le | E( \hat{X}_{0,n} \hat{X}_{n+p,n+p+m} ) | + \beta_2( n ) \sigma_n \sigma_m + \beta_2( m ) \sigma_n \sigma_m + \beta_2 ( n ) \beta_2( m ) \sigma_n \sigma_m .$$

Now let $0 < \varepsilon < 1$, then there is a $N \in \mathbb{N}$ such that $n,m > N$ implies:

$$\beta_2( n ) \sigma_n \sigma_m + \beta_2( m ) \sigma_n \sigma_m + \beta_2( n ) \beta_2( m ) \sigma_n \sigma_m \le ( \varepsilon / 2 ) \sigma_n( \mathfrak{Y} ) \sigma_m( \mathfrak{Y} ).$$

So it remains to prove:

$$| E( \hat{X}_{0,n} \hat{X}_{n+p,n+p+m} ) | \le ( \varepsilon / 2 ) \sigma_n( \mathfrak{Y} ) \sigma_m( \mathfrak{Y} ).$$

By Theorem 17. 2. 3. in [ I-L 71 ] we obtain in the $\varphi'$ - mixing case:

$$| E( \hat{X}_{0,n} \hat{X}_{n+p,n+p+m} ) | \le 2 \varphi'( p ) \sigma_n \sigma_m.$$

For large enough p we have : $2 \varphi'( p ) \sigma_n \sigma_m \le ( \varepsilon / 2 ) \sigma_n( \mathfrak{Y} ) \sigma_m( \mathfrak{Y} )$. This finishes the proof in the $\varphi'$ - mixing case.

For the proof of the inequality above in the $\alpha'$ - mixing case we use a truncation argument. For positive $r \in \mathbb{R}$ and a random variable Z let $Z^{r-} := Z\, 1_{\{|Z| \le r\}}$ and $Z^{r+} := Z - Z^{r-}$. Then we obtain using Hölders - inequality :

$$| E((\hat{X}_{0,n} / \sigma_n)(\hat{X}_{n+p,n+p+m} / \sigma_m)) | \le | E((\hat{X}_{0,n} / \sigma_n)^{k-}(\hat{X}_{n+p,n+p+m} / \sigma_m)^{k-}) |$$
$$+ | E((\hat{X}_{0,n} / \sigma_n)^{k+}(\hat{X}_{n+p,n+p+m} / \sigma_m)) |$$
$$+ | E((\hat{X}_{0,n} / \sigma_n)^{k-}(\hat{X}_{n+p,n+p+m} / \sigma_m)^{k+}) |$$
$$\le | E((\hat{X}_{0,n} / \sigma_n)^{k-}(\hat{X}_{n+p,n+p+m} / \sigma_m)^{k-}) | +$$
$$\| (\hat{X}_{0,n} / \sigma_n)^{k+} \|_2 + \| (\hat{X}_{n+p,n+p+m} / \sigma_m)^{k+} \|_2$$

Let $C > 1$ be so large that $\sigma_n \sigma_m \le C\, \sigma_n(\mathfrak{Y})\, \sigma_m(\mathfrak{Y})$, for $n,m > N$. ( C 3 ) implies, that $\| (\hat{X}_{0,l} / \sigma_l)^{k+} \|_2 < (\varepsilon / 6C)$ for large enough $k \in \mathbb{R}$ and $l \ge N$. By Theorem 17. 2. 1. in [ I-L 71 ] we obtain :

$$| E((\hat{X}_{0,n} / \sigma_n)^{k-}(\hat{X}_{n+p,n+p+m} / \sigma_m)^{k-}) |$$
$$\le 4\, \alpha'(p)\, k^2 + | E((\hat{X}_{0,n} / \sigma_n)^{k-}) | + | E((\hat{X}_{0,m} / \sigma_m)^{k-}) | .$$

By $E(\hat{X}_{0,l}) = 0$ , for all $l \in \mathbb{N}$, we obtain:

$$| E((\hat{X}_{0,l} / \sigma_l)^{k-}) | = | E((\hat{X}_{0,l} / \sigma_l)^{k+}) |$$
$$\le \| (\hat{X}_{0,l} / \sigma_l)^{k+} \|_2 < (\varepsilon / 6C)$$

Now choose p , such that $4\, \alpha'(p)\, k^2 \le (\varepsilon / 12C)$ is satisfied, then:

$$| E((\hat{X}_{0,n} / \sigma_n)^{k-}(\hat{X}_{n+p,n+p+m} / \sigma_m)^{k-}) |$$
$$\le 4\, \alpha'(p)\, k^2 + | E((\hat{X}_{0,n} / \sigma_n)^{k-}) | + | E((\hat{X}_{0,m} / \sigma_m)^{k-}) |$$
$$\le (\varepsilon / 12C) + (\varepsilon / 6C).$$

This together with the inequalities above leads to:

$$| E(\hat{X}_{0,n} \hat{X}_{n+p,n+p+m}) | \le (\varepsilon / 2C)\, \sigma_n\, \sigma_m \le (\varepsilon / 2)\, \sigma_n(\mathfrak{Y})\, \sigma_m(\mathfrak{Y}).$$

So $\mathfrak{Y}$ satisfies ( I-L 2 ). This finishes the proof that $\mathfrak{Y}$ satisfies ( C 2 ). The condition ( C 2 ) carries over from $\mathfrak{Y}$ to $\mathfrak{X}$ and the theorem is proved.

q. e. d.

The next Theorem helps to verify ( C 3 ) in the $\varphi'$ mixing case. The additive analogon of this theorem can be found in [ I-L 71 , Lemma 118. 5. 1. ] .

**Theorem 3** : *Let $\mathfrak{X}$ satisfy ( C 1 ) and ( C 2 ). Assume $\mathfrak{X} \subset L_{2+\delta}$ for some $\delta$ with $0 < \delta < 1$. Let $\mathfrak{X}$ be $\varphi'$ - mixing and let the following condition be satisfied :*

*( 3. 1. )* $\| X_{0,N} - Y_{0,N} \|_{2+\delta} \le C_1 \sigma_N$ , *with* $C_1 < \infty$.

*Then there is a constant* $0 < C_0 < \infty$ , *such that*

*( 3. 2. )* $\| X_{0,N} - E( X_{0,N} ) \|_{2+\delta}^{2+\delta} \le C_0 \sigma_N^{2+\delta}$ , $N \in \mathbb{N}$

*is satisfied.*

**Proof of Theorem 3: We first define some constants. There is a constant $0 < C_2 < \infty$ with :**

**( 3. 3. )** $$\max\{ \| \hat{X}_{0,p} \|_{2+\delta} , \| X_{0,p} - Y_{0,p} \|_{2+\delta} \} \le C_1 C_2 p$$

**Now let $\varepsilon > 0$ be so small, that**

**( 3. 4. )** $$2 + \varepsilon \le 2^{1+(\delta/2)}$$

**Now let p be so large, that the following inequality is satisfied for $p' \ge p$ :**

**( 3. 5. )** $$12 \varphi'( p' )^{\delta/(2+\delta)} \le \varepsilon$$

**Let $N_0 \in \mathbb{N}$ with $N_0 \ge \alpha$ such that for $N \ge N_0$ the following inequality is satisfied :**

**( 3. 6. )** $$C_2 ( p + 1 ) \le \sigma_N.$$

**Let $\varepsilon( t )$ and $C( x )$ be the functions in the integral representation of the slowly varying function $h( x )$ with $\sigma_N^2 = N h( N )$. Let $N_0$ above be chosen in sucha way that the following two inequalities are satisfied :**

**( 3. 7. )** $$\sup_{t_1,t_2 \ge N_0} ( C( t_1 ) / C( t_2 ) ) \le 2^{\delta/32(1+\delta/2)} , \quad \sup_{t \ge N_0} | \varepsilon( t ) | \le \delta/32 .$$

**If $0 < C_3 < \infty$ is large enough we have**

( 3. 8. ) $$( 1 + ( 12\, C_1 / C_3 ))^{2+\delta} \le 2^{\delta/32} .$$

Now we define

( 3. 9. ) $$C_4 := \max_{1 \le n \le 4(N_0 + p)} \| \hat{X}_{0,n} \|_{2+\delta}^{2+\delta} .$$

Finally let $C_5$ be such that for $\sigma_N > 0$ the following inequality is satisfied :

( 3. 10. ) $$1 \le C_5\, \sigma_N .$$

Now we can define $C_0$. Let $C_0$ be a real number, which satisfies the following two inequalities :

( 3. 11. ) $$C_0 > \max\{ ( C_3 + 12\, C_1 )^{2+\delta} , C_4\, C_5^{2+\delta} \} \quad \text{and} \quad C_0 + 3 \le C_0\, 2^{\delta/32} .$$

We will prove the Theorem by induction.

1) Let $1 \le N \le 4( N_0 + p )$. For $\sigma_N = 0$ the inequality is trivially satisfied. Let $\sigma_N > 0$ then :
$$\| \hat{X}_{0,N} \|_{2+\delta}^{2+\delta} \le C_4 \le C_4\, C_5^{2+\delta}\, \sigma_N^{2+\delta} \le C_0\, \sigma_N^{2+\delta}$$

2) Let $N > 4 ( N_0 + p )$ and let the inequality be valid for $n < N$ . Define

$$m := \left[ ( N - p ) / 2 \right]$$

then , using ( 3. 6. ) , we obtain :

( 3. 12. ) $$\| \hat{X}_{0,N} - \hat{X}_{0,m} - \hat{X}_{m,N-m} - \hat{X}_{N-m,m} \|_{2+\delta} \le 2\, \| X_{0,N} - Y_{0,N} \|_{2+\delta} + 4\, \| X_{0,m} - Y_{0,m} \|_{2+\delta} + 2\, \| X_{0,N-2m} - Y_{0,N-2m} \|_{2+\delta} \le 2\, C_1 ( \sigma_N + 3\sigma_m )$$

Together with ( 3. 3. ) we obtain :

( 3. 13. ) $$\| \hat{X}_{0,N} \|_{2+\delta} \le \| \hat{X}_{0,m} + \hat{X}_{N-m,N} \|_{2+\delta} + 3\, C_1 ( \sigma_N + 3\, \sigma_m )$$

**First case :** $\| \hat{X}_{0,m} - \hat{X}_{N-m,N} \|_{2+\delta} \leq C_3 \sigma_N$ .

The following lemma implies $\sigma_m \leq \sigma_N$ . This together with ( 3. 13. ) implies:

$$\| \hat{X}_{0,N} \|_{2+\delta}^{2+\delta} \leq ( C_3 + 12 C_1 )^{2+\delta} \sigma_N^{2+\delta} \leq C_0 \sigma_N^{2+\delta}$$

and the theorem is proved in this case.

**Second case :** $\| \hat{X}_{0,m} - \hat{X}_{N-m,N} \|_{2+\delta} \geq C_3 \sigma_N$ .

Then ( 3. 13. ) implies:

$$( 3. 14. ) \qquad \| \hat{X}_{0,N} \|_{2+\delta}^{2+\delta} \leq \| \hat{X}_{0,m} + \hat{X}_{N-m,N} \|_{2+\delta}^{2+\delta} ( 1 + ( 12 C_1 / C_3 ) )^{2+\delta}$$
$$\leq 2^{\delta/32} \| \hat{X}_{0,m} + \hat{X}_{N-m,m} \|_{2+\delta}^{2+\delta}$$

Here we have used $\sigma_m \leq \sigma_N$ and ( 3. 8. ). Using the inequality $| a + b |^{\delta} \leq | a |^{\delta} + | b |^{\delta}$, valid for $0 < \delta \leq 1$, we obtain:

$$\| \hat{X}_{0,m} + \hat{X}_{N-m,N} \|_{2+\delta}^{2+\delta} \leq E( | \hat{X}_{0,m} + \hat{X}_{N-m,N} |^2 ( | \hat{X}_{0,m} |^{\delta} + | \hat{X}_{N-m,m} |^{\delta} ) )$$
$$\leq 2 E( | \hat{X}_{0,m} |^{2+\delta} ) + 2 E( | \hat{X}_{0,m} |^{1+\delta} | X_{N-m,N} | )$$
$$+ 2 E( | \hat{X}_{0,m} | | \hat{X}_{N-m,N} |^{1+\delta} ) + E( | \hat{X}_{0,m} |^2 | \hat{X}_{N-m,N} |^{\delta} ) + E( | \hat{X}_{0,m} |^{\delta} | \hat{X}_{N-m,N} |^2 )$$

Handling the last 4 terms as in Lemma 18. 5. 1. of [ I-L 71 ] and using ( 3. 4. ) we obtain :

$$( 3. 15. ) \qquad \| \hat{X}_{0,m} + \hat{X}_{N-m,N} \|_{2+\delta}^{2+\delta} \leq ( 2 + \varepsilon ) \| \hat{X}_{0,m} \|_{2+\delta}^{2+\delta} + 6 \sigma_m^{2+\delta}$$
$$\leq 2^{1+(\delta/32)} \| \hat{X}_{0,m} \|_{2+\delta}^{2+\delta} + 6 \sigma_m^{2+\delta} .$$

By induction we obtain :

$$\| \hat{X}_{0,m} + \hat{X}_{N-m,N} \|_{2+\delta}^{2+\delta} \leq 2^{1+(\delta/32)} C_0 \sigma_m^{2+\delta} + 6 \sigma_m^{2+\delta} .$$

This , together with ( 3. 11. ) implies :

$$\| \hat{X}_{0,N} \|_{2+\delta}^{2+\delta} \leq 2^{1+(\delta/32)} ( C_0 + 3 ) \sigma_m^{2+\delta} .$$

**Together with ( 3. 11. ) this implies :**

$$\| \hat{X}_{0,N} \|_{2+\delta}^{2+\delta} \le 2^{1+(3\delta/32)} C_0 \, \sigma_m^{2+\delta} .$$

**This together with the following lemma completes the proof of Theorem 3.**

**Lemma 3.1 :** $( \sigma_m / \sigma_N )^{2+\delta} \le 2^{-(1+(6\delta/16))}$

**Proof of Lemma 3.1 We have**

$$( \sigma_m / \sigma_N )^{2+\delta} \le$$

$$\le \sup_{t_1,\, t_2 \ge N_0} ( C( t_1 ) / C( t_2 ) )^{1+(\delta/2)} ( m / N )^{1+(\delta/2)} \exp( \int_m^N | \varepsilon( t ) / t | \, dt \, ( 1 + ( \delta / 2 ) ))$$

**By ( 3. 8. ) we obtain:**

$$\max_{t \ge N_0} | \varepsilon( t ) | \int_m^N ( 1 / t ) \, dt \le ( \delta / 32 ) \log( N / m )$$

**This implies :**

$$( \sigma_m / \sigma_N )^{2+\delta} \le 2^{(\delta/32)} ( m / N )^{1+(\delta/2)-(\delta/32)(1+(\delta/2))}$$

$$\le 2^{(\delta/32)} ( m / N )^{1+(\delta/2)-(\delta/16)} \le 2^{(\delta/32)} ( m / N )^{1+(7\delta/16)}$$

$$\le 2^{(\delta/32)-(1+(7\delta/16))} \le 2^{-1-(6\delta/16)} .$$

**The lemma is proved. q. e. d.**

**The next Theorem helps to verify ( C 3 ) in the α' - mixing case. The additive analogon of this theorem can be found in [ K-P 80, lemma ( 2. 5. ) ] .**

**Theorem 4** : *Let $\mathfrak{X}$ satisfy ( C 1 ) and let $\mathfrak{X} \subset L_{2+\delta}$ . Let the following conditions be satisfied :*

*( 4. 1. )* $\sigma_N^2 \sim \sigma^2 N$ *, for a positive real number* $\sigma$.

*( 4. 2. ) There are positive real numbers* $\rho$ *and* $C_1$ *, such that :*

$$\| X_{O,N} - Y_{O,N} \|_2 \leq C_1 \; N^{(1-\rho)/2} \quad \text{, for all } N \in \mathbb{N}.$$

*( 4. 3. )* $\mathfrak{X}$ *is* $\alpha'$ *- mixing with* $\alpha'(N) \leq C_2 N^{-(1+x)(1+(2/\delta))}$ *, with positive real numbers* $x$ *and* $C_2$ *.*

*( 4. 4. ) There is a positive real number* $C_3$ *, with* $\| X_{O,N} - Y_{O,N} \|_{2+\delta} \leq C_3 N^{(1/2)}$.

*Let* $\alpha := \min \{ \rho\delta / ( 4 + 2\rho + 2\delta ) , x\delta / ( 6 + 6x + 3\delta + 2x\delta ) \}$ *, then there is a positive real number* $C_O$ *, such that the following inequality is satisfied :*

*( 4. 5. )* $\| X_{O,N} - E( X_{O,N} ) \|_{2+\alpha} \leq C_O \, \sigma_N$ *, for all* $N \in \mathbb{N}$ *.*

**Proof of Theorem 4** : Without loss of generality we can suppose $\rho = 2x / ( 3 + 2x )$. ( 4. 1. ), ( 4. 2. ) and ( 4. 4. ) imply , that there is a positive real number A , such that the following two inequalities are satisfied :

$$\| \hat{Y}_{O,N} \|_2^2 \leq AN , \qquad \| Y_{0,1} \|_{2+\delta} \leq A .$$

**Lemma 4. 1.** : *Let* $\Theta = \rho / 2$. *Then there is a positive real number* $C_4$, *such that for all* $n,N \in \mathbb{N}$ *and all* $\sigma( I_{O,n} )$ *the following inequality is satisfied :*

$$E( | E( ( N^{-(1/2)} \hat{X}_{n,n+N} )^2 | \sigma( I_{O,n} ) ) - E( N^{-(1/2)} \hat{X}_{n,n+N} )^2 | ) \leq C_4 N^{-\Theta}.$$

**Proof of Lemma 4. 1. :** Let $m := m(N) := [N^{1-\rho}]$ and $C_5 := (4C_1 + 2A)^2$ , then :

$$\| \hat{X}_{m,m+N} - \hat{X}_{0,N} \|_2 \le 2 \| X_{m,m+N} - Y_{m,m+N} \|_2 + 2 \| X_{0,N} - Y_{0,N} \|_2$$
$$+ \| Y_{0,m} \|_2 + \| Y_{N,N+m} \|_2 \le 4C_1 N^{(1-\rho)/2} + 2AN^{(1-\rho)/2} \le C_5^{(1/2)} N^{(1-\rho)/2} .$$

Using $a^2 + b^2 = (a - b)^2 + 2b(a - b)$ together with the Hölder - inequality and the Minkowski - inequality we obtain:

$$E( | \hat{X}^2_{m,m+N} - \hat{X}^2_{0,N} | ) \le 2C_6 N^{(1/2)} C_5^{(1/2)} N^{(1-\rho)/2} \le C_7 N^{1-\Theta} .$$

Here $C_6$ and $C_7$ are positive real numbers , such that $\sigma_N \le C_6 N^{1/2}$ is satisfied for all $N \in \mathbb{N}$ . We also obtain :

$$(4.6.) \quad E( | E( \hat{X}^2_{n,n+N} | \sigma( I_{0,n} )) - E( \hat{X}^2_{n,n+N} ) | ) \le 2C_7 N^{(1-\Theta)}$$
$$+ E( | E( \hat{X}^2_{n+m,n+m+N} | \sigma( I_{0,n} )) - E( \hat{X}^2_{n+m,n+m+N} ) | ) =: a_1 + a_2 .$$

Using [ K-P 80 , lemma ( 2. 1. ) ] together with the argument on the same page in [ K-P 80 , lemma ( 2. 2. ) ] , we obtain :

$$a_2 \le 10 \| \hat{X}^2_{n+m,n+m+N} - E( \hat{X}^2_{n+m,n+m+N} ) \|_{(2+\delta)/2} \alpha'( m )^{\delta/(2+\delta)}$$
$$\le \tilde{A}N^2 + C_2^{\delta/(2+\delta)} m^{-(1+\varkappa)} \le C_8 N^{(1-\Theta)} .$$

Where $\tilde{A} := 20 ( A + 2C_3 )^2$, and $C_8$ is a positive real number independent of the choice of $\sigma( I_{0,N} )$. This together with ( 4. 6. ) finishes the proof of lemma 4. 1.
q. e. d.

**Lemma 4. 2. :** *Let $\beta := \alpha / 2$ , then there is a positive real number $C_9$ , such that for all $I_{0,n}$ , the following inequality is satisfied :*

$$E( | E( ( N^{-(1/2)} \hat{X}_{n,n+N} )^2 | \sigma( I_{0,n} ) ) - E( ( N^{-(1/2)} \hat{X}_{n,n+N} )^2 ) |^{1+\beta} ) \le C_9 .$$

**Proof of lemma 4. 2. :** The lemma can be proved in the same way as [ S 68 , lemma 2. 1 ]. For the proof of [ S 68 , ( 2. 5 ) ] use ( 4. 4 ). The nonadditive analogon of [ S 68 , condition ( 2. 3. ) ] is verified in lemma 4. 1. . q. e. d.

**Proof of Theorem 4 :** Let

$$(4.7.)\quad C_{10} := \max_{0 \le t \le 2} \left( A^{t/2} + C_9^{t/(2+\alpha)} \right)$$

Now let $0 < C_{11} < +\infty$ so large that the following inequality is satisfied :

$$(4.8)\quad \left(1 + (6C_3 + A)/C_{11}\right) \le 2^{\alpha/(4(2+\alpha))}$$

Finally let $\tilde{C}_0 \ge 1$ be such that

$$(4.9.)\qquad 6C_{10} \le \tilde{C}_0^{\alpha/2}\left(2^{\alpha/4} - 1\right)$$

The following inequality together with ( 4. 6. ) finishes the proof of the theorem.

$$\| \hat{X}_{0,N} \|_{2+\alpha} \le \tilde{C}_0 N^{1/2}$$

We shall prove this inequality by induction :

1. For $N = 1$ the inequality is true.
2. Suppose that the inequality is satisfied for $n < N$ with $N \ge 2$.

Let $m := [N/2]$ , then we obtain :

$$\| \hat{X}_{0,N} - \hat{X}_{0,m} - \hat{X}_{m,2m} \|_{2+\alpha} \le 2 \| X_{0,N} - Y_{0,N} \|_{2+\alpha} + 4 \| X_{0,m} - Y_{0,m} \|_{2+\alpha} +$$

$$\| Y_{N-1,N} \|_{2+\alpha} \le 6\,C_3\,N^{1/2} + A.$$

In the case $\| \hat{X}_{0,m} - \hat{X}_{m,2m} \|_{2+\alpha} \le C_{11} N^{1/2}$ we obtain :

$$\| \hat{X}_{0,N} \|_{2+\alpha} \le 6C_3 N^{1/2} + A + C_{11} N^{1/2} \le (C_{11} + A + 6C_3) N^{1/2} \le C_0 N^{1/2},$$

and the inequality is proved.

In the other case we, obtain using ( 4. 8. ) :

$$\| \hat{X}_{0,N} \|_{2+\alpha}^{2+\alpha} \le \| \hat{X}_{0,m} + \hat{X}_{m,2m} \|_{2+\alpha}^{2+\alpha} 2^{\alpha/4}$$

Now we shall estimate $\| \hat{X}_{0,m} + \hat{X}_{m,2m} \|_{2+\alpha}^{2+\alpha}$. The first estimate follows as in [ S 68 , ( 3. 4 ), ( 3. 5. ) , ( 3. 6. ) and ( 3. 7. ) ] . The only difference is, that lemma 4. 2. instead of [ S 68, ( 3. 2. ) ] is used.

$$\| \hat{X}_{0,m} + \hat{X}_{m,2m} \|_{2+\alpha}^{2+\alpha} \le 2 \| \hat{X}_{0,m} \|_{2+\alpha}^{2+\alpha} + 2\, C_{10}\, m^{(1+\alpha)/2} \| \hat{X}_{0,m} \|_{2+\alpha} +$$

$$+ 2\, C_{10}\, m^{1/2} \| \hat{X}_{0,m} \|_{2+\alpha}^{1+\alpha} C_{10}\, m^{\alpha/2} \| \hat{X}_{0,m} \|_{2+\alpha}^{2} + C_{10}\, m \| \hat{X}_{0,m} \|_{2+\alpha}^{\alpha}$$

By induction this is at most

$$2\, \tilde{C}_0^{2+\alpha}\, m^{1+(\alpha/2)} + 2\, C_{10}\, m^{(1+\alpha)/2}\, \tilde{C}_0\, m^{1/2} + 2\, C_{10}\, m^{1+(\alpha/2)}\, C_0^{1+\alpha} +$$

$$+ C_{10}\, m^{1+(\alpha/2)}\, \tilde{C}_0^{2} + C_{10}\, m^{1+(\alpha/2)}\, C_0^{\alpha} \le 2\, m^{1+(\alpha/2)} ( C_0^{2+\alpha} + 6\, \tilde{C}_0^{2}\, C_{10} )$$

$$\le ( 2\, m )^{1+(\alpha/2)}\, \tilde{C}_0^{2+\alpha}\, 2^{-(\alpha/2)} .$$

The last inequality is a consequence of ( 4. 9. ).

q. e. d.

## § 3 The Invariance Principles

In this section we shall prove an a. s. invariance principle and we shall state an other without a proof. Two further invariance principles will be given as corollaries.

**Theorem 5** : *Let $\mathfrak{X}$ satisfy ( C 1 ) and let $\mathfrak{X} \subset L_{2+\delta}$ , $0 < \delta \le 1$. Let the following conditions be satisfied :*

*( S. 1. ) Let $\mathfrak{X}$ be $\alpha'$ - mixing with $\alpha'( N ) = O( N^{-x})$, with $x > 2$.*

*( 5. 2. )* *There are constants $\varepsilon$ and $\sigma$ with $0 < \varepsilon < 1/10$ and $0 < \sigma < \infty$, such that*

$$\sigma_N^2 = \sigma^2 N + O( N^{1-\varepsilon} ) .$$

*( 5. 3. )* $\| X_{0,N} - Y_{0,N} \|_1 = o( N^{(1/2) - 4\varepsilon} )$ .

*( 5. 4. )* *There are constants $\alpha$, $\Theta$, $\rho$ with $0 < \alpha < \delta$ , $0 \le \Theta \le \alpha/88$ and $0 \le \rho \le \alpha/( \beta + 1 )8( 1 + \alpha )$, where $\beta := \max\{ 2/\varepsilon, 16/\alpha x, 16/( x - 2 ) \}$, such that*

$$\sup_{k\ge 0} ( \sup_{N\ge k^{\rho}} ( N^{-1-(\alpha/2)-\Theta} \| \hat{X}_{0,N+k} - \hat{X}_{0,k} \|_{2+\alpha}^{2+\alpha} ) ) < \infty .$$

*Then there is a probability space with a process $\mathfrak{X}' \overset{\mathcal{D}}{=} \mathfrak{X}$, and a standard Brownian motion $\mathfrak{B} = \{ B( t ), t\ge 0 \}$ on it, such that*

$$\hat{X}'_{0,N} - B( \sigma^2 N ) = O( N^{(1/2) - \lambda} ) \qquad a. s.$$

*with $\lambda > 0$.*

**Proof of Theorem 5** : Let $n_k := [ k^{\beta} ]$ and $m_k := [ k^{\beta/4} ]$. Let $N_0 = 0$ and, for $k \ge 1$,

$$N_k := \sum_{i=1}^{k} ( m_i + n_i )$$

Then, for $k \ge 1$ : $k^{\beta+1} \ll N_k \ll k^{\beta+1}$ , i.e. , $k^{\beta+1} = O( N_k )$ and $N_k = O( k^{\beta+1} )$.

**Lemma 5.1 :** *There is a $\gamma > 0$, such that*

$$\sum_{i=1}^{k+1} ( \hat{X}_{N_{i-1}, N_{i-1}+m_i} + \hat{X}_{N_{i-1}+n_i, N_i} ) - \hat{X}_{0,N} \ll N^{(1/2) - \gamma} \qquad a. s.$$

**Proof of Lemma 5.1 :** There is a $\Theta_1 > 1$ , such that

$$P( | \hat{X}_{N_{k-1}, N_{k-1}+n_k} - \hat{Y}_{N_{k-1}, N_{k-1}+n_k} | \ge k^{(\beta/2) - 1} ) \le$$

$$\le 2\, k^{-(\beta/2)+1}\; k^{\beta((1/2)-4\varepsilon)}\; \| X_{0,n_k} - Y_{0,n_k} \|_1 \ll k^{-\Theta_1}.$$

Also, there is a constant $\Theta_2 > 1$, such that

$$P(\,|\, \hat{X}_{N_{k-1}+n_k, N_k} - \hat{Y}_{N_{k-1}+n_k, N_k} \,| \ge k^{(\beta/2)-1}\,) \le k^{-\Theta_2}.$$

Finally, there is a constant $\Theta_3 > 1$, such that

$$P(\,|\, \hat{X}_{0,N_k} - \hat{Y}_{0,N_k} \,| \ge k^{(\beta/2)}\,) \le k^{-\Theta_3}.$$

This together with

$$\sum_{j=1}^{k+1} j^{(\beta/2)} \ll k^{(\beta/2)} \ll N_k^{(1/2)-(1/2(\beta+1))}$$

and the Borel - Cantelli - Lemma proves lemma 5. 1.

q. e. d.

**Lemma 5.2** : *Let $\mathfrak{X}$ be a stationary process, which satisfies ( 5. 4. ), then there is a constant $\lambda > 0$, such that the following inequality is satisfied :*

$$\max_{N_k \le N \le N_{k-1}} |\, \hat{X}_{0,N} - \hat{X}_{0,N_k} \,| = O(\, N_k^{(1/2)-\lambda}\,) \quad a.\, s.$$

**Proof of lemma 5. 2.** : There is a constant $C < \infty$, such that for all $\nu$, k, l with $0 \le l \le \log_2(\, n_k + m_k\,)$ and $0 \le \nu \le m_k + n_k - 2^l$, the following inequality is satisfied :

$$\| \hat{X}_{0,N_k+\nu+2^l} - \hat{X}_{0,N_k+\nu} \|_{2+\alpha}^{2+\alpha} \le C\, 2^{l(1+(\alpha/2)+\Theta)}\, k^{\rho(\beta+1)(1+\alpha)}.$$

If $2^l \ge N_{k+1}^{\rho}$ the inequality is trivially implied by ( 5. 4. ). Otherwise we obtain, using $N_{k+1} = O(\, k^{\beta+1}\,)$ and $\Theta \le \alpha/2$ ,

$$\| \hat{X}_{0,N_k+\nu} - \hat{X}_{0,N_k+\nu} \|_{2+\alpha}^{2+\alpha} \leq$$

$$( \| \hat{X}_{0,N_k+\nu+2^l+[N_{k+1}^{\rho}+1]} - \hat{X}_{0,N_k+\nu+2^l} \|_{2+\alpha} +$$

$$\| \hat{X}_{0,N_k+\nu+2^l+[N_{k+1}^{\rho}+1]} - \hat{X}_{0,N_k+\nu} \|_{2+\alpha} )^{2+\alpha}$$

$$\leq C^{*} k^{\rho(\beta+1)(1+(\alpha/2)+\Theta)} \leq C^{**} k^{\rho(\beta+1)(1+\alpha)} .$$

The remaining part of the proof follows as in the proofs of [ K-P 80, proposition 2.2 and lemma 2.9.]. Let $N_k < N \leq N_{k+1}$ , and let $n = n(N)$ be the largest integer with $2 \leq N - N_k$ . Finally let $\sum_{l=1}^{n} \varepsilon_l 2 = N - N_k$ be the dyadic expansion of $N - N_k$ .

Let $F(r, s) := | \hat{X}_{0,N_k+r} - \hat{X}_{0,N_k+r+s} |$ and let $m_l := \sum_{i=l+1}^{n} \varepsilon_i 2^{i-l-1}$ , then

$$F(0, N-N_k) \leq \sum_{l=0}^{n} F(m_l 2^{l+1}, 2^l) .$$ Put $\gamma := \alpha/8(\beta+1)(1+\alpha)$,

and define the events $G_k(m, l) = \{ F(m2^{l+1}, 2^l) \geq N_k^{(1-\gamma)/2} \}$ and

$G_k := \bigcup_{l \leq N_k} \bigcup_{m \leq 2^{n_k}-1} G_k(m, l)$ , where $n_k := n(N_k)$. By the inequality above, we

obtain : $$P(G_k(m, l)) \ll 2^{l(1+(\alpha/2)+\Theta)} k^{\rho(\beta+1)(1+\alpha)} k^{-(\beta+1)(1-\gamma)(1+(\alpha/2))} ,$$

and therefore

$$P(G_k) \ll k^{\rho(\beta+1)(1+\alpha)} k^{-(\beta+1)(1-\gamma)(1+(\alpha/2))} \sum_{l=0}^{n_k} 2^{l(1+(\alpha/2)+\Theta)} 2^{n_k-l}$$

$$\ll k^{\rho(\beta+1)(1+\alpha)} k^{-(\beta+1)(1-\gamma)(1+(\alpha/2))} 2^{n_k(1+(\alpha/2)+\Theta)}$$

$$\ll k^{\rho(\beta+1)(1+\alpha)} k^{-(\beta+1)(1-\gamma)(1+(\alpha/2))} k^{\beta(1+(\alpha/2)+\Theta)}$$

$$\ll k^{-(1+(\alpha/2)) + \gamma(\beta+1)(1+(\alpha/2)) + \rho(\beta+1)(1+\alpha) + \beta\Theta} \ll k^{-(1+(\alpha/2)) + 2\gamma(\beta+1)(1+\alpha) + \beta\Theta}$$

$$\ll k^{-(1+(\alpha/2))+\alpha/4+\beta\alpha/8\beta} \ll k^{-1-\alpha/8}$$

**This together with the Borel - Cantelli - Lemma implies the lemma.**

**q. e. d.**

**Lemma 5.3** : ***There is a probability space with a process $\mathfrak{X}' \stackrel{\mathcal{D}}{=} \mathfrak{X}$ and a standard Brownian motion $\mathfrak{B}$ on it, such that the following inequality is satisfied :***

$$\hat{X}'_{0,N_k} - B\Big(\sum_{l=1}^{k} \sigma_{n_l}^{\ 2}\Big) = O\big(N_k^{(1/2)-\lambda}\big) \qquad a.\ s.$$

***with $\lambda > 0$.***

**Proof of lemma 5. 3. : We define**

$$u_k := \begin{cases} 0 & \text{if } \sigma_{n_k} = 0 \\ \sigma_{n_k}^{-1} \hat{X}_{N_{k-1},\, N_{k-1}+n_k} & \text{if } \sigma_{n_k} > 0 \end{cases}$$

**and $\mathfrak{A}_k := \sigma(u_1, \ldots, u_k)$.**

**Using [ K-P 80 , lemma 2. 2. ] we obtain :**

$$E \mid E(\exp(itu_{k+1}) \mid \mathfrak{A}_k) - E(\exp(itu_{k+1}) \mid \le 2\pi\alpha'(m_k) \le k^{-(x\beta/4)}.$$

**We define $T_k := k^{-(\beta/4)}$, $\lambda_k := 2\pi\alpha'(m_k)$ and $\delta_k := 4/T_k$ . Then, using the notation in [ B-P 79, theorem 1 ] , we obtain $\alpha_k = O(k^{-2})$. [ B-P 79, theorem 1 ] implies, that there is a sequence $\{u'_k\}_{k\in N}$ of independent random variables, defined on an enlargement of the original probability space, with $u'_k \stackrel{\mathcal{D}}{=} u_k$ and**

$$P(\mid \hat{X}_{N_{l-1},\, N_{l-1}+n_l} - \sigma_{n_l} u_l' \mid \ge \sigma_{n_l} \alpha_l) \le \alpha_l .$$

**Now we apply [ S 65 , theorem ( 4. 4. ) ] to $\{\sigma_n u'_k\}_{k\in N}$ with $f(t) = t^{1-8\lambda}$ and small enough $\lambda$ , where the sum [ S 65 , ( 138 ) ] will be estimated by ( 5. 4. ) :**

$$\int\limits_{\{(\sigma_{n_k} u_k^{\cdot})^2 > f(V_k)\}} (\sigma_{n_k} u_k^{\cdot})^2 \, dP \le f(V_k)^{-(\alpha/2)} \| \sigma_{n_k} u_k^{\cdot} \|_{2+\alpha}^{2+\alpha} , \qquad \text{where } V_k := \sum_{i=1}^{k} \sigma_{n_i}^2 .$$

**Now we apply [ B-P 79, lemma A1 ] and obtain a process $\mathfrak{X}' \overset{\mathcal{D}}{=} \mathfrak{X}$ and a standard Brownian motion $\mathfrak{B}$, on a common probability space with :**

$$\sum_{i=1}^{k} \hat{X}_{N_{i-1},\, N_{i-1}+n_i} - B\Big(\sum_{i=1}^{k} \sigma_{n_i}^2\Big) = O\big(N_k^{(1/2)-\lambda}\big) \qquad \text{a. s.}$$

**Here we have used $\sum_{i=1}^{k} \sigma_{n_i} i^{-2} = O(N_k^{(1/2)-\lambda})$. This, lemma 5. 1. and the following lemma 5. 4. completes the proof of lemma 5. 3.**

**q. e. d.**

**Lemma 5. 4. :** ***The following inequality is satisfied***

$$\sum_{i=1}^{k+1} \hat{X}_{N_{i-1}+n_i,\, N_i} = O\big(N_k^{(1/4)}\big) \qquad \text{a. s.}$$

**Proof of lemma 5. 4. : Using the Chebyscheff - inequality and the Minkowski - inequality we obtain :**

$$P\Big( \Big| \sum_{i=1}^{k+1} \hat{X}_{N_{i-1}+n_i,\, N_i} \Big| \ge N_k^{(1/4)} \Big) \le N_k^{-(1/2)} \Big( \sum_{i=1}^{k+1} \| \hat{X}_{N_{i-1}+n_i,\, N_i} \|_2 \Big)^2$$

$$= O\Big( N_k^{-(1/2)} \Big( \sum_{i=1}^{k+1} m_i^{(1/2)} \Big)^2 \Big) = O(k^{-2})$$

**An application of the Borel - Cantelli - Lemma yields the desired result.**

**q.e.d.**

**Lemma 5. 5. :** *There is a constant $\lambda > 0$, such that the following inequality is satisfied :*

$$\max_{N_k < t < N_{k+1}} \left| B(t\sigma^2) - B\left(\sum_{i=1}^{k+1} \sigma_{n_i}^2\right) \right| = O\left(N_k^{(1/2)-\lambda}\right) \quad \text{a. s.}$$

This lemma can be proved as in [ B-P 79, pp. 44,45 ] .

Now we can complete the proof of the theorem.

Let $N \in \mathbb{N}$, then there is a $k \in \mathbb{N}$ with $N_k < N \le N_{k+1}$ . Moreover there is a $\lambda > 0$ with :

$$| \hat{X}'_{0,N} - B(N\sigma^2) | \le \max_{N_k < t \le N_{k+1}} \left| B(t\sigma^2) - B\left(\sum_{i=1}^{k} \sigma_{n_i}^2\right) \right| +$$

$$+ \max_{N_k < t \le N_{k+1}} | \hat{X}_{0,N} - \hat{X}_{0,N_k} | + \left| \hat{X}_{0,N_k} - B\left(\sum_{i=1}^{k} \sigma_{n_i}^2\right) \right| = O\left(N_k^{(1/2)-\lambda}\right) \quad \text{a. s.}$$

q. e. d.

**Remarks** ( a ) In the case of a subadditive or superadditive process $\mathfrak{X}$ , ( 5. 3. ) can be replaced by

$$E(X_{0,N}) - N\gamma = O(N^{(1/2)-4\varepsilon}) .$$

( b ) ( 5. 4 ) is only used in lemma 5. 2. . In the other parts of the proof it suffices to assume

$$(5.\ 4.) \qquad \sup_{N \in \mathbb{N}} \left( N^{-1-(\alpha/2)-\Theta} \, \| X_{0,N} - E(X_{0,N}) \|_{2+\delta}^{2+\delta} \right) < +\infty .$$

Often an additive process $\mathfrak{Y}$ is given , which satisfies no dependence condition but which is close to a nonadditive mixing process. The following two corollaries deal with this case.

**Corollary 6 :** *Let $\mathfrak{X}$ satisfy ( C 1 ) and let $\mathfrak{X} \subset L_{2+\delta}$ , for some $\delta > 0$. Let $\mathfrak{X}$ be $\varphi'$ - mixing , and let ( 5. 1. ), ( 5. 2. ), ( 5. 3. ) and*

$$\| X_{0,N} - Y_{0,N} \|_{2+\delta} = O(N^{(1/2)})$$

*be satisfied. Then there is a probability space with a process $\mathfrak{X}' \overset{\mathcal{D}}{=} \mathfrak{X}$, and a standard Brownian motion $\mathfrak{B} = \{ B(t), t \geq 0 \}$ on it, such that the following holds*

$$\hat{X}_{0,N} - B(\sigma^2 N) = O(N^{(1/2)-\lambda}) \quad \text{a. s.}$$

*with suitable $\lambda > 0$.*

**Corollary 7** : *Let $\mathfrak{X}$ satisfy ( C 1 ) and let $\mathfrak{X} \subset L_{2+\delta}$, for some $\delta > 0$. Let $\mathfrak{X}$ be $\alpha'$ - mixing , and assume ( 5. 1. ), ( 5. 2. ), ( 5. 3. ). Let, in addition, the conditions*

$$\alpha'(N) = O(N^{-(1-x)(1+(\delta/2))}), \text{ for a constant } x > 0,$$

$$\| X_{0,N} - Y_{0,N} \|_2 = O(N^{(1/2)-4\epsilon}) \quad \text{and}$$

$$\| X_{0,N} - Y_{0,N} \|_{2+\delta} = O(N^{(1/2)})$$

*be satisfied. Then there is a probability space with a process $\mathfrak{X}' \overset{\mathcal{D}}{=} \mathfrak{X}$, and a standard Brownian motion $\mathfrak{B} = \{ B(t), t \geq 0 \}$ on it, such that*

$$\hat{X}_{0,N} - B(\sigma^2 N) = O(N^{(1/2)-\lambda}) \quad \text{a. s.}$$

*with suitable $\lambda > 0$.*

**Remark** The corrollaries can be obtained in the following way. Using theorem 2 in the $\varphi'$ - mixing case and theorem 3 in the $\alpha'$ - mixing case we verify ( 5. 4. )' . So lemma 5. 3. is satisfied and it follows as in the proof of lemma 5. 1. , that lemma 5. 3. is satisfied with $\mathfrak{Y}$ instead of $\mathfrak{X}$. Of course, ( 5. 4. )' is satisfied for $\mathfrak{Y}$ instead of $\mathfrak{X}$ and the additivity of $\mathfrak{Y}$ now implies ( 5. 4. ). So lemma 5. 2. implies to $\mathfrak{Y}$. The proofs can now be completed in the same way as at the end of the proof of theorem 5.

**The next theorem 8 will be presented without a proof. The additive version of this theorem can be found in [ B-P 79, Theorem 4 ]. A proof of theorem 8 can be found in [ W 83 ]. The proof of this theorem is an extension of the proof of [ B-P 79, Theorem 4 ].**

**Theorem 8** : *Let $\mathfrak{X} \subset L_{2+\delta}$, for some $0 < \delta \le 1$. Let $\mathfrak{X}$ be $\varphi$ - mixing, an let the following conditions be satisfied :*

( 8. 1. ) $\qquad \| X_{0,N} - X_{1,N} - Y_{0,1} \|_{2+\delta} = O( q^N )$, *with a constant* $0 < q < 1$.

( 8. 2. ) $\qquad \varphi( N ) = O( ( \log( N ) )^{-(160/\delta)} )$.

( 8. 3. ) $\qquad V( X_{0,N} ) \to \infty$ , *for* $N \to \infty$.

*Then there is a probability space with a process* $\mathfrak{X}' \overset{\mathcal{D}}{=} \mathfrak{X}$ *and a standard Brownian motion* $\mathfrak{B}$ *on it , such that*

$$\hat{X}'_{0,N} - B( a_N ) = a_N^{(1/2)} ( \log( a_N ) )^{-(1/4)} \qquad a.\ s.$$

*where* $a_N$ *is an increasing sequence of real numbers with* $a_N \sim V( X_{0,N} )$.

**Finally we shall give an application of the central limit theorem.**

**Example** : **Let $\mathfrak{Y}$ be an additive process. Let $\tau : \Omega \to \Omega$ be a measure preserving transformation with $Y_{N,N+1} = Y_{0,1} \circ \tau^N$ . Let $\mathfrak{M} = ( M_{m,n} )_{m,n}$ be a $\varphi'$ - mixing family of $\sigma$ - algebras which satisfies $M_{n,n+m} = \tau^{-n+k} M_{k,k+m}$ for all m and $n \ge k$ . Then the following theorem holds**

**Theorem E** : *Let, in addition, $\mathfrak{Y} \subset L_{2+\delta}$ , with $\delta > 0$, and assume that the following conditions are satisfied :*

$$(E.\ 1.)\qquad V(Y_{0,N}) \to \infty.$$

$$(E.\ 2.)\qquad V(Y_{0,N})^{-1/2} \sum_{n=1}^{N} \| E(Y_{0,1} \mid M_{-n,n}) - Y_{0,1} \|_2 \to 0\ .$$

$$(E.\ 3.)\qquad \| E(Y_{0,1} \mid M_{-n,n}) - Y_{0,1} \|_{2+\delta} = O(N^{-(1/2)-\varepsilon}),\ \textit{for a constant}\ \varepsilon > 0.$$

*Then the central limit theorem holds, i.e., we have*

$$V(Y_{0,N})^{-(1/2)} (Y_{0,N} - E(Y_{0,N})) \xrightarrow{\mathcal{D}} N(0,1).$$

We shall prove the theorem by an application of theorem 1. First we construct a nonadditive process $\mathfrak{X}$, which satisfies the conditions of theorem 1. These will be verified by means of theorem 2 and theorem 3. The use of theorem 2 and theorem 3 to verify the conditions of theorem 1 and, hence to prove a central limit theorem, is possible in many other cases ( for example for the the range of random walk and products of random matrices ). Let the three parameter process $\Psi = (\psi_{n,i,m})_{n\le i<m}$ be defined by $\psi_{n,i,m} = E(Y_{i,i+1} \mid M_{i-\min\{i-n,m-i\},\ i+\min\{i-n,m-i\}})$. Let the nonadditive process $\mathfrak{X} = (X_{m,n})$ be defined by :

$$X_{n,m} = \sum_{i=n}^{m-1} \psi_{n,i,m}\quad ,\ \text{for } 0 \le n < m\ .$$

$\mathfrak{X}$ is $\varphi'$ - mixing. Without loss of generality we can suppose $\varepsilon < 1/2$. ( E. 2. ) implies :

$$\| X_{0,N} - Y_{0,N} \|_2 \le 2 \sum_{i=0}^{[N/2]} \| E(Y_{0,1} \mid M_{-i,i}) - Y_{0,1} \|_2 = O(V(Y_{0,N})^{1/2})$$

This implies $V(X_{0,N}) \to \infty$, and theorem 2 can be applied, so ( C 2 ) is satisfied. In the same way as above we obtain

$$\| X_{0,N} - Y_{0,N} \|_{2+\delta} = O(N^{(1/2)-\varepsilon}).$$

So ( 3. 1. ) and ( C 1 ) are satisfied. Theorem 3 yields

$$\| X_{0,N} - E(X_{0,N}) \|_{2+\delta} = O( V(X_{0,N})^{1/2} ).$$

This implies ( C 3 ). By theorem 1 we obtain

$$V( X_{0,N} )^{-(1/2)} ( X_{0,N} - E( X_{0,N} ) ) \xrightarrow{\mathcal{D}} N( 0 , 1 ).$$

Theorem E now follows from [ B 68, theorem 4. 1. ] together with the following calculation :

$$\| V( X_{0,N} )^{-1/2} ( X_{0,N} - E( X_{0,N} ) ) - V( Y_{0,N} )^{-1/2} ( Y_{0,N} - E( Y_{0,N} ) ) \|_2$$

$$\leq | 1 - ( V( X_{0,N} )/V( Y_{0,N} ) )^{1/2} | \| V( X_{0,N} )^{-1/2} ( X_{0,N} - E( X_{0,N} ) ) \|_2 +$$

$$+ V( Y_{0,N} )^{-1/2} 2 \| X_{0,N} - Y_{0,N} \|_2 \to 0 , \text{ as } N \to \infty .$$

**References**

[ A-K 81 ] Akcoglu, M. A. and U. Krengel ( 1981 ) *Ergodic Theorems for Superadditive Processes*, J. Reine Ang. Math. 323,53-67

[ A-S 78 ] Akcoglu, M. A. and L. Sucheston (1978) *A Ratio Ergodic Theorem for Superadditive Processes*, Z. Wahrscheinlichkeitstheorie verw. Gebiete **44**, 269-278

[ B-P 79 ] Berkes, I. and W. Philipp (1979) *Approximation Theorems for Independent and weakly Dependent Random Vectors*, Ann. Prob. 7,29-54

[ B 68 ] Billingsley P. (1968) *Convergence of Probability Measures* Wiley. New York

[ G-S 77 ] Gänssler, P. and W. Stute (1977) *Wahrscheinlichkeitstheorie* Springer Verlag Berlin - Heidelberg - New York

[ F-K 60 ] Furstenberg, H. and H. Kesten (1960) *Products of Random Matrices*, Ann. Math. Statist. 39, 457-496

[ I 77 ] Ishitani, I. (1977) *A Central Limit Theorem for the Subadditive Process and its Application to Products of Random Matrices* , Publ. Rims. Kyoto Univ. 12,565-575

[ I-L 71 ] Ibragimov, I. A. and Linnik Y. a. V. (1971) *Independent and Stationary Sequences of Random Variables* Wolters-Nordhoff. Groningen

[ J-P 73 ] Jain, N. C. and W. E. Pruitt (1973) *The range of random walk*, Proc. Sixth. Berkeley Symp. Math. Statist. 3, 31 - 50

[ K 68 ] J. F. C. Kingman (1968) *The ergodic theory of subadditive stochastic processes*, J. Roy. Statist. Soc. Ser. B, 30, 499-510

[ K-P 80 ] Kuelbs , J. and W. Philipp (1980) *Almost sure invariance principles for partial sums of mixing B-valued random variables*, Ann. Prob. 8, 1003-1036

[ S 68 ] R. J. Serfling (1968) *Contributions of Central Limit Theory for Dependent Variables*, Ann. Math. Statist. 39, 1158-1175

[ S 65 ] V. Strassen (1965) *Almost sure behavior of sums of independent random variables and martingales* , Proc. Fifth. Berkeley Symp. Math. Statist. 2 , 315-343

[ W 83 ] U. Wacker (1983) *Grenzwertsätze für nichtadditive, schwach abhängige Prozesse* Dissertation. Göttingen

# Fixed point rays of nonexpansive mappings

RAINER WITTMANN*
Institut für Mathematische Stochastik der
Universität Göttingen

**Abstract.** Let $T$ be a nonexpansive mapping on a strictly convex and smooth Banach space $X$. It is shown, that $T(ty) = ty$ for any $t \in \mathbf{R}$ implies $T(x+y) = (Tx) + y$ for any $x \in X$.

## 1. INTRODUCTION

Throughout the sequel, $T$ will be a nonexpansive mapping on a Banach space $X$ with $T0 = 0$. In Theorem 5.7 of Lin, Wittmann[2] it was stated that for certain mappings $T$, which are simultaneously nonexpansive on all $L^p$ spaces, $Ty = y$ implies $T(x+y) = (Tx)+y$ for any $x \in X$. The basic assumption (DIS) of this theorem is equivalent to the property that the set of fixed points is a linear space. The proof was omitted, because it was too long and because [2] was primarily dedicated to ergodic theorems. While the original proof was heavily based on order properties and $L^1$ theory, which required special additional assumptions, we show this property for general nonexpansive mappings with $T0 = 0$, if $X$ is merely strictly convex. For smooth spaces, remark 3.5 outlines a fairly simple approach to this result. In the general case the proof depends on an ergodic theorem for the resolvent of $T$.

To introduce the resolvent, let $0 < \alpha < 1$ be given and define inductively for any $x \in X$

$$V_{\alpha,0} = \alpha x, \quad V_{\alpha,n+1}x = \alpha x + (1-\alpha)TV_{\alpha,n}x$$

Since $\|Tx\| \leq \|x\|$ the $V_{\alpha,n}x$ are norm convergent to an element $V_\alpha x$ in $X$ with $\|V_\alpha x\| \leq \|x\|$. Since $T$ is nonexpansive $V_\alpha$ is also nonexpansive. Passing to the limit in the induction equation we obtain the resolvent equation :

$$V_\alpha x = \alpha x + (1-\alpha)TV_\alpha x$$

## 2. A CONVERGENCE THEOREM FOR THE RESOLVENT

PROPOSITION 2.1. *Let $N = N_T$ be the closure of the set $\{x - Tx : x \in X\}$ with respect to the norm topology. Then we have*

(i) $N = \{x \in X : \lim_{\alpha \to 0} \|\alpha V_\alpha(x/\alpha)\| = 0\}$

(ii) $x - \frac{\alpha}{1-\alpha} V_\alpha(\frac{1-\alpha}{\alpha}x) \in N \quad (x \in X)$

(iii) $\limsup_{\alpha \to 0} \|\alpha V_\alpha(x/\alpha)\| = \liminf_{\alpha \to 0} \|\alpha V_\alpha(x/\alpha)\| = \inf\{\|x - y\| : y \in N\} \quad (x \in X)$

PROOF: From the resolvent equation we get

$$x = (V_\alpha(\tfrac{1-\alpha}{\alpha}x) - T(V_\alpha(\tfrac{1-\alpha}{\alpha}x))) + \tfrac{\alpha}{1-\alpha}V_\alpha(\tfrac{1-\alpha}{\alpha}x) \tag{1}$$

*Heisenberg Fellow of Deutsche Forschungsgemeinschaft

and (ii) follows. Since $T$ is nonexpansive and since $\lim_{\alpha\to 0}(\frac{1-\alpha}{\alpha})/\frac{1}{\alpha} = 1$, this implies also

(2) $$\liminf_{\alpha\to 0} \|\tfrac{\alpha}{1-\alpha} V_\alpha(\tfrac{\alpha}{1-\alpha}x)\| = \liminf_{\alpha\to 0} \|\alpha V_\alpha(x/\alpha)\| \geq \inf\{\|x-y\| : y \in N\}.$$

In particular, we have

(3) $$\{x \in X : \lim_{\alpha\to 0} \alpha V_\alpha(x/\alpha) = 0\} \subset N.$$

Again from the resolvent equation we get $(\frac{1}{\alpha}I - \frac{1-\alpha}{\alpha}T) \circ V_\alpha = I$, where $I$ is the identity mapping. In particular $G_\alpha := (\frac{1}{\alpha}I - \frac{1-\alpha}{\alpha}T)$ is surjective. But $G_\alpha$ is also injective, because for any $x, y \in X$ we have

$$\begin{aligned} 0 = \|G_\alpha x - G_\alpha y\| &\geq \tfrac{1}{\alpha}\|x-y\| - \tfrac{1-\alpha}{\alpha}\|Tx - Ty\| \\ &\geq \tfrac{1}{\alpha}\|x-y\| - \tfrac{1-\alpha}{\alpha}\|x-y\| = \|x-y\| \end{aligned}$$

For a bijective mapping any right inverse is also a left inverse. Thus we obtain

(4) $$V_\alpha \circ (\tfrac{1}{\alpha}I - \tfrac{1-\alpha}{\alpha}T) = I = (\tfrac{1}{\alpha}I - \tfrac{1-\alpha}{\alpha}T) \circ V_\alpha.$$

For any $x \in X$ we have $\|\frac{1}{\alpha}(x - Tx) - G_\alpha x\| = \|Tx\|$ and therefore, since $V_\alpha$ is nonexpansive,

$$\alpha\|Tx\| \geq \alpha\|V_\alpha(\tfrac{1}{\alpha}(x - Tx)) - V_\alpha(G_\alpha x)\| = \|\alpha V_\alpha(\tfrac{1}{\alpha}(x - Tx)) - \alpha x\|$$

Thus we obtain

$$\lim_{\alpha\to 0} \|\alpha V_\alpha(\tfrac{1}{\alpha}(x - Tx))\| = 0$$

Putting this together with (3) we obtain (i).

Let now $x \in X$ and $\varepsilon > 0$ be given. Then there exists $y \in N$ with $\|x-y\| \leq \inf\{\|x-z\| : z \in N\} + \varepsilon$. Since $V_\alpha$ is nonexpansive, we have $\|\alpha V_\alpha(x/\alpha) - \alpha V_\alpha(y/\alpha)\| \leq \|x-y\|$. Since $\lim_{\alpha\to 0} \|\alpha V_\alpha(y/\alpha)\| = 0$ by (i), this implies $\limsup_{\alpha\to 0} \|\alpha V_\alpha(x/\alpha)\| \leq \|x-y\| \leq \inf\{\|x-z\| : z \in N\} + \varepsilon$. Letting $\varepsilon > 0$ tend to 0 and combining this with (2) we obtain (iii). ∎

## 3. Fixed point rays

Proposition 2.1 becomes important for fixed point rays through the following result.

**Proposition 3.1.** *Let $y \in X$ be such that $T(ty) = ty$ for any $t \geq 0$. Then we have*

$$\|ty\| \leq \|ty - x\| \quad (x \in N_T)$$

**Proof:** From the definition of $V_\alpha$ it follows that $\alpha V_\alpha(ty/\alpha) = ty$ and the assertion follows from 2.1(iii). ∎

**Proposition 3.2.** *Assume that $X$ is strictly convex. Then for any $y \in X\setminus\{0\}$ the following properties are equivalent.*

(i) $$T(ty) = ty \quad (t \geq 0).$$
(ii) $$\|ty\| \leq \|ty - x\| \quad (t \geq 0, x \in N_T).$$
(iii) $$\|y\| \leq \|y - x\| \quad (x \in N_T).$$

PROOF: (i)$\Longrightarrow$(ii) follows from proposition 3.1 and (ii)$\Longrightarrow$(iii) is trivial.

Assume now that (iii) holds. By (1) of the proof of proposition 2.1 we have $y - \frac{\alpha}{1-\alpha}V_\alpha(\frac{1-\alpha}{\alpha}y) \in N_T$ and therefore (iii) implies

$$\|y\| \leq \|y - (y - \tfrac{\alpha}{1-\alpha}V_\alpha(\tfrac{1-\alpha}{\alpha}y))\| = \|\tfrac{\alpha}{1-\alpha}V_\alpha(\tfrac{1-\alpha}{\alpha}y)\|.$$

Since also $\|V_\alpha x\| \leq \|x\|$ for any $x \in X$, we obtain

$$\|\tfrac{\alpha}{1-\alpha}V_\alpha(\tfrac{1-\alpha}{\alpha}y)\| = \|y\|. \tag{1}$$

By the resolvent equation, we have

$$\tfrac{\alpha}{1-\alpha}V_\alpha(\tfrac{1-\alpha}{\alpha}y) = \alpha y + \alpha T V_\alpha(\tfrac{1-\alpha}{\alpha}y). \tag{2}$$

Since $\|\alpha T V_\alpha(\frac{1-\alpha}{\alpha}y)\| \leq (1-\alpha)\|y\|$, (1), (2) and strict convexity imply $\frac{\alpha}{1-\alpha}V_\alpha(\frac{1-\alpha}{\alpha}y) = y$ Inserting this into (2) we obtain $y = \alpha y + \alpha T(\frac{1-\alpha}{\alpha}y)$ and therefore $T(\frac{1-\alpha}{\alpha}y) = \frac{1-\alpha}{\alpha}y$. Since $\alpha > 0$ is arbitrary, (i) follows. ∎

We now recall that a Banach space is said to be *smooth*, if for any $y \in X \setminus \{0\}$ and $x \in X$ $\lim_{\varepsilon\to 0} \frac{1}{\varepsilon}(\|y+\varepsilon x\| - \|y\|)$ exists. Note that the right and left derivative of the norm exists always. The point is, that both derivatives are the same. Smoothness is a very weak property of a Banach space. For instance, if $X^*$ is strictly convex, then $X$ is smooth.

PROPOSITION 3.3. *Assume that $X$ is smooth and let $y \in X \setminus \{0\}$. If*

$$\|ty\| \leq \|ty - x\| \quad (x \in N_T) \tag{i}$$

*holds for all $t \geq 0$, then there exists $\ell \in X^*$ such that $\|\ell\| = 1$, $\ell(y) = \|y\|$ and $\ell(x - Tx) \leq 0$ for any $x \in X$.*

*If (i) holds for all $t \in \mathbf{R}$, then there exists $\ell \in X^*$ such that $\|\ell\| = 1$, $\ell(y) = \|y\|$ and $\ell(x - Tx) = 0$ for any $x \in X$.*

PROOF: If $N_T$ would be convex, then the above assertion follows from a well known result of approximation theory. The convexity of $N_T$ is known for certain Banach spaces including uniformly smooth Banach spaces (cf. Reich [3]). In the general case we consider the functional

$$p_y(x) = \lim_{0<\varepsilon\to 0} \tfrac{1}{\varepsilon}(\|y+\varepsilon x\| - \|y\|)$$

By an elementary result about convex functions on the line, the above onesided limits exist. Moreover, by the triangle inequality, $p_y$ is sublinear and by the smoothness of $X$ it must even be linear (in fact smoothness is equivalent to linearity of $p_y$ for any $y \in X \setminus \{0\}$. Now, if (i) holds for any $t \geq 0$, then we have $\varepsilon^{-1}(\|y - \varepsilon(x - Tx)\| - \|y\|) = \|\varepsilon^{-1}y - (x - Tx)\| - \|\varepsilon^{-1}y\| \geq 0$ and therefore $p_y(Tx - x) \geq 0$ for any $x \in X$. Thus $\ell = p_y$ has the required properties. If (i) holds for any $t \in \mathbf{R}$, then we have also $p_{-y}(x - Tx) \leq 0$. Since $p_{-y}(x) = p_y(-x)$ also the second part of the assertion follows. ∎

Combining propositions 3.1 and 3.3 we obtain

**COROLLARY 3.4.** *Assume that $X$ is smooth. and let $y \in X$ be such that $T(ty) = ty$ for any $t \geq 0$. Then there exists $\ell \in X^*$ such that $\|\ell\| = 1$, $\ell(y) = \|y\|$ and $\ell(x - Tx) \leq 0$ for any $x \in X$.*

*If even $T(ty) = ty$ for all $t \in \mathbf{R}$, then there exists $\ell \in X^*$ such that $\|\ell\| = 1$, $\ell(y) = \|y\|$ and $\ell(x - Tx) = 0$ for any $x \in X$.*

**REMARK 3.5:** (a) It was pointed out by the referee, that an argument, which can be found in the unpublished thesis of Ronald Bruck, may be used to give a direct and very simple proof of corollary 3.4 avoiding proposition 2.1. To this end, let $y \in X$ with $T(ty) = ty$ for any $t \geq 0$. Then for any $\varepsilon > 0$ we have $\varepsilon^{-1}(\|y - \varepsilon Tx\| - \|y\|) = \|\varepsilon^{-1}y - Tx\| - \|\varepsilon^{-1}y\| = \|T(\varepsilon^{-1}y) - Tx\| - \|\varepsilon^{-1}y\| \leq \|\varepsilon^{-1}y - x\| - \|\varepsilon^{-1}y\| = \varepsilon^{-1}(\|y + \varepsilon x\| - \|y\|)$. Letting $\varepsilon > 0$ tend to 0 we obtain $p_y(-Tx) \leq p_y(-x)$ for any $x \in X$. Since $p_y$ is linear the assertion follows.
(b) corollary 3.4 and proposition 3.1 can be regarded as abstract versions of lemma 3.7 and formula (2) in the proof of theorem 4.2 of [4].

We are now in position to prove our main result.

**THEOREM 3.6.** *Assume that $X$ is strictly convex and smooth. Let further $y \in X$ with $T(ty) = ty$ for any $t \in \mathbf{R}$. Then we have*

$$T(x + y) = (Tx) + y \quad (x \in X)$$

**PROOF:** By corollary 3.4 there exists $\ell \in X^*$ with $\|\ell\| = 1$, $\ell(y) = \|y\|$ and

$$\ell(x - Tx) = 0 \quad (x \in X) \tag{1}$$

Setting $M := \{x \in X : \ell(x) = 0\}$ we have

$$\inf\{\|ty - x\| : x \in M\} \geq \inf\{\ell(ty - x) : x \in M\} = \ell(ty) = \|ty\| \quad (t \in \mathbf{R}). \tag{2}$$

From (1) we get

$$T(M) \subset M. \tag{3}$$

We are now going to prove

$$T(x + ty) = (Tx) + ty \quad (x \in X, t \in \mathbf{R}). \tag{4}$$

Clearly it suffices to prove (4) for $x \in M$. By (1) we then have $\ell(T(x+ty)) = \ell(x+ty) = t\|y\|$ and therefore $x' := T(x + ty) - ty \in M$. On the other hand we have $\|T(x + ty) - Tx\| \leq \|(x + ty) - x\| = \|ty\|$ and therefore $\|ty + (x' - Tx)\| \leq \|ty\|$. By (3) $x' - Tx \in M$ whence $x' - Tx$ is also a best approximation of $ty$. Since $X$ is strict convex, the best approximation is unique, whence $x' - Tx = 0$ and (4) follows. ∎

**REMARK:** If $X = L^1$ then the above result is heavily false. In fact, it is false for any truely nonlinear disjointly additive mapping (cf. Krengel, Lin [1]).

We conclude with an analogue of the Kakutani–Yosida decomposition for nonexpansive mappings.

THEOREM 3.7. *Assume that $X$ is smooth, strictly convex and reflexive. Let $R_T$ be the closure of the linear space generated by $\{x - Tx : x \in X\}$. Then for any $x \in X$ there exists a unique $z \in R_T$ such that $T(t(x-z)) = t(x-z)$ for any $t \in \mathbf{R}$.*

PROOF: By our assumptions the best approximation $z$ of $x$ with respect to $N_T$ exists. Setting $y = x - z$ we see that 0 is the best approximation of $ty$ with repect to $R_T$ and therefore also with respect to $N_T$. By proposition 3.2 we have $T(ty) = ty$ for any $t \in \mathbf{R}$.

Let now $z' \in R_T$ such that $T(t(x-z')) = t(z'-x)$ for any $t \in \mathbf{R}$. Then, by corollary 3.4, there exists $\ell \in X^*$ with $\|\ell\| = 1$, $\ell(x-z') = \|x-z'\|$ and $\ell(y - Ty) = 0$ for any $y \in X$. Thus we have also $\ell(y) = 0$ for any $y \in R_T$ and therefore $\|x - z\| \geq \ell(x-z) = \ell(x-z') + \ell(z'-z) = \ell(x-z') = \|x-z'\|$. Thus $z'$ is also a best approximation of $x$ with respect to $R_T$. Since the best approximation is unique, the uniqueness part of the assertion follows. ∎

*Acknowledgement.* I am indebted to the referee for the simplified proof in the case of smooth spaces. In a first version of this paper most of the above results were only given for spaces with a uniformly convex dual, because we needed the convexity of $N_T$.

## REFERENCES

1. U. Krengel, M. Lin, *An integral representation of disjointly additive order preserving operators in $L^1$*, Stochastic Anal. Appl. 6 (1988), 289–304.
2. M. Lin, R. Wittmann, *Pointwise ergodic theorems for certain order preserving mappings in $L^1$*, Proceedings Conference A.E. Convergence, Evanston, Il. 1989 (to appear).
3. S. Reich, *Asymptotic behavior of contractions in Banach spaces*, J. Math. Anal. Appl. 44 (1973), 57–70.
4. R. Wittmann, *Hopf's ergodic theorem for nonlinear operators*, Math. Ann. 289 (1981), 239–253.

Lotzestr. 13, D-3400 Göttingen, Germany

# LIST OF PARTICIPANTS

| | |
|---|---|
| AFRAIMOVICH, V.S. | Gorky State University, Radiophysics Department, GSP-20, ul. Gorkogo 23, SU-603600 Gorky |
| ARBEITER, M. | Friedrich Schiller Universität, Institut für Mathematik, UHH 17. OG, D-O-6900 Jena |
| BAN, J. | Department of Mathematics, Comenius University, CS-84215 Bratislava |
| BANDT, C. | Universität Greifswald, Institut für Mathematik, Jahnstr. 15a, D-O-2200 Greifswald |
| BLOKH, A. | All-Union Hematological Scientific Centre, Nowozykovski pr. 4a, SU-125167 Moscow |
| BOGENSCHÜTZ, T. | Universität Bremen, Institut für Dynamische Systeme, Postfach 330440, D-W-2800 Bremen 33 |
| BOTHE, H.G. | Karl-Weierstraß-Institut für Mathematik, Mohrenstr. 39, D-O-1089 Berlin |
| BUNIMOVICH, L.A. | P.P. Shirshov Institute of Oceanology, Academy of Sciences, ul. Krasikova 23, SU-117218 Moscow |
| DENKER, M. | Universität Göttingen, Institut für Mathematische Stochastik, Lotzestr. 13, D-W-3400 Göttingen |
| FERENCZI, S. | CNRS, URA 225, Case 901, 163, av. de Luminy, F-13288 Marseille Cedex 9 |
| FLACHSMEYER, J. | Universität Greifswald, Institut für Mathematik, Jahnstr. 15a, D-O-2200 Greifswald |
| GEBEL, M. | Martin-Luther-Universiät Halle-Wittenberg, FB Mathematik und Informatik, Institut für Analysis, Postfach, D-O-4010 Halle |
| IWANIK, A. | Technical University, Institute of Mathematics, Wyb. Wyspianskiego 27, PL-50-370 Wroclaw |
| JACOBS, K. | Universität Erlangen, Mathematisches Institut, Bismarckstr. 1 1/2, D-W-8520 Erlangen |

KAMINSKI, B. Uniwersytet M. Kopernika, Instytut Matematyki, ul. Chopina 12/18, PL-87-100 Toruń

KELLER, K. Universität Greifswald, Institut für Mathematik, Jahnstr. 15a, D-O-2200 Greifswald

KOWALSKI, Z. Technical University, Institute of Mathematics, Wyb. Wyspianskiego 27, PL-50-370 Wroclaw

KRENGEL, U. Universität Göttingen, Institut für Mathematische Stochastik, Lotzestr. 13, D-W-3400 Göttingen

KRÜGER, T. Universität Bielefeld, Fakultät für Physik, BIBOS, D-W-4800 Bielefeld

KWIATKOWSKI, J. Uniwersytet M. Kopernika, Instytut Matematyki, ul. Chopina 12/18, PL-87-100 Toruń

LEDRAPPIER, F. Université Paris IV, Laboratoire de Probabilités, F-75253 Paris Cedex 05

LEMANCZYK, M. Uniwersytet M. Kopernika, Instytut Matematyki, ul. Chopina 12/18, PL-87-100 Toruń

LESIGNE, E. Université de Bretagne Occidentale, 52, rue Adolphe Leray, F-35000 Rennes

LIARDET, P. Université de Province, Case 96, 3, Place V. Hugo, F-13331 Marseille Cedex 3

MALCZAK, J. Uniwersytet Jagiellonski, Institut Informatyki, ul. Kopernika 27, PL-31-501 Krakow

MALICKY, P. Jablonova 618/2, CS-03104 Liptovski Mikulas

MAYER, D. Max Planck Institut für Mathematik, Gottfried Claren Str. 26, D-W-5300 Bonn 3

MIEBACH, J. Martin-Luther-Universiät Halle-Wittenberg, FB Mathematik und Informatik, Institut für Analysis, Postfach, D-O-4010 Halle

MISERA, J. Department of Mathematics, Comenius University, CS-84215 Bratislava

PATZSCHKE, N. Friedrich Schiller Universität, Institut für Mathematik, UHH 17. OG, D-O-6900 Jena

RICHTER, K. Martin-Luther-Universiät Halle-Wittenberg, FB Mathematik und Informatik, Institut für Analysis, Postfach, D-O-4010 Halle

SCHMELING, J. Karl-Weierstraß-Institut für Mathematik, Mohrenstr. 39, D-O-1089 Berlin

SIEGMUND-SCHULZE, R. Karl-Weierstraß-Institut für Mathematik, Mohrenstr. 39, D-O-1089 Berlin

SIPOS, J. SVST, Stavebná faculta, Radlinskeho 11, CS-81368 Bratislava

STEPIN, A.M. Moscow State University, Mech. math. faculty, SU-117234 Moscow

SUCHESTON, L. Department of Mathematics, Ohio State University, 231 W 18th Avenue, Columbus, Ohio 43210, USA

TOK, P. Universität Leipzig, FB Mathematik, Augustusplatz 10, D-O-7010 Leipzig

TRUBEZKOJ, S. Universität Bielefeld, Fakultät für Physik, BIBOS, D-W-4800 Bielefeld

URBANSKI, M. Uniwersytet M. Kopernika, Instytut Matematyki, ul. Chopina 12/18, PL-87-100 Toruń

DE VRIES, J. Stichting Mathematisch Centrum, Centrum voor Wiskude en Informatica, Kruislaan 413, NL-1098 SJ Amsterdam

WACKER, U. Universität Göttingen, Institut für Mathematische Stochastik, Lotzestr. 13, D-W-3400 Göttingen

WARSTAT, V. Martin-Luther-Universiät Halle-Wittenberg, FB Mathematik und Informatik, Institut für Analysis, Postfach, D-O-4010 Halle

WITTMANN, R. Universität Göttingen, Institut für Mathematische Stochastik, Lotzestr. 13, D-W-3400 Göttingen